Philosophy of Technology deepens especially my appreciation for the work of the authors Verkerk, van der Stoep, Hoogland, and de Vries. Their insightful contributions to critical philosophical reflection on technology fully deserve the wider audience this publication will promote

Carl Mitcham, PhD, Hans Jonas Chair at the European Graduate School EGS and Professor of Liberal Arts and International Studies, Colorado School of Mines, USA

This book is an interesting read for anyone interested in how philosophers have analyzed and interpreted the influence of technology on the modern human condition

Peter Kroes, Delft University of Technology, The Netherlands

PHILOSOPHY OF TECHNOLOGY

Philosophy of Technology: An Introduction for Technology and Business Students is an accessible guide to technology's changes, their ubiquitousness, and the many questions these raise. Designed for those with no philosophical background in mind, it is ideal for technology and engineering students or specialists who want to learn to think critically about how their work influences society and our daily lives.

The technological, business environment and daily experiences are the starting point of the book and the authors' reflect upon these practices from a philosophical point of view. The text goes on to present a critical analysis of the subject including development, manufacturing, sales and marketing and the use of technological products and services. The abstract ideas are made easier to grasp with a story-telling approach: a vivid history of the discipline and colourful portraits of the core thinkers in this domain, as well as four case studies drawing from various engineering disciplines to demonstrate how philosophy can and should influence technology in practice.

Maarten J. Verkerk is Affiliate Professor in Reformational Philosophy at Eindhoven University of Technology and Maastricht University, the Netherlands.

Jan Hoogland is Affiliate Professor in Reformational Philosophy at the University of Twente and Professor of Formative Education at Viaa University of Applied Sciences, the Netherlands.

Jan van der Stoep holds the chair for Media, Religion and Culture at the Ede Christian University of Applied Science, the Netherlands.

Marc J. de Vries is Professor of Science and Technology Education and Affiliate Professor of Christian Philosophy at Delft University of Technology, the Netherlands.

PHILOSOPHY OF TECHNOLOGY

An introduction for technology and business students

Maarten J. Verkerk, Jan Hoogland, Jan van der Stoep and Marc J. de Vries

Translated by Dr M. Nelson

LONDON AND NEW YORK

First published 2007 in Dutch
by Uitgeverij Boom, Amsterdam

First published in English 2016
by Routledge
2 Park Square, Milton Park, Abingdon, Oxon OX14 4RN

and by Routledge
711 Third Avenue, New York, NY 10017

Routledge is an imprint of the Taylor & Francis Group, an informa business

British Library Cataloguing in Publication Data
A catalogue record for this book is available from the British Library

Library of Congress Cataloging in Publication Data
Verkerk, Maarten Johannes, 1953-
[Denken, ontwerpen, maken. English]
Philosophy of technology : an introduction for technology and business students / Maarten J. Verkerk, Jan Hoogland, Jan van der Stoep and Marc J. de Vries. – English edition.
pages cm
Translation of the author's Denken, ontwerpen, maken.
Includes bibliographical references and index.
1. Technology–Philosophy. 2. Engineering–Philosophy. 3. Science–Philosophy. 4. Design–Philosophy. I. Title.
T14.V4913 2016
601–dc23
2015007992

ISBN: 978-1-138-90438-5 (hbk)
ISBN: 978-1-138-90439-2 (pbk)
ISBN: 978-1-315-69636-2 (ebk)

Typeset in Bembo
by Taylor & Francis Books

CONTENTS

List of illustrations *ix*
List of contributors *xi*
Foreword: philosophy, technology, and glocalisation *xiii*
Preface *xvi*

PART I
Thinking and making **1**

1 Thinking and technology: Between analysis and criticism 3

Portrait 1: Carl Mitcham (1941) 15

2 Speaking in a two-sided way: The meaning of disclosure and the disclosure of meaning 18

Portrait 2: Martin Heidegger (1889–1976) 35

PART II
Making and designing **39**

3 The world of technology: Three kinds of complexity 41

Portrait 3: Lewis Mumford (1895–1990) 59

4 The artefact [I]: Diversity and coherence 62

Portrait 4: Alasdair MacIntyre (1929) 86

5 The artefact [II]: Identity, function and structure 89

Portrait 5: Gilbert Simondon (1924–1989) 106

Case study I: Nanotechnology 109

6 Knowledge of designing: The role of the engineer 115

Portrait 6: Herbert Simon (1916–2001) 131

7 Design and reality: Methodological obstinacy 134

Portrait 7: Bruno Latour (1947) 153

8 Technology and production: From dehumanisation to human measure 156

Portrait 8: Larry Hickman (1942) 182

Case study II: A new factory 185

PART III
Designing and thinking **193**

9 The rules of the game: Technology as a social practice 195

Portrait 9: Langdon Winner (1944) 211

10 Symmetries: Between pessimists and optimists 214

Portrait 10: Jacques Ellul (1912–1994) 236

11 Clashing worlds: Globalisation and cultural diversity 239

Portrait 11: Albert Borgmann (1937) 258

Case study III: Network enabled military operations 261

12 Homo technicus: From device to cyborg 267

Portrait 12: Don Ihde (1934) 285

13 'Good' technology?: Normative artefacts and the web of responsibilities 288

Portrait 13: Egbert Schuurman (1937) 306

Case study IV: Innovation in health care 309

14 Expectations for the future: The secular sacred and the limits of technology 315

Portrait 14: Andrew Feenberg (1943) 330

Index *333*

LIST OF ILLUSTRATIONS

Figures

2.1	5¼ inch floppy diskette	19
2.2	A radu	26
3.1	A vidicon	44
3.2	The Plumbicon in a colour television camera	44
3.3	The old clinic at Vijverdal	49
3.4	The new clinic at Vijverdal	49
3.5	A sketch of a section of the new clinic at Vijverdal	51
4.1a	An industrial robot: mechanical manipulator	64
4.1b	An industrial robot: power supply	64
4.1c	An industrial robot: control system	65
4.2	A schematic drawing of a six-axised robot arm	65
6.1	The LOCOS process	117
6.2	Modal aspects and the dual nature of artefacts	123
8.1	Design sequence	170
8.2	Reduction of complexity by parallelisation and segmentation	171
9.1	Cross section through an actuator (20 mm x 8 mm x 1.5 mm)	200
C3.1	One of the soldiers involved in the Afghanistan mission working with the network technology	262

Portraits

1	Carl Mitcham	15
2	Martin Heidegger	35
3	Lewis Mumford	59

4 Alasdair MacIntyre 86
5 Gilbert Simondon 106
6 Herbert Simon 131
7 Bruno Latour 153
8 Larry Hickman 182
9 Langdon Winner 211
11 Albert Borgmann 258
12 Don Ihde 285
13 Egbert Schuurman 306
14 Andrew Feenberg 330

Tables

4.1 Overview of the modal aspects and their meaning nucleus 75
5.1 Overview of the diverse functions for the different stages of technological artefacts 96
9.1 Examples of ISO norms (4.1 t/m 4.4 and 4.9) and ISO procedures for the various disciplines 202
13.1 Overview of different ethical perspectives in relation to the practice model 296

Boxes

2.1 The radu 25
8.1 *Modern Times* 165
9.1 Ceramic multilayer actuator 199
9.2 The Philips Way – this is how we work 208
13.1 Framework 292

LIST OF CONTRIBUTORS

Jan Hoogland (1959) studied sociology and philosophy at the Erasmus University Rotterdam. In 1992 he received a PhD at the same university on a doctoral thesis on *T.W. Adorno: Autonomie en antinomie. Adorno's ambivalente verhouding tot de metafysica* (Autonomy and antinomy. Adorno's ambivalent relationship to metaphysics). Thereafter he worked for some years in the faculties of Philosophy of the Erasmus University and the VU University (Amsterdam). Since 1997 he has been affiliate professor in Reformational Philosophy at the University of Twente and professor in Society Issues (since 2010) and Formative Education (since 2014) at Viaa (University of Applied Sciences in Zwolle). He is editor in chief of the journal *Soφie* (formerly *Beweging*).

Jan van der Stoep (1968) studied biology at Wageningen University and philosophy at the VU University Amsterdam. In 2005 he received a PhD at the VU University Amsterdam on a doctoral thesis *Pierre Bourdieu en de politieke filosofie van het multiculturalisme* (Pierre Bourdieu and the political philosophy of multiculturalism) (Kok, Kampen). From 1993 to 1995 he did research, commissioned by the Dutch Railways, on the ethics of traffic and transport. From 1996 to 2010 Van der Stoep was attached to the Institute for Cultural Ethics in Amersfoort. For this institute he conducted research in the field of media and postmodern culture. Nowadays Van der Stoep holds the chair in Media, Religion and Culture at the Ede Christian University of Applied Science. Van der Stoep is a member of the editorial board of *Philosophia Reformata.*

Maarten J. Verkerk (1953) studied chemistry, theoretical physics and philosophy at the State University of Utrecht. He received a PhD in 1982 at the University of Twente on a doctoral thesis titled *Electrical Conductivity and Interface Properties of Oxygen Ion Conducting Materials*. In 2004 he received a second PhD on a doctoral thesis *Trust and Power on the Shop Floor* at the University of Maastricht. For some years he worked as a senior researcher in the Philip's Laboratory for Physics in Eindhoven. In the period 1986–2002 he worked in the management of diverse factories and developmental groups in the Netherlands as well as abroad. From 2003–2007 he was chairman of the board of management of the psychiatric hospital Vijverdal in Maastricht. Since 2008 he has worked on the Management Board of VitaValley, an innovation network in health care. Since September 2004 he has been an affiliate professor in Reformational Philosophy at the University of Technology of Eindhoven and since July 2008 of the University of Maastricht. He has published amongst others in the fields of materials science, feminism, business ethics, organisation science, and philosophy of technology.

Marc J. de Vries (1958) studied experimental physics at the VU University in Amsterdam. He was awarded a PhD in 1988 at the University of Technology Eindhoven on the doctoral thesis *Techniek in het natuurkundeonderwijs* (Technology in the teaching of physics). For a number of years he was responsible for the part-time training Technology of the Pedagogisch Technische Hogeschool (Educational Technological College) in Eindhoven. Since 1989 he has been a university lecturer in Philosophy of Technology at the University of Technology in Eindhoven and since 2003 he has been affiliate professor in Reformational Philosophy at the University of Technology in Delft. Since 2008 he has also been ordinary professor of Science Education at the latter institution. He is editor in chief of the *International Journal for Technology and Design Education* (Springer). He has published on the teaching of technology, design methodology, philosophy and ethics of technology, and on the history of the Philips Research Laboratory.

FOREWORD

Philosophy, Technology, and Glocalisation

Carl Mitcham

Since the 1970s philosophical literature on technology has grown considerably. In addition to leading monographs on specific themes and issues there now exist entries on philosophy of technology in standard reference works as well as a number of useful introductory texts and handbooks. This volume is a new and welcome addition to this body of literature, one that highlights some distinctive Dutch contributions while providing a broad conceptual framework that engages an important spectrum of European and American thinkers.

The history of the development of philosophical reflection on technology may be conveniently parsed into three stages of growth. After a long gestational pre-history, an explicitly denominated philosophy of technology emerged in the late nineteenth century through the work of Ernst Kapp (1808–1896) and Karl Marx (1818–1883). During its initial stage, technology was seen primarily as an extension of the human organism (as interpreted by Kapp) and a means to human liberation that (in the argument of Marx) needs only to be freed from societal deformations and cultural constraints. This view of the fundamentally liberating character of technology was given metaphysical expression in the thought of engineer-inventor Friedrich Dessauer (1881–1963) – and has been reiterated since by engineer philosophers such as Buckminster Fuller (1895–1983) and Samuel Florman. It remains a persistent ideology of contemporary culture.

During Dessauer's lifetime, however, there emerged a second stage in philosophical reflection on technology represented by such figures as José Ortega y Gasset (1883–1955), Martin Heidegger (1889–1976), and Jacques Ellul (1912–1994). Reflecting a philosophical shift manifested more generally in the existentialism of Søren Kierkegaard (1813–1855) and Friedrich Nietzsche (1844–1900), second stage philosophers stressed the importance of personal lived experience. In contrast with the first stage, these philosophers shifted the focus from economics and politics to that of daily life and living in a techno-material culture. To state the new perspective succinctly: whereas the first stage Marx saw technology being used by the bourgeois or capitalist class to oppress the working class, Ortega and associates argued that technology could oppress even the members of the bourgeoisie who were its ostensible masters and possessors. The problem presented by technology was not one of economics or of politics but of culture.

A third stage in philosophy of technology was reached through the work of such thinkers as Don Ihde, Albert Borgmann, Andrew Feenberg, and Kristin Shrader-Frechette. The distinguishing feature of this stage is some measure of acceptance of technology – together with a movement toward de-generalisation. Its distinctive feature, which remains a living presence even while there are hints of an emerging fourth stage, has been to try to learn to live with advancing technology, perhaps in piecemeal fashion.

The third stage approach to the philosophy of technology has also been described by Dutch thinkers as an empirical turn. It can also be read as exemplifying the applied turn in philosophy as a whole. In place of any attempt at general ethics of technology, third stage philosophers have advanced multiple regionalisations of applied ethics: nuclear ethics, bioethics, environmental ethics, computer ethics, engineering ethics, agricultural ethics, and more. Confronted with the holistic affirmations of the first stage and the comprehensive but impotent criticisms of the second, there was a spontaneous shift toward an emphasis on education and the kind of piecemeal, academic-based social activism associated with John Dewey (1859–1952). More than any previous stage, the third stage took philosophy and technology studies into the classroom by teaching and writing textbooks for courses especially in bioethics, computer ethics, and engineering ethics. Because of its interest in particulars, this stage has often made common cause with science, technology, and society (STS) studies.

In English the first self-defined textbook in philosophy and technology was Carl Mitcham and Robert Mackey's edited collection *Philosophy and Technology: Readings in the Philosophical Problems of Technology* (1972). Since then there have been other similar texts, among them *Controlling Technology: Contemporary Issues* edited by William Thompson (1991; second edition with additional editors Eric Katz and Andrew Light, 2003), *Readings in the Philosophy of Technology* edited by David Kaplan (2004 and 2009), and *Philosophy of Technology: The Technological Condition, An Anthology* edited by Robert Scharff and Val Dusek (2003 and 2014). The first monographic textbook was Don Ihde's *Philosophy of Technology: An Introduction* (1993), followed by Frederick Ferré's *Philosophy of Technology* (1995), and Val Dusek's *Philosophy of Technology: An Introduction* (2006). Complementing such publications is *A Companion to the Philosophy of Technology* edited by Jan Kyrre Berg Olsen, Stig Andur Pedersen, and Vincent Hendricks (2009).

Against such a background, *Philosophy of Technology: An Introduction for Technology and Business Students* is distinctive in at least three important respects: it is monographic yet in a form that highlights seminal figures from both the first and second stages in the philosophy of technology. More than any other it calls attention to modern engineering as key to understanding technology. Indeed, its structure emphasises the centrality of design within technology and engineering experience rather than philosophical categories as epistemology, metaphysics, and ethics. And it engages otherwise three almost completely ignored second and third stage contributors from Dutch traditions in philosophy of technology: Herman Dooyeweerd, Hendrik van Riessen, and Egbert Schuurman.

In this third respect, *Philosophy of Technology* may be described as a local introduction to the philosophy of technology for a globalising world – a glocal text, as it were. In a world where justice requires developing respect for local ways of life and culture in a progressively networked globality, this volume bears philosophical witness to a dialectic between particularity and universality as well as a resistance to the hegemonic tendencies of English speaking discourse.

Acknowledgment: in most cases books are honoured by the writing of forewords. In the present instance, however, given the special achievements of this book, I am equally if not more honoured by having been invited to write – especially insofar as the volume favorably profiles my own work. Additionally, *Philosophy of Technology* deepens especially my appreciation for the work of authors Verkerk, van der Stoep, Hoogland, and de Vries. Their insightful contributions to critical philosophical reflection on technology fully deserve the wider audience this publication will promote.

PREFACE

The existing introductions to philosophy of technology are either collections of independent essays or 'classic' texts or they deal with only a limited part of this fascinating and ever-expanding subject field. With this book we are attempting to fill this void. In this introduction we have deliberately tried to establish a connection between classical philosophy of technology (Heidegger, Ellul), the analytical tendency within philosophy of technology (Simondon, *Dual nature*-approach) and the tendency of the social-empirical reversal (Bijker, Latour). In our choice of subjects we have set it as our goal to deal with as many subfields of philosophy of technology as possible, like the analysis of technological artefacts, the status of technological knowledge, the methodology of designing, the organisation of manufacture, the analysis of technological practices, the anthropology of technology, the ethics of technology and also the relationship between technology and religion. In addition we have also dealt with subject fields which traditionally do not belong to philosophy of technology but which are cognate and are to a great extent relevant for people working in technology daily, such as the positioning of engineers, the ethics of organisation and the sociology of post-industrial society. Moreover, this book differs from other books in that consideration is intentionally devoted to the way human beings intervene not only in their physical but also in their biotic environment: agriculture, nature conservation, bioengineering, medical technology.

The book was born in practice. Three of the four authors lecture in philosophy of technology at one of the universities of technology in the Netherlands. The fourth author, too, has conducted varied research projects in this field. Yet this book intends to be more than a textbook. It offers an introduction to anybody who wants to dig down deeply into the subject field of the philosophy of technology later. In addition it is a reference book for people who have some acquaintance with philosophy of technology but who do not yet have an overview of the whole field or are on the lookout for an outline of this field. Furthermore this book is intended to link up directly with everyday engineering practice. For this reason much room has been devoted to practical examples and case studies.

In the book our philosophical orientation is on a Dutch 'school' of which the foundations were laid by Herman Dooyeweerd and his ideas were applied to technology by Hendrik van Riessen. Later Egbert Schuurman took up their line of thinking and worked it out into a critical perspective on technology in which the normativity that is inherent in technology plays an important role. The book deals with all existing 'schools' in the philosophy of technology, but the way in which we have arranged the material bears the marks of their work.

Writing a book with four authors is a complicated matter. It entails much coordination and harmonisation. But most of all it entails intensive discussions on the diverging approaches in philosophy of technology and its relevance to technological practice. The past two years we met regularly in the station restaurant at Eindhoven. In particular it was a fascinating and enriching undertaking because each one of us is working within philosophy of technology from a different background so that we were able to hone one another with examples from, for example, the history of technology, the practice of designing, the practice of manufacture, medical practice and the world of agriculture.

This book was originally published in Dutch by Boom publishers in Amsterdam. We thank Boom's Wouter van Gils for his tremendous input and involvement and Niels Cornelissen and Cyril Lansink for editing the text and asking numerous critical questions. We thank Marietjie Nelson for the excellent cooperation during the translation of this book and Rika Fliek for her translation advice. For this English edition, we want to thank Routledge's Amy Laurens for her confidence in the quality of the text. We also want to thank the two anonymous reviewers for their positive response and their valuable comments for revision. We want to thank Nicola Cupit for guiding us through the process of preparing the manuscript for Routledge and all the other Routledge people who worked hard to turn our manuscript into the book that it is now. Finally, we want to thank Carl Mitcham for writing a Foreword for the book. Given his status in the field, this was a great encouragement for us.

The thinking has been done, the book has been designed and finished. Now it is up to the reader! We hope that reading this book will lead to many people looking at the multi-coloured world of technology with new insights and new enthusiasm!

Eindhoven, January 2015
Maarten J. Verkerk
Jan Hoogland
Jan van der Stoep
Marc J. de Vries

PART I

Thinking and making

1

THINKING AND TECHNOLOGY

Between analysis and criticism

Summary

Philosophy of technology is a subject in which one reflects on technological thinking and acting with regard to technology. Philosophy regarding technology has three functions. The analytical function means that one attempts to make good definitions of concepts and in this way to create a conceptual framework. The critical function is directed at having a discussion on the issue whether the working of technology is harmful or beneficial. The directional function attempts to determine what would be a good development of technology. Philosophy of technology is a relatively new subject field. In the past decades four themes have emerged: technology as artefacts, as knowledge, as processes and as a part of our being human.

Engineers are often practically minded people. After all, they are busy with things that are focused on everyday practice. They make things, maintain and repair technical apparatus, and design sophisticated systems. To be sure, technology is not merely a matter of dexterous hands, but also a matter of knowledge. One can study technology. But often that knowledge is of a practical nature and focused directly on its application in daily life. To most people philosophy is the direct opposite. It sometimes seems as if there is nothing practical at all in philosophy. This is a subject field generally regarded as a kind of unworldly activity. Philosophers ask bizarre questions, like: What is 'being'?, What is 'knowing'?, What is 'time'?, What is 'truth'? and What is 'reality'? These questions sound particularly strange since the answers seem to be so obvious. Do we not all know the meaning of 'being', 'knowing', 'time', 'truth' and 'reality'? What is the point in asking these questions? And, moreover, engineers think: how can the answer to such questions help me in my practical technical work? Thom Morris's book *Philosophy for Dummies* (1999) was intended for a wide audience. In it he jokingly quotes Voltaire who is poking fun at himself while examining the (apparently) not so practical focus of philosophers. 'If the listener does not comprehend the intention of the speaker, and if the speaker himself does not comprehend his intention, then you are dealing with philosophy' (Morris, 1999: 14).

Indeed, the question of practical usefulness is not the most obvious one when philosophy is concerned. One might ask oneself whether the question of practical usefulness should indeed always be raised. For there are many areas in life where one would not ask such a question. Imagine, if in the choice of a partner one were led by the question of practical usefulness. However, these apparently very impractical questions become more relevant for engineers when we define them more closely. Thus one can, for instance, ask: What actually *is* a technical thing (which then really is the question of being, applied to technology)? When does one call something 'natural' and when does one call it 'technical'? Or: What is technical knowledge and in what respects does it differ from scientific knowledge (or the question of knowing, but then applied to technology)? Is technology applied physical science or is it something different? For engineers who consciously strive to contribute to a healthy society, a philosophical reflection on technology is particularly useful, for it can help them to define their own technical thinking and conduct. Therefore, engineers would also have to count among the *'reflective practitioners'* as Donald Schön (1983) calls them.

In this chapter we will be looking at three ways in which philosophy can help them in their reflection. Subsequently, we give consideration to the development of philosophy of technology and the themes which it lays on the table. We also discuss the specific perspective from which this book was written.

1.1 Functions of philosophy

How can philosophical reflection be useful to an engineer? To answer this we have to differentiate between three functions of philosophy. In the first instance, philosophy has an analytical function. Philosophy helps with the development of conceptual frames. Often discussions are fruitless because there is no proper consensus on the meaning of the terms that are used. A distinct example in technology is the discussion about whether technology is or is not applied physical science. Supporters of the statement that technology is nothing more than applied physical science point out examples like the laser and the transistor which were indeed more or less direct consequences of scientific research. A somewhat older example, the atom bomb, was the stimulus for Vannevar Bush, scientific advisor to President Roosevelt, after the end of the Second World War, to write a report, entitled *Science: the Endless Frontier* in which it was recommended that the technology policy of the government be geared to basic research only since it would always in one way or another lead to industrial applications. The report had great influence. Companies like Bell in the USA and Philips in the Netherlands tailored their research policies to it and stimulated the kind of research advised by Vannevar Bush. However, over the following two decades they observed that this strategy did not always work.

Philips had frustrating experiences with its work on hot air engines. Although the researchers consistently made progress in capturing the behaviour of the hot gas in computer models, they did not succeed in making this type of engine commercially viable.

Opponents of the statement that technology is applied physical science would point out that the history of these engines shows clearly that Robert Stirling designed them before the thermodynamic theories on the behaviour of hot gases had been developed. Stirling, who was by no means an engineer, but a church minister, for the greater part worked by intuition. He used the idea of heat as a kind of fluid that has long since been discarded. The answer to the question whether technology is or is not applied physical science therefore depends very much on what one understands by the term technology. If one presupposes that technology is characterised by developments like the laser, transistor and atom bomb, one might conclude that technology indeed is applied physical science. But if one supposes, rather, that technology is characterized by developments like a hot air engine, one comes to the opposite conclusion. Clarity on the meaning of the concept 'technology' therefore is essential for having a fruitful discussion on the question of whether technology is or is not applied physical science.

The above example is used to indicate that clarity on concepts is important in allowing a good discussion. We want to emphasise that it is more than a game with language. There are philosophers who are of the opinion that the analytical function of philosophy is exclusively a matter of language. Analysis would then only concern the reality of the language and not the reality represented by means of language. According to these philosophers what we mean by a speech act is nothing more than an agreement between human beings. Speech acts do have meanings but according to these philosophers these are not to be found in their reference to an underlying reality. We, however, are convinced that the primary goal of language is to enable us to speak about reality. Hence with regard to concepts one can indeed raise the question of whether they give an adequate representation of reality.

A second role of philosophy is to look critically at reality. In the philosophy of technology, this function emerges in the critical analysis of the role that technology has played, still plays and will have to play in culture and society. In this critical analysis value judgements are expressed that encompass the total development of science and technology. The main issue is whether technology constitutes a threat to human beings or actually contributes to their well-being. The critical function of philosophy is connected to its analytical function. If it is set up effectively the critical view uses the conceptual framework developed by means of the analytical function. Without analysis no satisfactory criticism is possible. But there is also a dependency that works the other way round: establishing a good conceptual framework becomes more or less meaningless if that framework is never used for having a good critical discussion. Thus, in critical views on technology a *horror scenario* is sometimes sketched in which technology as an autonomous phenomenon jeopardises the freedom of Western humanity. Technology then gets the same function as 'big brother' in the novel *1984* by George Orwell that to a great extent determines the life of individual citizens. If such a view is based on a hazy understanding of what technology really is, it could be equally biased as the very positive view that sees only blessings in technology. This view is also found in philosophy of technology. On the one hand we can only evaluate such views if we approach the nature and character of modern technology analytically. On the other hand this kind of view makes us more sensitive to possible blind spots in our analytical investigation of technology.

A third function of philosophy is to point direction. A critical analysis of culture and society automatically leads to the question 'Well, then how …' (should it be seen)? The directional function of philosophy is of great significance for technology. It concerns questions like 'What is a good apparatus?', 'What is the influence of technology on human beings and society?', 'When may we apply a certain technique and when not?', 'What are the effects of new techniques on the environment?', and 'To what extent can technology solve problems in our society?' All these types of questions are encountered in the ethics of technology. The directional function of philosophy is closely linked to the analytical and critical functions. The ethics of technology can only be properly developed on the basis of the analysis of technology and in dialogue with the criticism of technology. Thus, the ethics of technology will also have to consider seriously the question to what extent modern technology endangers the freedom of Western humanity. It can only have a meaningful reflection by extensively analysing modern technology and identifying the (possible) social implications in a critical manner. In this way only will the ethics of technology be able to express itself normatively on the development of modern technology.

Technology has long been a neglected topic in philosophy. And that is strange when one realises how important technology is in our culture. Perhaps this is because most philosophers did not pursue any technical studies (and still do not). Or perhaps it is a remnant of the old platonic idea that science as a spiritual activity is much more important than the crude technical activity of engineers. Whatever the case, it is only since the late twentieth century that philosophy of technology has been spoken of. The fact that most philosophers who expressed themselves on technology had no technical background could certainly be detected in their ideas. They often spoke in very general terms about technology without really sensing that there are significant differences between the various technical disciplines. An architect is a completely different kind of engineer to a mechanical engineer or a chemical engineer. Therefore, in more recent philosophy of technology increasing use is made of historical and sociological case studies in order to understand the practice of various technologies before making generalisations about technology.

Until recently, literature in the subject field was almost exclusively written by German and French authors who focused mostly on the critical function of philosophy. In this regard we speak of 'Continental' philosophers since they lived and worked mostly on the European continent. This approach led to very general statements about technology which often did not see much good in technology.

Technology was assumed to be the cause of a deprived experience of reality, of an exploitation of nature, of all kinds of environmental problems, of 'ordinary men and women' experiencing a strong feeling of powerlessness and of being technically dominated. Other 'Continental' authors have highlighted the blessings of modern technology. They thought – often without any critical notes – that every problem in society could be solved with the help of technology. The latter view probably had even greater influence in practice than the more critically focused view. In Part III of this book a number of these philosophers and their ideas are reviewed. This form of philosophy of technology was mostly practised by people who themselves stood outside of technology. In a book that has become a classic, *Thinking Through Technology* (1994), Carl Mitcham, an American philosopher of technology, refers to this as a 'humanities philosophy of technology' (Mitcham, 1994: 39 et seq.).

The second approach, practising the analytical function of philosophy with regard to technology has only recently begun and philosophers of technology have started concerning themselves with questions like 'What do we actually mean when we say "technology" or "technical artefact", or "technical knowledge"?' These are concepts used by engineers, and their exact meanings are not always clear. So, for instance, much is said and written about knowledge management in technological industries but what exactly is understood by 'knowledge' often is not clear. This can lead to misunderstanding and confusion, for instance when a certain view of knowledge leads to too high expectations regarding the possible effects of knowledge management. A thorough reflection on what we understand by knowledge can then lead to a meaningful analysis of knowledge management. This is merely an example of the usefulness of the analytical function of philosophy and this is often practised by people who have also studied technology themselves. So Mitcham calls it an 'engineering philosophy of technology' (Mitcham, 1994: 19 et seq.).

1.2 Growing a philosophy of technology

How philosophy of technology has grown over time can be illustrated in a concise overview of some important persons in the history of this discipline. It is difficult to know where such an overview should start. In any case, the first philosopher to use the expression 'Philosophy of technology' was the German Ernst Kapp (1808–1896). In 1877 he published a book with the title *Grundlinien einer Philosophie der Technik* (Basics of a Philosophy of Technology). In that book he claims that one could view technology as a complement to our human body. By means of technological aids we support and strengthen the functioning of our human limbs and organs. A pair of spectacles means a support for our eyes, while binoculars are a strengthening thereof. Pots and pans are extensions of our hands in respect of what objects we can hold. So are a knife and an axe, but then with respect to the ability of our hands to separate things. In his approach Kapp also makes use of metaphors. Thus he viewed the railway network as an externalisation of our blood circulatory system and the telegraph as an extension of our nervous system. Kapp was inspired by the German philosopher Georg W.F. Hegel. So was Karl Marx. And this is the one person we can definitely not omit among philosophers who reflected on technology. Marx had high expectations that technology would help in realising a classless society. If technology could come into the hands of the workers, an ideal society could come into being. Kapp and Marx were both philosophers without a technological background. However, a few decades later there also were engineers who began using the term 'philosophy of technology'. One of them was the Russian Peter K. Engelmeier (1855–1941). In 1911 he published an article for a philosophy conference entitled *Philosophie der Techniek* (Philosophy of technology). In this article he emphasised the role of the human will in technology. For a while he would be one of the few engineers to help in establishing the philosophy of technology. Another well-known name from the initial years of philosophy of technology was Friedrich Dessauer, another German. He published several books reflecting on technology: *Technische Kultur?* (Technological culture?) in 1908, *Philosophie der Techniek* (Philosophy of technology) in 1927, *Seele im Bannkreis der Technik* (The soul under the spell of technology) in 1945 and *Streit um die Technik* (Battle about technology) in 1956. From some of these titles one senses that he was not altogether happy with the role of technology in our culture. Dessauer sees technology as a high vocation, but also as a danger to society when work is done with the wrong motives.

In France, too, philosophy of technology has started developing. One of the first persons involved in this was the philosopher Gilbert Simondon (1923–1989). In his book *Du mode d'existence des objets techniques* (About the nature of existence of technical objects) of 1958 he gives an analysis of the nature of technical artefacts. With hindsight it is a noteworthy book since for a long time thereafter no further philosophical analyses of technical artefacts would appear. Together with the Dutch reformational philosopher Hendrik van Riessen, to whom we will return later, Simondon in that respect was far ahead of his times, for the next attempt to characterise the nature of technical artefacts was the research programme *Dual Nature of Technical Artifacts* that was carried out from 2000–2004 at the Delft University of Technology.

In the meantime, philosophy of technology was dominated by culture-critical opinions which mostly emerged in the 1950s and 1960s. This was an era when a general distrust in society of the dominance of science and technology came to the fore, partly because at that time the far-reaching consequences of technology for the environment had begun to dawn on the broader public. Even before that time there had been a few philosophers making critical noises, such as the American Lewis Mumford and the Spanish philosopher José Ortega y Gasset. In his book *Technics and Civilization* (1934), Mumford gives a historical and culture-philosophical analysis of the role of technology in society. On the basis of this analysis in his two-volume work *The Myth of the Machine* (1967, 1970) he characterises modern technology as a 'megamachine' that leads to the muzzling and dehumanisation of the human being. José Ortega y Gasset in 1933 held a series of lectures on technology. He elaborated on these lectures in various writings. Ortega y Gasset emphasises that technology plays an important role in the way in which humans creatively give form to their existence in this world. He points out the danger of the perfection that characterises modern science and technology leading to a loss of people's creative capacity.

Among the philosophers who wrote critically about technology in the 1950s and 1960s were the Frenchman Jacques Ellul and the German Martin Heidegger. Ellul in his books (amongst others in *La technique*, 1954) emphasises that technology has taken on the character of an autonomous system which implies that there remains hardly any possibility of its being directed by society. Heidegger warns against the danger that technology leads to reality merely being seen as something that has to be processed in order to meet our needs, and that we no longer see the value of reality as such. In this way we are estranged from reality. Like Ellul he hardly sees a way out. One of his last remarks about the almost inevitable consequences of technology is that 'only a God' can still save us. By this Heidegger did not mean that according to him the Christian faith offers a way out but that it is (almost) impossible to escape the rule of technology.

In the midst of this flood of cultural criticism of technology the person who did point at the way out offered by the Christian faith was Egbert Schuurman, a reformational philosopher. A group of philosophers representing a different trend from the 1950s and 1960s and later was made up by the neo-Marxist *Frankfurter Schule* where names like Marcuse, Habermas, Horkheimer and Adorno are to be found. Karl Marx's idea that history would end up in a classless utopia was not taken over by these philosophers, but they still found a distinct role for technology in effecting the social reversal. Different representatives of the *Frankfurter Schule* however, also pointed out the dangers of technology as a potential means of reinforcing the *status quo*.

Giving a good description of technology is almost impossible without saying something about the relations between science and technology. One of the first philosophers to reflect on this was the Argentinian Canadian Mario Bunge who in 1966 published an article entitled 'Toward a Philosophy of Technology' in the journal *Technology and Culture* which is a magazine for the history of technology. He works with a very broad concept of technology and he sees a great role for science in giving direction to technology. In the course of the history of philosophy of technology the debate would be precisely about the idea that technology is an independent activity and is not completely dependent on science.

Apart from philosophers there also are organisations that contributed to the development of the philosophy of technology. Two especially deserve mention in this overview. The *Verein Deutscher Ingenieure* (Society of German Engineers), established in 1947, had a division occupying itself with philosophy of technology from as long ago as 1960. Almost all well-known names in the German philosophy of technology of the Second World War belong to this division: Alois Huning, Hans Lenk, Simon Moser, Friedrich Rapp, Günther Ropohl, Hans Sachsse, Klaus Tüchel and Walther Christoph Zimmerli. Rapp was one of the few among them who practised an analytical philosophy of technology (he became known in particular for his book *Analytische Technikphilosophie* (Analytical philosophy of technology) published in 1978). The second organisation is the *Society for Philosophy of Technology*, established towards the end of the 1990s. The core of this organisation was located in the USA, where the philosophy of technology was dominated by pragmatists like Paul T. Durbin and Larry Hickman, phenomenologists like Don Ihde and Alberg Borgmann, and Andrew Feenberg, who initially was particularly oriented towards the *Frankfurter Schule*.

So we see how philosophy of technology has developed from an initial approach within philosophy that was interested in analysis to a trend in which quite soon the emphasis was laid on culture-critical views on technology. Over the last ten years there has been a renewed quest for philosophical analysis of concepts related to technology. We also note that philosophy of technology was initially more of a German–French matter and that more recently interest in the subject has also been increasing in the USA.

1.3 A classification into focus areas

Although systematic consideration of technology in philosophy is therefore still something recent (about a century old), a considerable number of books and articles have been published on it. In his book *Thinking Through Technology*, Carl Mitcham gives a more extensive overview of the development in philosophy of technology up to the present. He structures that overview by using four ways in which to look at technology. These four ways are related to the most important divisions of philosophy. In Part II of this book we will make further use of his categorisation.

In the first instance one can see technology as a collection of products and this is the way many people speak about technology. If you ask a fourteen-year-old girl or boy what technology is, s/he will at once start naming all kinds of products. Journalists, too, regularly use the word 'technology' in this way. But in philosophy of technology consideration is given to the question: What actually is a technical artefact (product)? Here one is concerned with the question of 'being': ontology is the subfield of philosophy that occupies itself with this in particular. Carl Mitcham shows that discussions in this field are mainly concerned with the question of how one can categorise technical artefacts according to their nature, and how one can distinguish them from non-technical artefacts, like objects of art. So, for instance, he discusses the difference between a tool, a machine and an automatic device. One important element in the discussion is the understanding that artefacts have a social dimension.

A second view, according to Mitcham, is technology as a field of knowledge. Technology is something one can study. If one considers it from a philosophical angle, one finds oneself in the subfield of epistemology (the theory of knowledge). Mitcham raises as the main point of discussion the question in which respect technological knowledge differs from scientific knowledge. On reading the relevant chapter in Mitcham's book one notices that it is even more concerned with the question in which respect technology differs from science; little is said about knowledge. But consideration is given to the way in which knowledge comes into being and that is the domain of the philosophy of science. Still, philosophy of science focuses on scientific knowledge and the question is whether all forms of technical knowledge should be called scientific. Engineers sometimes know something intuitively which causes a philosopher of science to hesitate to accept it readily as scientific knowledge. Yet that sort of knowledge often is very important to engineers. Discussions on this issue are still at an early stage. In this book we will attempt to say some more about this. A second point of discussion that Mitcham mentions in relation to technology as a field of knowledge, is that of *knowhow*. To what extent can technical skills also be called knowledge? And how does *knowhow* come into being? The discussion on this is also in its early stages.

The third way in which, according to Mitcham, we can think and speak about technology is as a collection of activities and processes. In technology a great number of things are done: designing, making, using, assessing. In philosophy the subfield that relates to such issues is methodology. In philosophy methodology is much more than the 'theory of methods'. Literally it is the study of 'the way in which' (in Greek: *meta hodos*) things come into being. These can be beautifully paved roads but also tortuous, bumpy roads. Every experienced engineer knows examples of projects that went off smoothly but not because of a beautifully paved way (a method). Methodology also occupies itself with these less beautiful ways. One of the most important processes in technology is design, which is the subject of the methodology of design. Initially this was mostly practised by designers themselves who wanted to reflect systematically on their work and thus learn how it could be improved. Gradually non-designers also, among whom there were many philosophers, started applying themselves to this. Another important process in technology is *production*. Originally this was the field of the tradesman. Later, engineers also became involved in the field, and then began the division of labour within the production process.

Finally, there is technology as an aspect of our being human, as something related to our will. Mitcham shows that there is a whole range of 'forms of the will'. Human beings have a will to survive, a will to control, a will to be free, etc. All of this concerns the values we employ while being technically busy. Fundamentally it concerns our view of the world and reality. Metaphysics is the philosophical subfield that focuses on this, and ethics and aesthetics are the subfields in which values take a central position (values such as 'good' and 'bad', or 'beautiful' and 'ugly'). Most of what has been written up to the present on technology as a part of being human was written in the framework of this last-mentioned approach of Mitcham. It was mainly the philosophers in the German–French tradition who were engaged in this. The first three of the approaches discussed above are still more or less virgin territories, and it is mostly the philosophers of technology interested in the analytical function of philosophy who have been occupying themselves with these. In this respect the development of philosophy of technology is lopsided, at least in the light of the four approaches described by Mitcham.

For that matter, Mitcham's four approaches and the most important sub-disciplines of philosophy do not tally completely. Technology as a collection of technical artefacts is not only found in ontological questions but also in our reflection on technical knowledge. For this knowledge pertains mainly to the features of technical artefacts. Neither do we reflect on design processes only in the framework of methodology. Here, too, epistemological questions are significant because designing not only renders products but also knowledge. The links we have laid between Mitcham's four approaches and the philosophical sub-disciplines should be taken in a flexible way so as not to miss all kinds of interesting questions.

In conjunction with the new attention that philosophers interested in analysis are giving to technology, a so-called 'empirical turn' has taken place in the philosophy of technology.[1] By this is meant that increased consideration is given to empirical studies of technological developments as a source of inspiration for philosophical reflections. Contemporary philosophers of technology, who are mainly occupied with the analytical function, generally find that culture-critical philosophers of technology tend to make very broad statements on technology without first considering carefully whether technology in all its variety actually does answer these statements. In order to reach a more differentiated argument that does better justice to the fact that all technologies are not alike, philosophers of technology have begun to pay careful attention to what has been written on technological advances by historians and sociologists. It is true that these writings cannot be used as a basis for normative, culture-critical statements, but they can indeed contribute to our knowledgeability when we make culture-critical statements on technology. This could spare us from statements on technology that are too bold and general.

1.4 Structure of this book

Chapter 2 ends Part I with an exploration of the relation between technology, cohesion and disclosure. We show that technology can be defined as a search for sense. Part II of this book deals in particular with the analytical approach in philosophy of technology, Part III with the critical approach. A connection between the analytical and the critical function assists us in developing a language and conceptual framework in which the critical reflection on technology can be practised productively. In Part III, the directional function of philosophy will also be dealt with in detail. Philosophical analysis takes place in close connection with the practice of technology. We develop our view in the various chapters with reference to concrete examples. In addition, four case studies of technologies and their relation to society are discussed on the basis of empirical data. In this the book joins up with what we called above the 'empirical turn' in philosophy of technology. In this way philosophy of technology takes on a concrete form. Creativity plays a huge role in this. Philosophy not only helps reaching an appraisal of technology with reference to an analysis of concepts but also to pointing out alternative routes. Actually in this respect the work of philosophers links up closely with the work of engineers. In her book *Ethics in Engineering Practice and Research* (1998) Caroline Whitbeck also indicated this in the framework of the ethics of engineering. According to her, ethicists should actually go about their work like engineers.

Instead of merely analysing what the elements of an ethical dilemma are and pointing out the possible choices, they should reflect in a *creative* way on the possible ways of breaking out of that dilemma. She is the first to see an ethical problem as a problem of *design*. Engineers, too, often encounter dilemmas in their business of designing. Like a new iron which should in fact be both light and heavy: light to be easy to pick up, and heavy to press down properly on the fabric. Instead of choosing exclusively between two options or looking for an average as a compromise, engineers try to break out of the dilemma in a creative manner. In this way the steam iron as a household device came into being. Without extra weight from the iron the fabric does become smoother due to the effect of the steam. A similar directional function is found in philosophy of technology when it concerns the issue of how we should handle technology or contribute to its development. That can entail more than mere analysis and critique. We can also reflect on creative methods of breaking out of a dilemma.

Note

1 The term 'empirical turn' is used in various ways. Earlier in the history of philosophy people also spoke of an empirical turn, namely in the philosophy of science (see Chapter 10). That implied that people started using empirical studies, especially sociological and historical ones. The 'turn' caused scientific knowledge to no longer be seen as something flowing directly from observations, but as something that was determined in particular by social actors. The empirical turn mentioned in this chapter, likewise means the use of empirical studies, but in this case the 'turn' focuses particularly on a deeper insight into the nature and character of technology. That has not (yet) led to such a pronounced view on the role of social actors. It has formerly demonstrated that the influence of social actors is limited by what is physically and technically possible.

References

Bush, V. (1945) *Science The Endless Frontier*. A Report to the President by Vannevar Bush, Director of the Office of Scientific Research and Development, July, Washington, DC: United States Government Printing Office.

Dessauer, F. (1908) *Technische Kultur?* (Technological culture), Kempten and Munich: Kösel.

Dessauer, F. (1927) *Philosophie der Techniek* (Philosophy of technology), Bonn: Cohen.

Dessauer, F. (1945) *Seele im Bannkreis der Technik* (The soul under the spell of technology), Freiburg i. Breisgau: Otto Walter AG Olten.

Dessauer, F. (1956) *Streit um die Technik* (Battle about technology), Frankfurt am Main: Knecht.

Mitcham, C. (1994) *Thinking Through Technology. The Path between Engineering and Philosophy*, Chicago, IL: Chicago University Press.

Morris, T. (1999) *Philosophy for Dummies*, New York: Wiley.

Schön, D.A. (1983) *The Reflective Practitioner*, New York: Basic Books.

Further reading

Fearn, N. (2005) *Philosophy. The Latest Answers to the Oldest Questions*, New York: Grove Press.

Hospers, J. (1996) *An Introduction to Philosophical Analysis*, 4th edn, London: Prentice Hall.
Kroes, P.A. and Meijers, A. (eds) (2000) *The Empirical Turn in the Philosophy of Technology*, Amsterdam: Elsevier.
Scharff, R.C. and Dusek, V. (eds) (2002) *Philosophy of Technology. The Technical Condition. An Anthology*, London: Blackwell Publishers.
Schuurman, E. (2009) *Technology and the Future. A Philosophical Challenge*, Grand Rapids, MI: Paideia Press. Ttranslation of *Techniek en Toekomst. Confrontatie met wijsgerige beschouwingen* (1972).
Scruton, R. (1996) *Modern Philosophy. An Introduction and Survey*, New York: Penguin Books.
Staudenmaier, J. (1985) *Technology's Storytellers. Reweaving the Human Fabric*, London: MIT Press.
Vries, M.J. de (2005) *Teaching About Technology. An Introduction to Philosophy of Technology for Non-philosophers*, Dordrecht: Springer.

CARL MITCHAM

(1941)

PORTRAIT 1 Carl Mitcham

Carl Mitcham will probably always be best known for his overview *Thinking Through Technology* (1994) on the development of philosophy of technology. Since 1999 he has been lecturing at the Colorado School of Mines in the USA. Before that he was a professor at Brooklyn Polytechnic University and Pennsylvania State University. Mitcham studied philosophy at the University of Colorado and Fordham University and was the editor of several series of books on philosophy of technology, in particular the series *Philosophy and Technology*. More recently he edited the *Encyclopaedia of Science, Technology and Ethics* (2005). This is the first multivolume encyclopaedia on philosophy of science and philosophy of technology. Mitcham is regarded as one of the most prominent living philosophers of technology.

Although Mitcham did not develop a philosophy of technology of his own he has had great significance for the subject by creating an overview of what many others do and have done. In his book *Thinking Through Technology* he demonstrates that in practice there are two kinds of philosophy of technology (see also Chapter 1): one that looks at the consequences of technology as if from the outside (practised mostly by philosophers without a technological background) and another that attempts more from the inside to clarify what we should understand by technology (practised only by philosophers with some technological training). Furthermore he shows that technology can be described in four ways: as a collection of artefacts and systems, as a field of knowledge, as a process (for instance designing or making) and as something belonging to our being human (in particular to our will).

Mitcham has always shown particular interest in ethical issues in technology and in the relation between technology and religion. In 1984 he worked with Jim Grote as they jointly edited a volume of essays titled *Theology and Technology: Essays in Christian Analysis and Exegesis*. He is currently working on a new book on religion and technology. Mitcham is of Roman Catholic origin but has developed a special interest in Buddhism. According to him, Buddhism, as a complement to a Christian approach, offers useful gateways to the development of peaceful and lasting technology.

Finally here are some characteristic quotations:

> My thesis, in Aristotelian language, is that 'technology' is not a univocal term; it does not mean exactly the same thing in all contexts. It is often, and in significant ways, context-dependent – both in speech and in the world. But neither is it an equivocal like 'date' (on a calendar and on a tree). There is a primacy of reference to the making of material artifacts especially as this making has been modified and influenced by modern natural science, and from this is derived a loose, analogous set of other references. An initial need in the philosophy of technology is for clarification of this conceptual one and many, a conceptual one and many which evidently exists because it reflects a real diversity of technologies with various interrelationships and levels of unity.
>
> *(Mitcham, 1978: 230)*

> One historical complication in the birth of the philosophy of technology is that not only was it somewhat overdue, it was not even the outgrowth of a single conception. The philosophy of technology gestated as fraternal twins exhibiting sibling rivalry even in the womb. 'Philosophy of technology' can mean two quite different things. When 'of technology' is taken as a subjective genitive, indicating the subject or agent, philosophy of technology is an

attempt by technologists or engineers to elaborate a technological philosophy. When 'of technology' is taken as an objective genitive, indicating a theme being dealt with, then philosophy of technology refers to an effort by scholars from the humanities, especially philosophers, to take technology seriously as a theme for disciplined reflection. The first child tends to be more pro-technology and analytic, the second somewhat more critical and interpretative. Before trying to decide which is more closely affiliated with philosophy itself, it is appropriate simply to observe some differences in character.

(Mitcham, 1994: 17)

References

Mitcham, C. (1978) 'Types of technology', *Research in Philosophy & Technology*, 1: 229–294.

Mitcham, C. (1994) *Thinking Through Technology*, Chicago, IL: University of Chicago Press.

Mitcham, C. (ed.) (2005) *Encyclopedia of Science, Technology, and Ethics*, Detroit, MI: Thomson & Gale.

Mitcham, C. and Grote, J. (1989) *Theology and Technology*, Lanham, MD: University of America Press.

2

SPEAKING IN A TWO-SIDED WAY

The meaning of disclosure and the disclosure of meaning

Summary

In this chapter we discuss the concept of 'meaning'. Technology is a part of culture. By culture we understand everything brought about by human activity. In this technology plays an important part. For many things made by human beings are created with the aid of technical means. Acting is concerned with meaning: in their activity people unlock the meaning of what they encounter. In this chapter a distinction is made between 'meaning' as a concept referring to the relationships in everyday experience on the one hand and 'meaning' as a metaphysical or worldviewish concept that refers to the foundation of being or the essence of life. Eventually the conclusion emerges that both concepts are mutually connected. Human activity is subject to the directional or regulatory guidance of worldviewish ideas about reality.

We begin this chapter with the introduction of the concept 'meaning'. Subsequently we proceed with a closer analysis of this concept and illustrate that there are two sides to it. On the one hand 'meaning' is a given: human beings interpret the 'meaning' of things. On the other hand 'meaning' has to be unlocked by human activity. Thus both passivity and activity come into play. On this basis we describe how the terms 'meaning', 'activity', 'culture', 'history' and 'technology' are mutually connected. This chapter concludes with a view on the 'metaphysical' or 'worldviewish' concept of meaning.

2.1 Introduction: meaning as a question

If one asks a random number of people on the street what philosophy is, there is a great chance that at least some of them will say that philosophy is concerned with questions of meaning, in particular the meaning of life. Philosophy and meaning have a lot in common. This also applies to the philosophy of technology. Anyone who raises the question of technology for a philosophical discussion can hardly circumvent the question about the *meaning of technology*. Yet it is to be doubted whether there are many philosophers who would phrase it thus. For the concept of *meaning* is extremely abstract. What are we talking about when we talk about *meaning*? Let us speak in somewhat more concrete terms by referring to an example.

When using a sophisticated electronic apparatus one does not always grasp straightaway how it works. When *Personal Computers* first came into use (in the 1980s) there was much uncertainty attached to working with them. No user-friendly interface like Windows was available yet. Moreover, the use of such an apparatus was not common at that stage. Nowadays even a young child knows what can be done on a computer. Initially this was different. If you had bought such a thing and were a novice in the field of computers it was the beginning of a long and tiresome voyage of discovery. You were the owner of a kind of box of synthetic fibre or metal (the actual computer), a monitor and a keyboard. In the box there were disc drives (5¼ inch, soft floppys; see Figure 2.1), the processor and sometimes – for people who could pay more – a hard drive. When you switched on the computer you did not see much more than what was called the 'prompt': a

FIGURE 2.1 5¼ inch floppy diskette

black screen and in the top line: C:\>. Following this, one had to type commands to which the *Disk Operating System* then responded. One learned that if one had particular programmes at one's disposal one could also type after the prompt the command for starting that one specific programme. Only after having done all this could one really start doing something useful with the newly acquired apparatus. One can imagine that real novices sometimes must have felt quite desperate as they sat behind their sophisticated apparatus not knowing how to go about doing something significant with it.

It is quite conceivable that such an apparatus might engender a feeling of dissatisfaction, helplessness or even estrangement. You do not know how it operates. You do not understand its function. You ask yourself what you can do with it. In short, you cannot find a *meaningful place* for it in your life with the possible result that you yourself begin feeling unhappy. It is as if you are confronted with a confusing situation: the experience that there are specific matters you cannot fit into the self-evident coherence of meaning of your life. Usually this problem can be easily overcome by familiarising yourself with the world of computers. However, it is different for someone who lacks the ability or the self-confidence to learn to master a computer. For such a person is increasingly confronted with an environment in which competence in working a computer has become more and more a matter of course. This might cause such a person to develop a sense of no longer being able to keep up with the world around him or her. This could lead to a deeper crisis: the feeling of being excluded or of not counting. By this example we want to illustrate that there actually is an important difference between the questions regarding the function and the working of things on the one hand and questions of meaning (questions regarding the meaning of one's existence) on the other hand, but that at the same time there also is a kind of relation between the two. To our minds the reality in which we live forms a coherence of meaning that can be disturbed on various levels with minor or major consequences.

Generally one is confronted with questions of meaning particularly when the self-evidence of meaning falls away. Here are some diverse events at which the 'why-question' presents itself. For instance, you have to catch a train, get on your bicycle to ride to the station and then observe that you have a flat tyre. Your first reaction is 'Why should this happen to me?' This is a matter of a relatively minor event. It is something completely different when you lose a loved one through illness or an accident. Or when you hear of enormous disasters, like the tsunami on Christmas Day 2004. Asking the question as to the meaning in such situations is seen by many people as meaningless. For it is something that happens to one outside of one's will: illness, a natural disaster. Others will indeed ask the question even though they cannot answer it (adequately). For instance, those who believe that behind the events in our experience of reality there lurks a God holding world events in his hands. This is a belief that inadequately answers the question of meaning, if at all. Much doubting of faith arises precisely from this kind of event: how can God allow such horrific events as a tsunami that cost thousands of lives?

The word 'meaning' may be an important philosophical concept but should an elaborate view of this concept be necessary in a book on philosophy of technology? Indeed, even in the context of technology the word 'meaning' is very important. Technology can be seen as a *search for meaning*. This may seem a strange and perhaps even bold statement. Is this not to allot too significant a meaning to technology? After all, technology is just a means of rendering life a little more convenient. Would people who are in search of meaning not better apply themselves to philosophy or theology, for those are the fields of study concerned with meaning? However obvious these objections may be, in this chapter we stand by our opinion that technology can be seen as a *search for meaning*.

2.2 Technology and meaning

According to commonly held prejudice technology and meaning form two separate worlds. 'Technology' belongs with the *natural sciences*. On the basis of 'hard' scientific research the latter brings to light functional relationships and regularities which are subsequently applied in technology to control nature. The concept of 'meaning', however, is placed in the *humanities*. This discipline is not concerned with hard scientific research but for instance with the interpretation of old texts or foreign cultures. The results of research in humanities are therefore often 'less hard' than those in the natural sciences.

One important reason for doing away with this prejudice is the fact that technology itself forms part of culture. It is true that many technological developments are based on scientific research but technology is something that is brought about entirely by human beings. Behind technology lie the intentions of people and interpretations of reality. In other words, there is much *meaning* in technology. A second question raised by this prejudice regards the purpose of technology itself. Why do people actually develop technology? Why is so much money invested in *research and development?* Very broadly one could answer: to make the world more predictable, more controllable, safer or more comfortable. Or in somewhat more abstract terms: to make the world more meaningful. Here is an example. We spoke about the tsunami of 2004, a terrible disaster. And a meaningless event: more than 225,000 people lost their lives. It would be fantastic if people could succeed in installing an alarm system in the ocean that could warn in time of a threatening tsunami. It would then be possible to prevent so many people from dying in a flood. It could make life in vulnerable coastal regions a lot more predictable and safer. This increase in predictability and accompanying safety of one's life can be regarded as one of the developments that make life more 'meaningful'. Not that 'safety' and 'meaningfulness' are identical concepts. But they are in this respect that where human beings have better control of their living environment their vulnerability to events that rupture the meaningful coherence of their lives is diminished.

Why then do we call technology a *search for meaning?* Is it not odd to specify technology like this? Technology is entirely concerned with the way in which human beings attempt to order and control reality with the purpose of leading a better, more satisfying existence. In other words: with the help of technology human beings attempt to gain control of the incidental facts of existence. People utilise the latent possibilities of their surrounding reality and thereby also enable themselves to make life more meaningful. Technology is very often concerned with gaining control of the coincidences and threats of daily life so that we can direct our own lives. And when subsequently we have become less vulnerable to the whims of nature by means of technology we can plan our way of life better and organise it according to the points of departure we ourselves have chosen. Possibilities are 'opened up' by technology which can unfold, develop and give more meaning to our own lives and those of others.

However, the preceding statement asks for a normative framework. When does technology contribute to the 'opening up' of possibilities hidden in reality and thus to a more meaningful reality? In this chapter we would like to start with that normative framework which will be developed further in this book.

2.3 The concept 'meaning'

Describing a normative framework demands an exact definition of the concept 'meaning'. 'Meaning' is an ambiguous concept which, amongst other things stands for 'sense' or 'reason for existence'. We speak of 'meaning', for instance, when talking about the sense of a specific word or expression: 'I use the word "line" in the meaning of a row of words from left to right on a printed page'. A second meaning is closely connected to the above, namely the word 'meaning' in the sense of use, utility, reason for existence or *telos* (goal): 'What is the meaning of disasters?' It would seem as if this is a metaphorical use of the word 'meaning' in the first sense mentioned above. In the same way one could ask about the meaning of a word or a linguistic expression, one could also ask about the meaning of an event or a specific fact.[1]

We have now differentiated between two senses of 'meaning': 1) the word 'meaning' in the more linguistic sense as an indication of the sense of symbolic expressions, the meaning of an expression; and 2) the word 'meaning' as an indication of the use, the utility, the reason for existence or even the goal of something. The latter meaning also approaches a 'metaphysical' use of the term in, for instance, the question about the meaning of suffering or the meaning of life. Although one is tempted to look for a connection between these different meanings, we will not do so here. However, we do want to point out a certain parallel between the use of the word 'meaning' in the first and the second sense.

It is clear that the word 'meaning' in the second sense is all about the human capacity to read and interpret the world and express this in language. Human beings are capable of communicating about reality by means of symbols. That is to say, human beings can express reality in speech in a conversation, dialogue or an exchange of views with other human beings. In doing so they make use of language that is intelligible to the communication partners. Much can be said about the use of language, communication and the underlying human competence. In this framework we want to highlight one element: the meaning of letters, words, sentences, expressions and speech acts are also fully concerned with a so-called 'referential coherence'. Words refer to things, to events, to acts and to one another. People who express themselves in words often want to say something about the world in which they find themselves or about the experiences they have. In such a case the signs of language and words are in a mutual relationship called 'language'. There has been much reflection on the question of how exactly this is possible and in which way human beings can give meaning to linguistic forms of expression. Language plays an extremely important part in the way in which human beings experience the world around them. Human beings cannot live otherwise than continuously giving meaning to the things, phenomena and events in their environment. Often they do not consciously do so, but instinctively. A human being cannot help but go through life reading and interpreting.

If it goes without saying (note the word!) that human beings read, interpret and talk about the world, it is not strange that they also regard reality as a *readable text.* The interpretations that human beings give to reality and express in language have to do with a kind of 'readability' of reality itself: a meaning that presents itself. In this connection we would like to refer to an expression that has become popular in the Christian tradition, namely 'the book of nature' (Jorink, 2010). Behind this expression lies the belief that nature should be seen as a creation of God. He created the world and had a certain intention with it. Everything He created has some destination. Calvin emphasises in his writings that human beings can get to know God from two sources: from the Bible as his revelation as well as from nature as his creation. The Bible was given to people because as a result of the fall human beings are no longer capable of getting to know God from creation on the basis of their own ability to know and gain insight. In the first instance they have to rely on the Bible, which is a readable message in the literal sense. The second source for getting to know God is his creation. According to this view people can acquire some knowledge of God by just looking at reality.

We are not concerned here with whether this view is justifiable or not. What we are concerned with is that in this manner of speaking 'observation of reality' is specifically compared to 'reading a book'. And in our opinion this comparison is most meaningful. For this expresses the idea that human observation of reality always instinctively amounts to the giving of meaning or reading. Someone who is reading a book does not only see black spots on a white background (the letters on the page), he sees meaningful units like letters, words, sentences, stories. And immediately – without any conscious decisions – these take on meaning in the mind and imagination of the reader. But this does not apply to reading only. It applies to everything that people observe. Human beings do not observe any meaningless '*sense-experiences*' as was once supposed. Observation instantly also becomes reading. Though this does not say that people always succeed in also reading efficiently everything that they observe. For 'reading' is a human activity which relies on fallible powers of the human being: someone can give an incorrect or faulty interpretation of the facts s/he observes.

Since we are concerned with the word 'meaning' in the first and second sense (that of words or lingual expressions and that of use, utility, reason of existence or goal) we note that it is a concept which refers to reference itself: if one asks about the meaning of something, one is asking for a context of references. Take for instance a hammer. If you do not know what a hammer is, the object says nothing to you. What you can see is that it was made (hammers do not grow in the wild) and thereby you have received a first reference, namely to its producer. And then the question spontaneously arises as to the purpose for which s/he made it. What was his/her intention or goal with it? If you ask yourself these questions, you are actually asking about *the context of meaning* in which the object 'hammer' functions. Soon it will become apparent that the hammer plays an important part in the lives of people, especially in the production of objects. This was the purpose for which the hammer was made and this function also qualifies the hammer.

A philosopher who said a lot about this 'coherence of meaning' is Martin Heidegger (see Chapter 10). He particularly tried to show in what way people move around in this context. For this is how the world presents itself to us. The *coherence of meaning* is therefore one of our everyday experiences. Normally people would not ask as many questions about a hammer as we did above. In our reality a hammer is an obvious instrument. Even young children know what it is and what one can do with it. The meaning of the hammer is fully clear in our frame of reference (the way in which we see and understand the world) and therefore hardly evokes any questions.

This obvious coherence of meaning can be ruptured in several ways. In the first place by relating directly to experience. We regularly experience reality not as a meaningful coherence. Oftentimes certain parts of reality around us cannot be read or only with difficulty. For instance when fate strikes as in the case of an accident, a natural disaster or unexpected illness. Other events can also have the same effect. Think for instance of people coming home to find that their house has been burgled and that everything is in a shambles. Often it takes time before everything is back in its place. Such happenings can shock our experience of meaning to such an extent that we temporarily lose our feeling of having a grip on reality.

We live in a reality that is far from perfect. Many things happen in our lives and in those of others that shock us and can profoundly confuse us. This also applies to matters that in themselves are 'normal', like the fact that life is finite and that every human being will die. Dying may be normal but it can have drastic effects when one has to face it in one's own life or in one's immediate environment. And even to people who see death as a 'normal' part of life, death is usually an unwelcome guest. Often death is experienced as something that disrupts the natural course of things. On being confronted with death many will even think of evil. About evil (both in the moral meaning and in the non-moral meaning) one could also say that it ruptures the coherence of meaning in our lives and – worse still – destroys it. Evil makes things aimless and meaningless. It can also shock our experience of meaning so badly that we begin to experience all of life as meaningless and aimless.[2] The power of evil can therefore be unsettling.

When we meet people, habits, customs or instruments from a culture completely different to our own, we can go through a similar experience of alienation and rupture. And to people from another culture in which technological development is not as advanced as in our culture, our technical artefacts are completely unknown. Would they understand what a computer is? Or an mp3-player? Chances are that they will be at a loss with these artefacts within the frame of reference of their everyday world of experience. Conversely we could ask the question whether we would be able to grasp without further explanation the meaning of the technical means by which they provide for their livelihood. Here is an example.

BOX 2.1: THE RADU

To illustrate the importance of the everyday world and the coherence of meaning we live in let us look at a technical artefact from a completely different culture that we will find strange at first sight and that we cannot fit into our interpretation of reality. We refer to a radu, an instrument used in both Guinea-Bissau and in the Casamance in Senegal where this object is called a kayendo (see Figure 2.2).

At first sight the radu looks like a huge oar with a metal part that resembles a horseshoe at the end of the blade. We understand better what a radu is when we know that this implement is used for making small dikes and earthen embankments for growing rice in former mangrove forests. Let us go one step further and 'try' to use the radu. It then proves not to be so simple for inexperienced Westerners to 'spoon' the wet mud onto the dikes with this tool. After 10 minutes one already has a serious pain in the back. Because the radu is cut from a tree trunk in a special way and has a certain thickness at various points, it also has some elasticity which should be used when shovelling the mud. One digs the blade into the mud and pushes down on the handle. Next one lets go, at which point the blade comes up due to its own flexibility. The trick then is to support the upward movement somewhat and with the wrist turn it into a swinging movement so that the mud is deposited on the dike or bank. But even this knowledge of how the implement works is not in itself sufficient to enable one to work properly with the radu. One also needs detailed knowledge of the terrain because the dikes and banks should have a specific height depending on the unevenness of the land and the gradient of the surface. And for building the dikes and banks knowledge is also needed of the physical and chemical quality of the soil. This knowledge always depends on practical experience and familiarity with the land and not on abstract scientific knowledge. For instance by the colour, the smell and the taste of the soil (saltiness) one recognises the kind of soil at hand.

To comprehend what a radu is and especially in order to be able to work effectively with it one has to know in which context it is used. So as an implement the radu only gets meaning when we know what it is for and why it is made the way it is. We can understand the meaning of it when we know the social context. It becomes clearer that a radu is not just a piece of wood but functions within an intricate system of activities when we discover that not everybody is competent in making and using a radu. One first has to get permission for this from the elders of the village. Besides, working with the radu is associated with power and vitality and it may only be used by men. The younger and stronger the men are, the bigger their radu. When the iron shoe is worn away it is placed in the hiding-place of the gods to show them that the owner of the radu has performed his duty well. So the radu is also a reflexion of a complex and intricate social order. As an outsider one can only learn this step by step. And actually also only by gaining experience in the field and not from information obtained from books. Things that are a matter of course to the farmers in Guinea Bissau and Senegal, because they are familiar with the radu and its use, are completely incomprehensible to an outsider since the coherence of meaning within which the radu functions can only be understood in a sustained and intensive process of trying and trying again. Imagine what will happen when a manufacturer discovers the radu and starts making horseshoe plates via mass production. As can be expected a set of standard sizes will be offered which will cause the farmers to feel that they can no longer make the radu according to their own size or insight. Either the product of the manufacturer will not catch on, or the practice of 'spooning' and the social order associated with it will sooner or later undergo drastic changes.

Scientific and technological research concerns a breach with the coherence of meaning of our everyday world of experience. This is something deliberately sought after, for the scientist attempts to strip her/his subject of research from the matter of course of daily experience and turn it into a theoretical question or

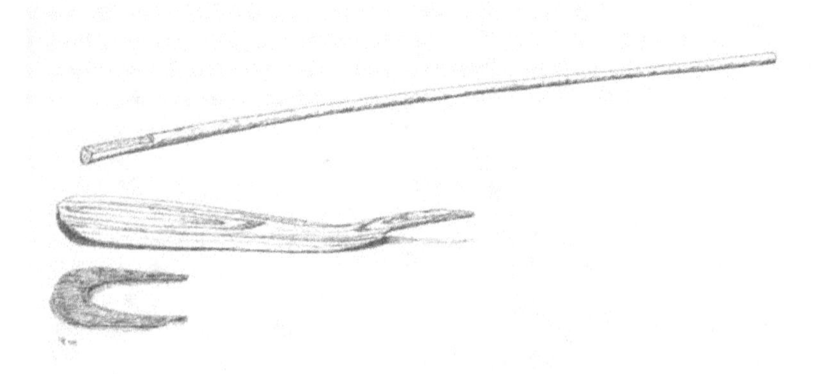

FIGURE 2.2 A radu
Source: van der Ploeg (1991: 189–199)

problem. S/he does this by separating a particular kind of experience or one aspect of reality from the integral coherence of all the kinds of experience. This process of analysis can be called abstraction, literally 'setting free' or 'drawing away from'. In other words science is based on 'theoretical abstraction'. Something that is a matter of course is stated as a problem.

Exactly what the concept 'abstraction' entails will be dealt with in a later chapter. Here we want to elucidate one facet of it only, namely that the scientific approach of reality has an estranging effect. The philosopher Martin Heidegger resisted the dominant idea that the scientific way of looking at things was supposedly the most 'objective' way of approaching reality. By this he opposed a view that has had great influence in modern times. For it was often thought that in order to know how reality is put together one actually had to research it scientifically. What Heidegger highlighted is that science is a specific way of looking at reality. And one that looks at reality according to a specific *design*. Think, for example, of the way mathematics is applied in the physical sciences. Mathematics plays an important structuring part enabling us to research the natural phenomena and to visualise functional relationships. Science differs from the normal way in which we focus on reality in its analysing and abstracting procedure. In normal life we see things in their interconnection, without making them into a theoretical problem. In science the subject of research is as though isolated from its context.

We have discussed two ways in which the everyday or the intuitively accessible coherence of meaning in our reality of experience can be ruptured. First there are phenomena or occurrences that we cannot read because we do not know their meaning or do not understand it: a completely foreign language, instruments of which we do not know the function, nasty experiences or serious events that affect or impair our confidence in our environment. In this framework we also mentioned the role of evil in the reality of our experience. Secondly by a scientific view within which we deliberately detach ourselves from (abstract) the coherence of meaning in our everyday experience. It is important to realise that experiencing meaning and a coherence of meaning is not a matter of course. However much we focus intuitively on reality as a coherence of meaning, contradictory experiences are always possible and can have a confusing effect on us. Furthermore it is important to realise that in a certain sense science is based on a 'violent' intervention in reality. That makes us careful in engaging with science. However much science can give insight and help us to a better understanding of reality, it is still based on a way of observation that simultaneously distorts or reduces reality. The image of reality in science is one-sided and works with models, and therefore has to be applied with the necessary precision and caution.

2.4 A 'two-sided' discussion of meaning

Apart from coherence the word 'meaning' also denotes something else. For not everything that exists asks the question about the meaning of existence. An ordinary stone does not ask the question. Stones just are and do not think about their existence. But people do. Heidegger therefore investigates in detail how a human being[3] thinks about his/her own existence. A human being always lives and experiences from the angle of a kind of *a priori* understanding of their being. A human being cannot lead their life without continuously reading and interpreting in a specific way the reality in which they find themselves. This is always a bilateral process, namely reacting to what is (reality precedes our reading), and simultaneously a creative process (in their reading of reality human beings reveal something of themselves).

The word 'meaning' in other words presupposes a human capacity to interpret meaning. And with this presupposition we find traces of a human trait that will be playing an important part as our argument develops. In the depth of their being humans are geared towards *answers*. In contrast to other beings humans lead their own existence in which they make choices and act on the basis of a particular attachment of meaning to reality.

Therefore speaking about meaning has two sides. On the one hand things or events either have meaning or they do not. This is the way it is expressed in everyday language: 'What you are doing there, is meaningless'. On the other hand the meaning of things can only be determined by someone who can also interpret them. And actually the two sides are not detached from each other, for the moment we use the word 'meaning' and ask about the sense of something, we are already interpreting. Even the way people observe things is a form of interpretation. When I look through the window, I not only experience an invasion of light particles on my retina, I also see trees, houses, clouds, people and lampposts. In other words, I see meaningful things belonging to my world of experience that should be there and that hardly cause me any amazement. Sometimes there are things or events that I cannot yet 'place'. I then start looking for their meaning until I can 'give them a place'. The fact that we speak of two sides illustrates again that the concept 'meaning' implies reference, relationship. Meaning has to be read. And the person who can interpret meaning, can open up meaning. In other words, 'meaning' has to do with a person as a *reading, speaking, interpreting* and *acting* being. In a certain sense one could say: being human is a dialogue with reality in its meaningful coherence.

Implicitly, the last sentence harbours an anthropology, a view of mankind. It is a view in which human identity is seen from the angle of being spoken to as the second person, as you, as somebody who is called to answer. Being able to answer is the basis for being responsible. People can respond to claims of meaning that they pick up from their environment. People are continuously challenged by their experiences to respond to the appeal made on them by these to read reality and open up its meaning.

It is our opinion that this capacity for taking responsibility is therefore closely linked to the capacity for reading, interpreting. The capacity for interpreting the meaning of something, or determining its purpose, supposes the capacity to act and set in motion processes. In order to be able to use the hammer I first have to be able to read the hammer as a hammer. With the act I open up the meaning hidden in things or – in the case of the hammer – I realise its function.

2.5 Acting as the opening up of meaning

A human is a responsible being who responds to claims of meaning s/he experiences. The reality in which s/he lives is not experienced just like that and as a matter of course as a coherence of meaning: it asks for human activity. All the time a human being is present in reality, functioning and acting. Eating and drinking, making or manufacturing things, formulating an industrial plan or designing machines, these actions not only interpret what is already there – the coherence of sense around us – but are continuously expanding further the meaning that is there. Thus the activity of human beings is an activity that *opens up* meaning. In this it differs from 'natural' or 'instinctive' reaction. By 'reaction' we here understand action without intention. A spontaneously uttered cry of fear, taking flight, yawning and nodding are various forms of reaction. They are not deliberate choices, but something that just happens. Does this mean that, for instance, nodding is a meaningless reaction? No, but its function does not depend on the intentions of those who show this reaction. For acting that is different. One could say that acting is not merely functional like a reaction but that the person who acts (the actor) opens up meaning.

Activity realises latent meaning. Therefore we call activity meaning-involved conduct. By activity stone that simply exists (without function) can become the raw material for iron. Iron-containing rock is chipped loose and treated in such a way that the iron can be isolated from it and processed into usable steel. It is a complicated process that requires a long period of learning. People must have learned that there are rocks that contain precious metals and that the metals can be isolated from that rock. Mining iron ore and extracting iron from the ore are therefore forms of acting, geared towards bringing to light the potential buried in the ore. In the ore there is a metal with functional characteristics that make it usable by people to manufacture objects. Many of these objects in turn have a function in the manufacture of other objects (instruments or tools).

It is exactly through this meaning-unlocking characteristic that activity forms the basis of everything we call culture and history. Human activity creates culture and where culture appears, history begins. Both of these, culture and history, refer to the opening up of meaning. Human activity brings to light the meaning present in reality. What takes place is an opening up of meaning. Objects previously simply present in the wild (stones, fruit, animals, water, iron ore) can obtain a place in the meaningful coherence of culture and history by means of human intervention (becoming building material, food, housemate, raw material). Sometimes these objects get this function almost immediately (like fruit from a tree), sometimes after intensive processing (like rock or iron ore). But in all cases something that formerly was latently present (and therefore was not directly visible) is opened up by human acts.

With culture and history the field of using means also comes into the picture. Precisely the fact that a human being acts instead of merely showing reactions, determines the capacity of human beings to use means (instruments, tools). In this, too, humans differ from animals. It is because human beings can set goals for themselves (intentions) that they can also choose the intentional employment of means. And if there is something that characterises technology, it is this 'means-feature' of it.

Thus it transpires that the concept 'activity' forms the link between the concepts 'culture', 'history' and 'technology'. And the particular aspect of the concept 'activity' is that it concerns 'meaning-unlocking activity': someone who acts, has an aim or intention. They knows what they are doing. This is in contrast to reactions that can also be meaningful but do not open up meaning because they are without intention.

2.6 Activity, meaning and worldview

Above we have spoken about the coherence of the concepts 'meaning', 'activity', 'culture', 'history' and 'technology'. To ensure a good understanding of what follows we want to add one more term, namely 'worldview'.

As we have already pointed out the concepts 'activity' and 'meaning' are closely connected. The fact that human beings can act has to do with their capacity for asking about the meaning of things and events. By acting, human beings construe their own reality: they hunt animals, cultivate vegetables and fruit, build houses, create art, etc. In other words, they open up the meaning locked up in reality. They organise a life for themselves in which they are less vulnerable to chance. Seen like this one can say that acting is unlocking meaning. But this is only possible if one can say that there is meaning that is latently locked up in reality. Also, activity is always based on a prior reading of the reality to which the act pertains.

Take for instance the sculptor who wants to make a sculpture from a particular block of stone. Such a person assigns meaning to a rock that in itself is 'meaningless'. Yet the latter statement is somewhat too easily said for two reasons. First, even the bare unexcavated stone is already part of a coherence of meaning, for it is an integral part of a particular scene and has its own place in it. If someone lives in that place or enjoys that landscape during a walk even that stone or rock mass belongs there. Also it emerges that the stone can at any moment be found suitable material for making a sculpture. This 'suitability for …' one could call the 'potential for meaning' that is present in the rock. It is a 'potential for meaning' that is subsequently realised by the fact that the sculptor makes a sculpture out of it. And this 'suitability for …' indicates that it is incorrect to speak about a 'meaningless' rock. The rock is always a part of a reality meant for meaning and its interpretation. This also becomes evident from the fact that for all these seemingly 'meaningless' elements there are very often words to name them. Second, the bare, 'meaningless' rock is the bearer of meaning in various ways. For instance when one assumes that the particular sculptor did not merely find the rock in his backyard, but had to buy it from a stone merchant. In that case the rock was dug out somewhere from a quarry where the stonecutters had to decide whether the piece of rock had a good-looking structure (aesthetical meaning) and would be suitable for sale (economic meaning). Moreover the stonecutters must have had the authority to acquire the rock or a legal mandate to do so (juridical meaning). Seen like this it is difficult to call a bare rock 'meaningless'.

At the same time it is clear that the rock that was dug out as material for making a sculpture still has something incomplete. It still has to reach its 'destination' in the final sculpture. Only then will it have become 'interlaced' in the constructed world of human art products. The (sculptured) rock is put on a pedestal and people look at it admiringly.

We have shown that activity opens up meaning but that it always happens on the basis of discovered meaning or of reference to meaning. Even the rock in the wild already forms part of a meaningful coherence, if only in the particular framework of a natural cycle. The sculptor uses a rock that was chosen earlier by a stonecutter and regarded as suitable to be used as raw material for a sculptor. And in order to be able to choose the rock the stonecutter must have had an eye for it. It is part of his expertise to choose the rocks according to their suitability.

Something (seemingly) meaningless like a stone can become more and more filled with meaning, or more accurately formulated, the latent meaning is opened up. The same stone or rock can become a marketable block, suitable for a sculptor. A marketable block of stone becomes a sculpture exhibited in a museum. The unlocking of meaning in practice always happens under the guidance of certain opinions or ideas that we will here call 'directional' or 'regulative' ideas. With these directional or regulative ideas we find ourselves in the field of the worldview. And here our discussion on technology, meaning and activity gains another dimension. For while above we spoke in a limited way about the concept 'meaning', we now reach what we formerly called the 'metaphysical' dimension.

So far we have connected the word 'meaning' with the word 'coherence'. Something is meaningful when it fits into a particular coherence, if it can be interpreted from the angle of a particular frame of reference. In this way the mobile phone refers to the network with which it is connected. Without that network the instrument would have no function. Solely by the fact that the network always 'knows' where my phone is located, it can make a connection between me and the person who phones me or is phoned by me. But what about all these coherences themselves? One can ask about the meaning of the mobile phone and then refer to the network, but one can also inquire further about the meaning of the network itself. And naturally one gets to the question of the meaning of mobile phones in general. Did the invention of a mobile telephone system enhance our lives? This naturally leads us to questions about communication, the meaning of it and the various ways (technologies) that facilitate communication. And questions like these cannot be answered unless one has some notions about what is good and bad or what is understood by a 'good life'.

This example shows that one can keep on asking questions. That is a very old philosophical experience, the problem of infinite regression. Again and again it becomes clear that one asks questions from the angle of a specific framework, but that it is likewise possible to query the framework itself. And so one can go on and on. The word 'metaphysics' denotes reflection about the ultimate questions or about the last basis (of being). In metaphysics philosophy searches for the answer to the last questions based on the conviction that reality is in essence rational and that the question about the meaning of being must eventually be answerable with the aid of human reason.

The enlightenment philosopher Immanuel Kant (1724–1804) in particular seriously criticised this metaphysical way of thinking. In his opinion the human capacity for knowing is too restricted to be able to answer the last questions. Therefore since the time of Kant metaphysics is suspected of overestimation of the powers of the human intellect. People have to give up the idea of an answer to their deepest questions. Ultimately the questions about the meaning of being and the meaning of existence cannot be answered, according to the modern idea. The philosopher Wilhelm Dilthey (1833–1911) agrees with Kant that it is not possible to formulate a rational answer to questions about the meaning of reality and of life. But, just like Kant, he is of the opinion that people will nevertheless keep on asking such questions. It simply is part of being human to reflect on the meaning of life. Therefore Dilthey thinks that putting the question is not wrong, but only the idea that one can answer the question in a scientific or philosophical manner. The fact that people ask this question and keep on doing so is taken by Dilthey as perfectly normal. He calls it the domain of worldview.[4]

We have spoken about meaning on two different levels: in detail about the word 'meaning' referring to the meaningful coherence of our everyday experience and subsequently and much more briefly about the metaphysical concept of 'meaning'. However, we are of the opinion that it is not quite possible to separate these two distinct levels when speaking about 'meaning'. The 'meaning' of everyday experience is inevitably connected to the more fundamental concept of meaning. We will return in more detail to the way in which this happens in later chapters.

As we have said, the opening up of meaning takes place under the guidance of 'directional' or 'regulative' ideas. These ideas lie at the level of worldview or religion. In our opinion such ideas always contribute to determining the way in which people are acting. Whether it is a doctor, a lecturer, a technical engineer or a carpenter: for all of them performing their work is partly determined by ideas about good and bad, about responsible or professional conduct and finally about 'the good life'. These ideas also harbour various views about the nature of reality. Views about the question where everything comes from, where everything is going, what the origin and destination is of all things and what the connection is behind the great variety of things, phenomena and events. Because of the enormous importance these ideas have for human activity we find that in discussing the meaning of science and technology for our contemporary culture we cannot avoid the issue of life and worldview. For 'regulative ideas' always have a religious or worldviewish origin. The way in which people by their activity 'add meaning' to reality is determined to a great extent by regulative ideas. In this regard we refer the reader to *The Religion of Technology* (1997) by David Noble. In Chapter 14 we will return to this in more detail.

Not all forms of human activity work towards opening up meaning. The activity of people can also have an adverse effect and shut out possibilities of meaning. A person is a responsible being. S/he can make correct or incorrect choices and take good and bad decisions. Especially where science and technology are concerned, it is urgent to keep in mind this capacity and make a careful analysis of where such a locking up of meaning could come in. For there is always the threat of reduction that lies in ambush in discussions about science and technology! How easy it is in the practice of scientific and technological enterprise to look at our reality in a completely one-sided way, thereby disregarding or ignoring numerous aspects of reality that can be detected. All these and related issues will be explored in detail in Parts 2 and 3.

2.7 Conclusion

Technology plays an important part in the way in which people process reality. In this the notion of 'acting in a way that opens up meaning' takes a central role. 'Meaning' here is a given of experience: people normally experience the reality in which they live as a meaningful coherence, in which things have their meaningful place. In the way in which people give form to this reality a directional role is played by worldviewish notions about the good life and about the origin, coherence and destination of reality.

Notes

1 In English the meanings mentioned here are usually denoted by two different words: *sense* and *meaning*. Yet even so the word *sense* can sometimes occur in the sense of meaning (for instance in an expression like: *in the proper sense of the word*).
2 There are indeed people who think that it is not meaningful to search for the meaning of life. By that they mean that it is not meaningful to search in a *metaphysical* sense: they are of the opinion that there is no higher, more profound intention behind life. When we speak here of the coherence of meaning of everyday experience we do not mean this *metaphysical* sense of the word meaning. It does not mean that we consider the discussion on this as unimportant. On the contrary. But what we are here talking about is the everyday experience of meaning and coherence in particular.
3 Heidegger consistently speaks about being human as 'Dasein' ('Being-There'). He does this because the being of human beings is different from the bare existence of other beings, since humans can ask themselves the question about being.
4 In German and in Dutch the words 'lifeview' and 'worldview' ('Lebensanschauung' and 'Weltanschauung') both occur. There is indeed a difference in meaning between the two. 'Lifeview' refers more to the question of the meaning of life, while 'worldview' seems more to denote a comprehensive view or theory about the world we live in. However, the words are often interchanged.

References

Heidegger, M. (1963) *Sein und Zeit* (Being and Time), 10th edn, Tübingen: Niemeyer.
Jorink, E. (2010) *Reading the Book of Nature in the Dutch Golden Age, 1575–1715*, Leiden: Brill.
Noble, D.F. (1997) *The Religion of Technology. The Divinity of Man and the Spirit of Invention*, New York: Knopf.
van der Ploeg, J.D. (1991) *Agriculture as Work of Man. Labor and Technology in the Agricultural Development* (in Dutch), Muiderberg: Coutinho.

MARTIN HEIDEGGER

(1889–1976)

PORTRAIT 2 Martin Heidegger

Martin Heidegger was born in 1889 in Messkirch and died in 1976 in Freiburg. He was a student of the Jewish philosopher Edmund Husserl. Without doubt Heidegger was one of the most influential but also most controversial philosophers of the twentieth century. Like no-one else he was able to express how many Europeans experienced the years between and after the two World Wars. For a short time he had a leading role in the National Socialist German Workers Party (National Sozialistische Deutsche Arbeiter Partei) perhaps hoping to become a leading philosopher of the national movement. However this was short lived and because he never left the party nor showed any remorse there were always doubts about his real attitude towards its ideology.

Heidegger's first important book was *Sein und Zeit* (1927). Although it was presented as Part 1 of a larger work, Part 2 was never published. Nevertheless this first part became one of the greatest classical works of the twentieth century. The essence of the book is the question about the sense of being. To penetrate this difficult question Heidegger sought to link up with one particular being: the human being. Human beings can ask themselves questions about being. A striking feature of *Dasein*, Heidegger's term for 'being', is its fundamental temporariness: being is after all a *being until death*. Mortality marks the complete existence of mankind. Against this two fundamental attitudes can be taken: one can undertake existence and actively take responsibility for it ('*Eigentlichkeit*') or one can be submerged in the masses and allow oneself to be lived ('*Uneigentlichkeit*').

Yet Heidegger's analysis of the *Dasein* provides no sufficient answer to the question of being. In his later work (after 'the turn', 'die Kehre') Heidegger indeed puts this question in a different way. He focuses on the fundamental temporariness of being as 'Ereignis' ('event', but with the connotation of 'appropriation'). People (*Dasein*) have a place in this 'Ereignis', but they do not 'make' it. In his later work, poetry and technology play a central role in his analysis of the 'sense of being'.

In the above discussion the word 'technology' has only once been used. In the strict sense Heidegger is indeed no philosopher of technology. But he did give much consideration to technology. According to him modern events of being are dominated by technology. The way in which being reveals itself in our times is dominated by technology. Man does not control this event, much rather he is controlled by it.

> What is modern technology? It too is a revealing. Only when we allow our attention to rest on this fundamental characteristic does that which is new in modern technology show itself to us. And yet the revealing that holds sway throughout modern technology does not unfold into a bringing-forth in the sense of poiesis. The revealing that rules in modern technology is a challenging [Herausfordern], which puts to nature the unreasonable demand that it supply energy that can be extracted and stored as such.
>
> *(Heidegger, 1977: 14)*

> The revealing that rules throughout modern technology has the character of a setting-upon, in the sense of a challenging forth. That challenging happens in that the energy concealed in nature is unlocked, what is unlocked is transformed, what is transformed is stored up, what is stored up is, in turn, distributed, and what is distributed is readjusted. Unlocking, transforming, storing, distributing, and readjusting are ways of revealing. But the revealing never simply comes to an end. Neither does it run off into the indeterminate. The revealing reveals to itself its own manifoldly interlocking paths, through regulating their course. This regulating itself is, for its part, everywhere secured. Regulating and securing even become the chief characteristics of the challenging revealing.
>
> *(Heidegger, 1977: 16)*

> What kind of unconcealment is it, then, that is peculiar to that which comes into being through this demand? Everywhere everything is ordered to stand by, to be immediately at hand, indeed to stand there just so that it may be on call for a further ordering. Whatever is ordered about in this way has its own standing. We call it the standing-reserve [Bestand]. The word here expresses something more, and something more essential, than mere 'stock'. The name 'standing reserve' assumes the dignity of a title. It characterises nothing less than the way in which everything presents that is wrought upon by the challenging revealing. Whatever stands by in the sense of standing-reserve no longer stands over against us as object.
>
> *(Heidegger, 1977: 17).*

References

Heidegger, M. (1963) *Sein und Zeit* (Being and Time), 10th edn, Tübingen: Niemeyer.

Heidegger, M. (1977) *The Question Concerning Technology – And Other Essays*, New York and London: Garland Publishing.

PART II

Making and designing

3

THE WORLD OF TECHNOLOGY

Three kinds of complexity

Summary

This chapter explores the world of the engineer. With reference to some case studies we will show that an engineer has to deal with three kinds of complexity. First, the engineer has to take into consideration the various aspects of the products that s/he designs. S/he is not only concerned with technological aspects like mathematical or physical problems, but also with non-technological aspects like social, economic, judicial or moral issues linked with a particular product. Second, technology has become an integral part of our society. One of the consequences of this is that an engineer has to consult with various parties whose interests are affected by the new product. In this regard we are not only thinking of the client but also of the users, the banks, the government, action groups, etc. Third, technology itself has also become complex. Technology has developed into a great number of specialities and sub-specialities. Nowadays much more than formerly, different specialities are integrated in new designs. These different kinds of complexity render the work of the engineer extremely challenging. Finally we demonstrate that the phenomenon 'abstraction', which plays an important part in technological research, is characterised by detachment from the various forms of complexity.

During the course of history the world of the engineer has become increasingly complex. On the one hand this has to do with the developments in technology itself. In the past technology comprised just a few specialities. For making technical artefacts knowledge of one speciality was often enough. Currently there are a great number of specialities and these are divided in turn into sub-specialities. We also see that in many cases different specialities are integrated in one technical artefact. On the other hand the complexity is caused by developments in society. When society becomes more complex this has direct repercussions for the various technologies that are developed. Therefore it is of the utmost importance that the engineer continuously enters into dialogue with the different groups of society.

In Part II of this book we will be exploring the complex world of the engineer by means of a philosophical analysis. Insight into this complexity is vital for comprehending the influence of technology on society. Moreover, this insight is essential for developing an ethics of technology. In this chapter we offer an introduction to the world of the engineer. We will show that every engineer has to deal with three kinds of complexity. We explore these different types of complexity further with reference to some case studies: the development of the Plumbicon television pickup tube, the design of a psychiatric hospital and the modification of a plant. In the subsequent chapters of Part II we discuss the numerous aspects of technology (Chapter 4), the structure of technical artefacts (Chapter 5), the nature of technical knowledge (Chapter 6), the process of designing (Chapter 7) and the production process (Chapter 8).

3.1 A complex world

The complexity of technology emerges first of all in the process of designing. An engineer not only has to consider the technical aspects of his or her design but also has to give attention to diverse non-technical aspects like the price of the design, the environmental impacts and the aesthetic appearance of the product. Students are trained to be engineers at a technological university. They occupy themselves with the technological sciences, a discipline with a particular theoretical basis and a particular methodology. The essence of the scientific approach is to detach oneself from the total reality to study separate phenomena or mechanisms. Thus a physicist would focus on physical phenomena without worrying about various economic regularities and an economist would focus on various economic phenomena without paying attention to various physical mechanisms. This essential trait we also find in the engineering sciences. Engineers in the first instance focus on a limited part of reality, namely the technological aspects of designing products, apparatus and machines. They employ theories to work out their designs and apply models to detect the critical aspects of their designs. However, engineers can never confine themselves exclusively to the technical aspects of their design. They will also have to take into consideration the ergonomic aspects of the design, the cost price of the product, liability in case of accidents, the aesthetic appearance of the entity, and at the end of the product's life span, the recycling of the materials used.

The complexity of technology emerges in the second place in its relation to society, a relation that cannot be described in just a few words. Technology is interwoven with all activities of our lives and has become an integral part of our society. Technology also is one of the key driving forces of changes in our society. The introduction of new products can have a very profound effect on people's lifestyle and the same is valid for the effect of the development of new technologies on the structure of societal institutions. However, developments in society also have a great influence on technology. Intensive education and training of the youth, the increase in wealth and the changes in our worldview enable all kinds of technical developments. The engineer should have insight into the way in which products and technologies are embedded in society. It is important that the engineer should keep in mind the various groups who will have to deal with the design. Thus, the client will set certain functional and economic requirements, the consumer will consider comfort and ergonomics, and the government will focus, amongst other things, on aspects of safety and the environment. In other cases consultation will be needed with financiers or various pressure groups. In short, the engineer has to deal with all kinds of requirements set by various parties or groups in society for their design. The success of the engineer will therefore be determined substantially by the extent to which s/he succeeds in realising a design that meets the diverse – sometimes even conflicting – demands.

In the third instance technology itself has also become more complex. Over time various specialities have been developed; beginning with the classical technologies like mining, the textile industry, architecture, the steel industry, chemical industry, foodstuff industry, transport technology and energy technology. Later additional technologies emerged, like information technology, communication technology, medical technology and micro-electronics. And the very newest specialities are biotechnology and nanotechnology. The first aspect of the complexity of technology is the way in which the various technologies are integrated with one another. The classical technologies actually are no longer 'classical' because they have changed as a result of the application of modern technologies. So for instance the processes in the 'old' industries are now all managed with the aid of modern electronic systems. But we also observe that the modern technologies themselves are closely interwoven. The most recent developments in medical technology are made possible by using the newest insights into information technology.

In this introduction we have discussed three different elements of the complexity of technology: the different aspects of the process of designing, the requirements set by the various groups involved in the design, and the complexity of the design itself. We will elaborate on these three elements by referring to case studies in which we will highlight one form of complexity.

3.2 The development of the Plumbicon

A first example in which the complexity of technology can be seen is the discovery of the so-called Plumbicon by researchers of N.V. Philips Gloeilampenfabrieken (Philips Company) in Eindhoven. The example aptly demonstrates the various aspects with which engineers are confronted in the design of a new product. The description of this case study is taken from the book *80 Years of Research at the Natuurkundig Laboratorium 1914–1994* (2005) by Marc J. de Vries.

The Plumbicon is a small glass tube for recording programmes for colour television. Figure 3.1 shows a tube and Figure 3.2 is a diagrammatic drawing of the Plumbicon in a colour television camera. For decades the Plumbicon could be found in professional television cameras in recording studios. That small tube forms

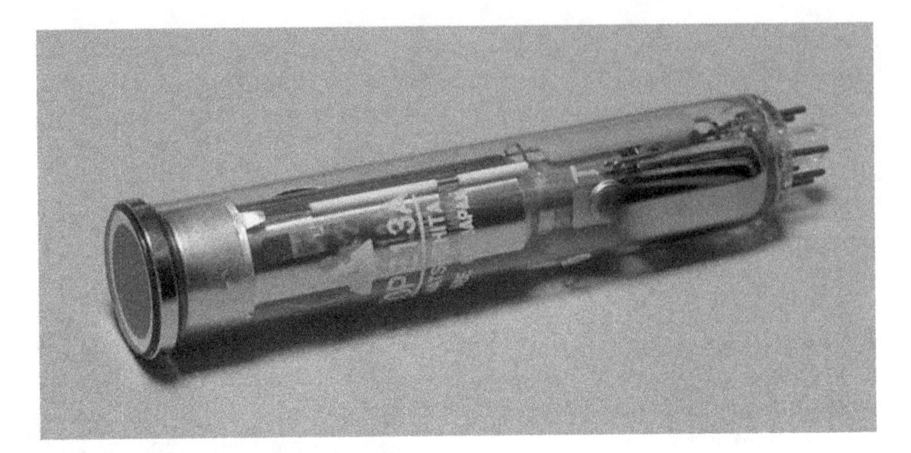

FIGURE 3.1 A vidicon
This was the direct predecessor of the Plumbicon with the same shape.

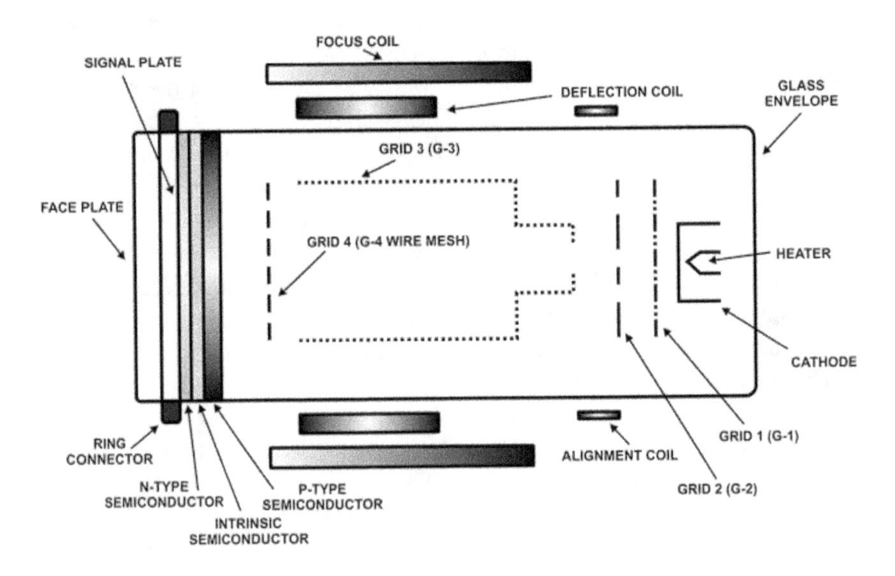

FIGURE 3.2 The Plumbicon in a colour television camera

the heart of the camera: it is responsible for transforming a light image into an electric signal. In the tube there is a plate with light sensitive material that converts differences in the light intensity into more or less electric charge on the surface of the plate. In this way the recorded image is 'saved' on the light sensitive plate. With an electron beam the surface is scanned and measured in every position to see whether there is more or less charge on the surface, and in this way it is determined whether much or little light fell on it.

The Plumbicon was developed during the 1950s in the Physics Laboratory of Philips in Eindhoven. This laboratory was referred to as the 'Natlab'. At the time the management of the laboratory meticulously followed the latest scientific developments. Whenever there was the slightest chance that such a development could lead to industrial application they started their own research. The development of the Plumbicon is a good example of the research at the time. The chief interest of the researchers focused on the fundamental aspects of the behaviour of electrons. They searched with dedication and soon found the most suitable material for the conversion of light to electricity in an application like the Plumbicon. This television pickup tube proved to be exceptionally successful. For more than 25 years it was applied in professional cameras.

The achievement by the Philips researchers is all the more impressive the more insight one gets into the number of problems with which they were confronted. A television image is made up of horizontal lines. To get an acceptable image there has to be a sufficient number of lines. When this number falls below a particular value, one gets a 'coarse' image. Moreover, an adequate number of images have to be projected per second to prevent the image from looking jerky. The last-mentioned fact has to do with the *biological* characteristics of one's eye and the way in which the nerves interpret the signals from the eye. In the beginning only stationary images could be recorded. However, from the outset it was clear that in this way television would never become a success. Stationary images do not captivate people's interest. The way we experience reality asks for more. Therefore an important part of the earlier phase in the development of camera tubes was the search for a technique to catch an adequate number of images per second, and a technique that could at the same time realise enough lines. There also was a *spatial* problem. The new camera tube was meant for colour television. But with colour television there are three tubes in a camera instead of one because a recording for each of the main colours (red, blue and green) has to be made. The tube therefore has to be very compact; otherwise the camera would be unmanageably large. The advent of colour television happened quite fast because an image with colours is *aesthetically* more attractive than a black and white image.

The researchers also struggled with the problem that there were two kinds of tubes that divided the market: one kind worked with a scanning beam with fast moving electrons and the other with a beam with slowly moving electrons. The choice for one type or the other was a decision taken by the governments of countries. It was mainly an issue of social or political preference, since whether the tube was suitable depended on the television standard used. In the USA they used NTSC and in Europe PAL (except in France where SECAM, yet another standard, was used). The difference in standards concerns amongst other things the number of lines per image: for NTSC there are 525 lines and for PAL 625. The American television image is therefore less well-defined than in Europe. This difference still exists. Television pickup tubes with fast scanning beams were more suitable for PAL and tubes with slower scanning beams for NTSC. The reason is that fast beams can be focused better, which is especially important when there are many lines in the image that have to be kept neatly separated. The most important question for the researchers was from which material the plate in the tube should be made. They knew it had to be a semiconductor, but which one? That is a *physical* problem that has to do with the energy of the electrons in the plate. The application of the knowledge naturally was coupled with extensive analysis of the tube and its behaviour. While doing this they not only made use of their own knowledge of the behaviour of electrons, but also cleverly made use of the knowledge of others.

For a long time Philips did little more than follow the developments of other companies. It was only when Radio Corporation of America (RCA) brought out the so-called vidicon tube that Philips saw a chance for making a substantial improvement with little effort, and large gain. They were well aware that new technical realisation created new opportunities. The reason the Physics Laboratory had not started developing a tube themselves at that time had to do with a *moral* aspect of television. The managing director of the laboratory, Gilles Holst, considered television a human-unfriendly medium. It forced the viewer if s/he wanted to see the news to sit down at 8 p.m. before the set, whether it suited her/him or not. Holst preferred the home news cinema which delivered the news on a film at people's homes. But the management of Philips more or less compelled Holst to have his people work on television. With hindsight, perhaps Holst with his moral considerations was not altogether wrong.

Once Philips had created the Plumbicon tube in 1958, a way had to be found of *legally* protecting the knowledge they had acquired. In order to prevent others from inferring from the patent how the tube was produced, Philips described in the patent document more than 20 different ways of possible production. The Philips researchers knew that only one of these would work. In this way they could legally protect the right way of producing the tube, whilst not betraying the relevant knowledge, as it was hidden among the description of ways of production that would not work.

The production of the tube initially progressed with great difficulty. Many tubes imploded or showed spots in the image. Only due to the perseverance and endless patience of one of the technical assistants could the tube eventually be produced. He was convinced that this tube would work and believed that by means of endless experimenting the problems of production would prove to be solvable. He even went to work in the factory himself for a while to see what was going wrong. The technical assistant displayed the aspect of *faith* (*trust*).

Finally in the patent itself careful consideration was given to the spelling of the name of the tube. Up until the Plumbicon all types of tubes were written in lower case: for instance iconoscoop, orthicon and vidicon. Philips insisted that their tube have a name and not an indication of its type. Therefore it became 'Plumbicon' and not 'plumbicon'. This is a small example of how a *language* aspect worked out in the finest detail can be coupled with a technological development. Philips did well with the tube, and for a long time the Plumbicon could be found in almost every studio camera. At first the Philips company refused to supply separate tubes to other producers of cameras (like RCA) but supplied only complete television cameras. They thought that from an *economic* point of view this was more advantageous. Later on they found out that it was more inviting to sell separate tubes as well. And so after a long struggle with numerous problems the Plumbicon was developed to the standard in colour television recordings.

In retrospect the development of the Plumbicon was a difficult assignment that the engineers carried out with great success. In this case study the first form of the complexity of technology emerges, with the researchers having to keep in mind many different aspects. There are nine key words connected to the development of this television pickup tube: the *physical* characteristics of the materials, the *spatial* characteristics of the tube, the *biological* characteristics of the human eye, the *lingual* denotation of the tube, the *economic* considerations in the sales strategy, the *aesthetic* characteristics of a colour version, the *judicial* (*legal*) protection of the design by means of a patent, the *moral* aspect of the design and the *faith* that making the tube would be feasible. These key words are not only connected with the technological aspects of the development of this tube, like the physical characteristics of the material and the size of the tube, but also with the non-technological aspects as considerations of marketing, legal protection of knowledge and the morality of the design. These are key words that each represents a particular domain of meaning.

In this example a second form of complexity also comes to the fore. The researchers had to consider the *different parties* whose interests were at stake: the clients who wanted a sharp and flowing image, the government that could set a certain standard, the management of Philips who wanted to market the colour television and the competition who was busy developing comparable tubes.

Finally it must also be noted that in the development of the Plumbicon the third form of complexity was also distinctly present: a multitude of *scientific theories* and more than one *technical discipline* were needed in order to realise this design.

3.3 Designing a new psychiatric hospital

In this section we will discuss the example of designing a clinic for the psychiatric hospital Vijverdal in Maastricht, a city in the southern part of the Netherlands. In particular we want to examine the complexity that emerges in the interests of the different parties with which the architect was confronted. One of the authors of this book (MJV) was managing director of this hospital from 2004 to 2007.[1]

In 1968 the psychiatric hospital Vijverdal went up in the hilly landscape on the outskirts of Maastricht. To conform with the views of the time the hospital had to have a function properly integrated in the *community* with differentiated care-giving services. The architecture was seen as a significant social factor with the potential of having a positive influence on the behaviour of *patients*. Various functions in everyday social life would therefore have to be realised in this complex. In the middle would be the clinic and the treatment centre, and on the outside of this various facilities: from shop to hair-dresser, from swimming pool to gym hall and from coffee shop to recreational space. The clinic and the treatment centre would be housed in an apartment building of nine stories. The underlying idea of the *architect* was that a psychiatric hospital predominantly geared towards patients from the city had to have an urban character, seen from an architectural point of view. An apartment building would be the best reflection of the living conditions in the urbanised Netherlands. In this new vision the idea that psychiatric patients had to be accommodated in an outbuilding in the woods far from normal society was abandoned. Formulated differently, the psychiatric institution is no longer 'alien to society' but 'committed to society'. In the press the new design was called 'revolutionary' because the classical idea of separate smaller buildings had been relinquished. We would like to point out that there is a remarkable parallel between the 'flat-optimism' of the architect and the utopian outlook of the designers of the Bijlmer in Amsterdam.[2] Finally it has to be stated that the high rise building fitted in with the contemporary town planning of the *municipality of Maastricht*.

Building the hospital proved problematic. There were differences of opinion concerning the desired size of the hospital and the number of beds needed. The *medical staff* for reasons of care-giving proposed that a smaller hospital be built. However, the *Limburger project developer* who was to build the hospital for economic reasons pleaded for a bigger design. The *management* eventually decided to follow the proposals of the project developer. Another complicating factor was that the building of the new psychiatric centre had been started without approval by the then Minister of Health in The Hague. This led to fierce discussions. The *officials of the ministry* thought that the building was both too expensive and too big.

The original intention – the hospital as a small society – was indeed realised in its physical form. But long before Vijverdal was *officially* opened the first critique was heard. The concrete buildings were felt to be depressing. The occupation rate of the beds lagged far behind the expectations, which led to serious financial problems. The *teams providing the treatment* proved to function not as well as was originally anticipated. In addition the societal climate regarding psychiatry had changed substantially. At the opening of the building, the book *Not Made of Wood. A Psychiatrist Discovers his Profession* by the anti-psychiatrist[3] Jean Foudraine was in its 25th edition in the Netherlands. Finally, it proved that the view of care-giving which formed the basis for the architecture – the institution as town or as urban centre – was already outdated. All in all, the hospital did not develop into an organisation that was firmly integrated in society, but into an inward-looking institute. The distance between the institute and society remained huge, with the result that it became more difficult for patients to return from the hospital (the 'small society') to their own situation at home (the 'real' society). At the same time it proved that the goal for an open social community had been set too high.

From about 1990 a new wind began blowing through care-giving institutions in the Netherlands. The call for care on demand became steadily stronger. The realisation that a psychiatric hospital should become a patient-oriented organisation began taking root. It also became steadily clearer that taking care of a psychiatric patient should not be concentrated in the hospital as a 'small society' but should be done in such a way that the patient could keep on living at home ('real' society).

FIGURE 3.3 The old clinic at Vijverdal

FIGURE 3.4 The new clinic at Vijverdal

Only when it was no longer possible to give care at home, for instance because the patient poses a threat to him/herself or to their environment, would admission to a psychiatric hospital be justified. Vijverdal also actively contributed to the development of this new vision. These ideas had great influence on so-called chronic psychiatry. Instead of accommodating these patients for many years in the clinic of Vijverdal, an attempt was made to let them live as independently as possible in a 'socio-home' on the grounds of the hospital or in a 'socio-home' in the city. The more independently the patient could live, the further the distance from the hospital.

In due course the various ideas developed into a coherent vision. The idea that care had to be concentrated in different sections that each offered their own particular treatment module was discarded. The main problem was that the different modules for a specific target group were not well coordinated, because each section had its own psychiatrist and its own practitioners. An entirely different structure was selected. All treatment modules needed for a certain target group were now brought together in a so-called clinical pathway. The main advantage for the patient was that s/he no longer needed to go from one section to another to take part in the various modules but that these were all represented in the pathway and linked up neatly with one another. Another advantage was that patients did not have to change psychiatrists or practitioners when they needed a different treatment module.

The problems attached to the 'old' hospital became ever more serious. The Vijverdal building was too big and too expensive, the increasing demand for single rooms could not be met and the absence of outside space for patients in closed sections was considered a fault. In addition the architecture of the hospital no longer fitted in with the newly developed vision of care-giving. The dominating high rise would have to make room for a modest low building that would harmonise with the surrounding environment. The functional giant would have to be replaced by accessible, patient-friendly sections on a smaller scale. Figures 3.3 and 3.4 give a picture of the old and the new clinic.

In developing the new clinic for Vijverdal much attention was given to the design. Admission to the clinic is given to adults and elderly people for whom the duration of their stay could vary from a few days to a few years. The clinic consists of nine units each having 10–12 beds. Every unit has its own lounge, smoking room and outdoor space. Figure 3.5 shows a sketch of such a unit. In the design the architect has had to take into consideration various legal requirements. The patient should have a single room of 12–15m^2 with their own toilet, shower and wash basin. And there are specific basic requirements for the so-called isolation rooms. Other guidelines relate to fire safety, energy economy and environmental demands. During the process of designing the architect regularly consulted with the *association of patients*. In some associations the *relatives* of the patient play an important part. In this way extensive talks were held with them about the privacy of the patient and the way the isolation rooms had to be furnished (a patient is admitted to an isolation room when he or she is stimulated too much in the section and isolation is needed to calm down). Another point of discussion was the impression the building would make: does the building emanate power over the patient or

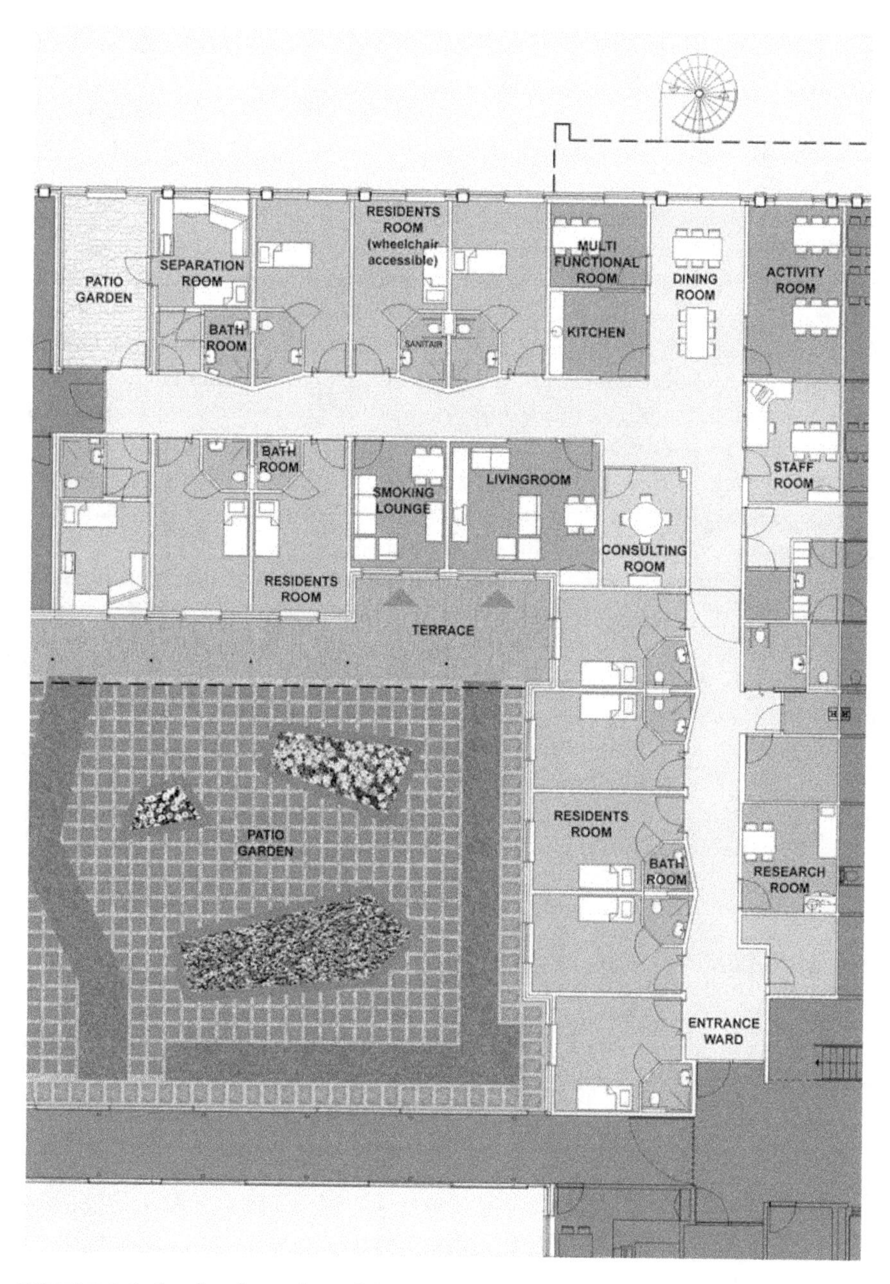

FIGURE 3.5 A sketch of a section of the new clinic at Vijverdal

does it actually make him/her feel at home. The associations also pleaded for outdoor facilities for each unit. During the designing phase there was much consultation with the *nurses* and the *practitioners*. In the interior design of the rooms it was assumed that lively colours would be used and that old and new materials would be used harmoniously. Moreover, much care was given to the safety of the patients, staff members and visitors. The nursing stations and the passages are designed in such a way that the staff have maximal supervision of the section. Technical aids like cameras are used to improve the supervision even further. All rooms and nursing stations are equipped with cables for electronic data access so that the Electronic Medical Record can be consulted in each of these locations. Finally, heat storage in the ground is utilised. This leads to a substantial reduction in the energy budget and at the same time contributes to a cleaner environment.

Once again, in the design of the new Vijverdal, the complexity of technology is highlighted. This is evident in the various aspects the architect has to deal with: the physical characteristics of the building materials, the spatial aspects of the design, the economically limiting conditions set by government, the legal rights of the patient and the moral issues relating to their care.

This case study shows that a great number of groups exert influence on a design: the different *managements*, the *builder*, the *architect*, the *ministry*, the *provincial government*, the *municipality*, the *patients*, the *relatives of the patients*, the *patient pressure groups*, the *nursing staff*, the *teams giving treatment*, and the *practitioners*. This leads to a complex dynamics between the interests of the various groups.

As a third form of complexity in the design of a building we see that a clinic is not merely a spatial facility but a building in which numerous different technologies are integrated. In conjunction with the different architectural technologies, energy technology, information and communication technology and safety technology all play essential roles.

3.4 Modification of soya beans

A technological design is not only complex because there are so many aspects to it, or because the design always takes shape in interaction with social processes, but also because the design in itself is an entity in which different scientific insights and technical applications are integrated. Tools like a hammer or a screw driver are relatively simple. Different materials are indeed used for them, like iron, wood and plastic, but they each have an unambiguous function. Therefore it is highly predictable how and for what purpose these tools will be used in practice, although a hammer can of course also be used to hit someone over the head or a screw driver for breaking open the lock of a door.

For many contemporary technical designs having a univocal function is no longer valid. Often knowledge from different disciplines is used and designs come into being in the interaction between numerous disciplines resulting in a complex network of insights and applications which constantly interact. A recent development aptly illustrating this is the convergence of nanotechnology, biotechnology, information technology and cognitive sciences (the so-called *NBIC convergence)*. Parallel developments take place in these fields and comparable concepts and metaphors are used. Think for instance of the concept 'information' and the notions 'encoding' and 'decoding'. Exchange between the different disciplines is promoted by this and thus numerous hybrid areas of knowledge and potential applications originate. As a direct consequence it is never entirely clear where a new design or a new application will eventually lead. When *Arpanet*, the forerunner of the Internet, was built in 1969 for American Defence no-one could predict what an enormous social revolution would be set in motion by it, nor how radically science and technology would be changed by it.

An excellent case for illustrating the complexity of technology itself is provided by the genetic modification of soya beans. This application from agriculture shows that technology does not stop with the design and manipulation of physical objects and structures. For the technical-scientific method is deliberately utilised to intervene in organic nature, for instance, to enhance food production or develop new medicine. Biotechnology in itself is not new. For centuries man has been manipulating nature by developing new races or by using bio-cultures in making cheese, yoghurt or wine. What is new, is the systematic application of technical-scientific methods and the use of technologies that make it possible to experiment with genetic material at DNA-level. These developments also demonstrate that the boundaries between agriculture and technology are becoming increasingly vague, just as are the boundaries between technological and medical science.[4]

In the cultivation of soya, but also maize, cotton and coleseed, multiple use is made of herbicides, or weed-killing products. However, many kinds of weeds have become resistant. A solution to this problem was found by the American biotechnology company Monsanto. With the help of genetic manipulation they developed plants that were resistant to the herbicide RoundUp. This herbicide contains the chemical compound glyphosate which inhibits the enzyme EPSP-synthase. This development is significant because RoundUp is a typical kill-all that would normally also affect the soya plants themselves. By inserting a specific gene derived from a bacterium in the soya plants this plant becomes resistant, and the use of RoundUp in combatting weeds becomes possible. In the first instance, it seems as if this amounts to nothing more than a *technological fix*, a technological solution to a technological problem without going to the root of the causes of this problem. On closer inspection, however, it seems that the matter is somewhat more subtle. First one could expect that resistance in weeds would not occur very soon because we are here dealing with a 'broad spectrum' herbicide that intervenes in the vital processes of a huge group of plants. Secondly, glyphosate is generally seen as a relatively environment-friendly weedkiller that is not or only slightly poisonous to animals and plants and can easily be broken down. Environment-unfriendly combatting agents can thus be replaced by a less harmful agent without losing efficiency.

This example of modified soya beans shows once more that in every design there is a diversity of aspects which the engineer should take into consideration:

physical-chemical, biotic, social, economic, judicial, etc. Likewise it transpires that we are concerned with a technical design that is always developed within a specific social constellation in which different groups are involved. While in the USA genetically modified soya is cultivated on a large scale, within the European Union it is met by significantly more resistance. Since 1996 protest actions have been regularly launched in Europe with the result that large food producers like Numico and Unilever have given up using modified soya. An important body who has to supervise the use of genetically modified strains in the Netherlands is the Commissie Genetische Modificatie (the Commission for Genetic Modification). Until now only a limited number of genetically modified strains have been allowed in the Netherlands.

The complexity of technical design itself is manifested in the diversity of subjects and disciplines involved in the development of genetically modified soya beans. First of all, laboratories are needed, equipped with microscopic and electronic apparatus and computer systems. All analysing techniques require thorough technical knowledge for manning the apparatus and adequate theoretical knowledge for interpreting the results. Secondly, intimate knowledge is needed of biochemistry, plant physiology, genetics and the refinement of strains. In order to make a good risk analysis and to investigate how the product can be marketed without meeting much opposition in society, disciplines like ecology, social sciences, judicial sciences (patent right) and ethics are indispensable. And, obviously, a sound knowledge of the agricultural sector is required. Moreover, it is striking that in this kind of development one can hardly discern any difference between pure science and applied science, and between science and technology. Business concerns often take the lead in generating new ground-breaking knowledge. Besides, scientific research and the development of technology go hand in hand and can hardly be distinguished from each other. The philosopher Bruno Latour who is discussed in detail in Chapter 10 therefore with good reason speaks about *technoscience*. What is happening in fact is that science, technology and business are getting ever more interwoven, thereby creating an intricate network of relations of which it seems government bodies often have hardly any understanding. In Chapter 11 we will discuss this problematic aspect further.

The complexity of disciplines and fields of application also means that the possible consequences of a certain technical application are hard to foresee. Regarding the genetically modified soya beans it is not clear what effects there are on the environment, biodiversity and national health. Although the risk attached to the use of glyphosate, the most important component of RoundUp, is said to be relatively slight, it cannot be excluded that in the long run it may lead to ecological damage. Thus it may happen that when the agent is used on a large scale and for some time resistance could appear in weeds. This risk is not altogether fictitious, as proved by the fact that glyphosate-resistant grass has been observed in Australia and the USA. Moreover it is possible that herbicide tolerance can be genetically transferred via cultured plants to wild congeners, or that a herbicide tolerant plant can develop into a weed in another location. Exactly because RoundUp seems to be an agent that causes little damage to the environment, farmers may find this a reason for using it unsparingly. Problems with biodiversity can occur because the use of RoundUp Ready plants on a large scale makes farmers more and more dependent on this herbicide. So when resistance does occur, or problems are encountered in other ways with the herbicide, the consequences of this cannot be foreseen, and the food supply can even be jeopardised. There are also possible effects on food safety. Although this threat at the moment seems to be relatively small, it cannot altogether be excluded that the accumulation of glyphosate in soya beans could cause health risks.

Nor is it clear what the consequences will be of glyphosate-resistant plants for current agricultural practice. An interesting issue in this regard is the litigation between the company Monsanto and the Canadian farmer Percy Schmeiser about the use of these plants. In 1997 glyphosate-resistant coleseed was found in Schmeiser's fields. Schmeiser stated that the coleseed had blown over from other fields. Subsequently he used part of the harvest of this 'blown over' coleseed to sow his fields the following year. Eventually Monsanto won the case and Schmeiser was not allowed to use the 'blown over' glyphosate-resistant coleseed as seed. The system of patents on genetically modified plants therefore means that farmers more and more become customers of the agricultural industry and that they can no longer freely decide what they do with the seed they harvest from their fields. Moreover the plants have to be meticulously checked by experts all the year round to prevent the use of genetic material by third parties. Another development that could have drastic consequences is the so-called terminator gene. This gene prevents genetically modified plants from reproducing. It has the advantage that it can help prevent the risk of spreading the genetic material in the ecosystem. At the same time, however, it also prevents farmers from cultivating their own plants. In particular this is a problem for the farmers in Third World countries who cannot afford to buy new seed every year. At the end of 1999 Monsanto announced that it was desisting from the production and sale of seeds with the terminator gene.

Also it is not clear how the development of genetically modified plants like the RoundUp Ready plants of Monsanto could impact on human beings and nature in the long term. What new technological-scientific developments are evoked by this? Some people claim that genetically modified plants could over a period contribute to the fight against hunger in the world since the production per hectare can be improved. Others point out the dangers to food safety, the possibility of ecological disasters and the formation of monopolies. Especially the violation of the boundaries between different species is often seen as controversial. The genetically modified soya plants that we have mentioned above also have a transgenic character. They were made resistant to glyphosate with the aid of a gene derived from the bacterium *Agrobacterium CP.4*. What is this eventually going to mean for technological-scientific development? Should this not be stopped? Some ethicists warn that human beings are taking on them the role of God when they ignore the existence of boundaries between different species. According to them permission should be asked for every matter and new applications must only be permissible when all risks have been properly forestalled. Other ethicists, however, coming from an evolutionist view, state that intrinsic characteristics of a kind do not exist. They are of the opinion that with technological-scientific development morals will in the long run be adapted too because hybrid forms of life will become more and more common. In this book we will be discussing this controversy in more detail in Chapter 12, in which we discuss the movement of transhumanism.

3.5 Engineers and complexity

We have now looked at three different cases in which engineers have played the leading role and we have shown that the work of an engineer is very complex.

To what extent is every designer confronted by this broad complexity? It is not easy to answer this question as it is so comprehensive. Amongst other things it depends on the nature of the product, the organisation of the business, and the experience of the engineer. While designing a relatively simple product a young engineer will soon take responsibility for a whole project and come up against all kinds of complexity. To design very complicated artefacts engineers often work together in teams. In such a team a young engineer will often focus on one aspect, one component or one sub-technology of the design, an experienced engineer will be responsible for the integration of the different aspects, components or technologies in the total design, and a manager will give particular attention to the relations with and among the different groups.

3.6 Technological research as abstraction

In Chapter 2 we characterised technology as a search for meaning. We related the meaning of technical artefacts to the meaning of the artefacts in a greater context. In this chapter we have specified the main elements of the context by means of an analysis of the various kinds of complexity.

However, there is one activity by which an engineer deliberately breaks the coherence: technological research (see Section 2.3). During research an engineer often focuses on one element or one phenomenon of a technical artefact, one particular aspect, the interests of one group or the features of one technology. What happens is that an engineer isolates one element or one phenomenon and subjects it to closer investigation. This isolation makes sense: for it is the only way of investigating that element or phenomenon in detail and recording the various regularities. However, we have to realise that the isolation of one element or phenomenon means that we abstract from the different forms of complexity of the technical artefact. Abstracting means that we disregard the numerous aspects, the different groups and the integral technical system.

A first example to illustrate the phenomenon of isolation and abstraction is taken from the doctoral thesis *Electrical Conductivity and Interface Properties of Oxygen Ion Conducting Materials* for which Maarten J. Verkerk was awarded his PhD at the Technical University of Twente in 1982. In it were reported the results of the research on the characteristics of materials conducting oxygen ions. These kinds of materials can be used for storing, transforming and saving energy. During the research attention was focused on physical characteristics: the electrical conductivity of the materials and the characteristics of interfaces as granular boundaries and electrodes. Only one of the many aspects was investigated: the physical-chemical aspect. Hardly any attention was given to the interests of the different groups in the development of these materials and to the technical systems in which these materials would be applied. This isolation made sense because detailed insight into the various physical-chemical mechanisms was obtained. But the consequence of this isolation was that the coherence of numerous aspects, the different groups and the various technologies were lost from sight.

A second example is found in the research *Trust in Systems: Effects of Direct and Indirect Information* by the engineer Peter de Vries for which he received his PhD at the Eindhoven University of Technology in 2005. This doctoral thesis was focused on the effect of direct and indirect information on the confidence that drivers have in the navigation system of their cars. De Vries considered one aspect (confidence), one party (the user), and one element of the total system (the 'box in the car'). In this research, too, we see that abstractions are made from the complexity of reality: the numerous aspects, the many parties and the complete technical system. In this research we once more see that the isolation made sense: insight into the various factors playing a role in the way people judge a system. But this was only possible because De Vries ruptured the different coherences.

We have mentioned three elements of abstraction in technological research: abstraction from the many aspects, abstraction from the numerous parties and abstraction from technical systems. For the sake of comprehensiveness we point out that abstraction also takes place in technological research from the concrete technical artefact itself (the issue was not confidence in this specific navigation system but confidence in navigation systems in general) and also from one individual researcher (the research has to be done in such a way that another researcher doing a comparable study would obtain the same results). These also are the reasons why often it is hard to use the results of technological research directly in practice: the *researcher* abstracts from the various forms of complexity, the *designer*, however, has to take into consideration all the various forms and can run into serious problems.

3.7 Conclusion

In this chapter we have seen that the world of technology is exceptionally complex. With reference to three case studies we have been able to distinguish different kinds of complexity: the many aspects of technology, the numerous groups who are stakeholders in technology, and the integration of technology in systems. Each of these kinds of complexity has its own mechanisms and phenomena. This analysis also gives a better insight into technological research: for in it there is a deliberate disregard for the various forms of complexity in order to be able to investigate only one mechanism or phenomenon.

Notes

1 We thank A. Klijn for the information about the history of Vijverdal.
2 In the same period a new district was built in Amsterdam named 'De Bijlmermeer'. This district consisted of flat buildings of 10 floors. The intention was to build a quiet and green area for modern people. However, in practice it proved to be quite different: 10 years later this area became known for its social problems.
3 Supporters of anti-psychiatry have fundamental critiques of western psychiatry. They consider that psychiatric illness should be seen as a natural reaction to society, and that this kind of disorder can best be explained from the angle of social, ethical or political factors. To their way of thinking admission to and treatment in a psychiatric hospital will generally lead to a deterioration of the illness.
4 The case study description is based on Bos (2008).

References

Bos, A.P. (2008) 'Instrumentalization theory and reflexive design in animal husbandry', *Social Epistemology* 22(1): 29–50.

Verkerk, M.J. (1982) *Electrical Conductivity and Interface Properties of Oxygen Ion Conducting Materials*, thesis, Enschede, the Netherlands.

Vries, M.J. de (2005) *80 Years of Research at the Natuurkundig Laboratorium 1914–1994*, Amsterdam: Pallas Publications.

Vries, P. De (2005) *Trust in Systems: Effects of Direct and Indirect Information*, thesis, Eindhoven, the Netherlands.

LEWIS MUMFORD

(1895–1990)

PORTRAIT 3 Lewis Mumford

Lewis Mumford was born in Flushing (New York) on 19 October 1895. After finishing high school he studied at the City College of New York and the New School for Social Research but left the university early due to illness. Mumford called himself a writer and not a scientist, historian or philosopher. His studies covered a wide field of research: the urban phenomenon, architecture, literature, cultural history and technology. He wrote 25 books of which almost all are still available. Mumford was substantially influenced by the work of Patrick Geddes, a Scottish botanist, sociologist and urban architect. Geddes inspired him to go on an exploration trip in his own city, New York. This resulted in the first instance in numerous articles and essays and later on led to the classics *Sticks and Stones* (1924), *The Brown Decades* (1931) and *The City in History* (1961).

After his first studies in cities and architecture Mumford expanded his field of research to the origin of modern society. According to him the face of modern Western society was formed to a great extent by the development of the 'machine'. He did extensive research on its development. This led to the publication of his great work *Technics and Civilization* in 1934. In this book Mumford shows that all kinds of cultural changes preceded the development and implementation of new technologies. He made a new division into periods of the history of technology and distinguished three waves: (1) the period of aeotechnology, from the tenth to the middle of the eighteen century, with water and wind as energy sources and wood and glass as the main materials, (2) the period of paleotechnology, from the middle of the eighteenth century to the second half of the nineteenth century, with coal as energy source and iron as the main material, and finally (3) the period of neotechnology that started about the second half of the nineteenth century and in which oil and electricity are the main energy sources and steel and other alloys are the main materials.

Well over twenty years later his follow-up study *The Myth of the Machine* was published in two volumes: *Technics and Human Development* (1967) and *The Pentagon of Power* (1971). In this work Mumford differentiated between two different types of technology. He demonstrates that technology was originally harmoniously interwoven with the life of human beings. This technology he calls 'poly- or biotechnology'. The development of modern technology, however, led to a change. The situation where it was harmoniously interwoven with human life was ruptured and gave way to a power struggle. Mumford places the origin of this 'mono- or megatechnology' some five thousand years ago in Egypt, Mesopotamia and the Far East. In these cultures hierarchical organisations were developed who could realise enormous works like the pyramids and the Chinese Wall. He calls these organisations 'megamachines'. With the development of modern science in the sixteenth and seventeenth centuries the idea of the megamachine was once more revived. The modern megamachine surpasses its forerunners in all aspects. In particular the combination of modern technology with a hierarchic organisation results in complete control of man and society. Eventually this control leads to death. However, Mumford was convinced that human beings can escape the hold of technology. According to him it is a myth that total control is inescapable.

Mumford was active in society in a variety of ways. As a journalist for *The New Yorker* he wrote fierce critiques of the town planning policy of 'his' New York. He was one of the first to criticise the use of nuclear weapons and voiced fierce resistance against America's involvement in Vietnam. In 1964 he was awarded the *Presidential Medal of Freedom.*

We cite a few quotations which clearly show the power of the megamachine and the ability of human beings to escape from it:

> Technics and civilization as a whole are the result of human choices and aptitudes and strivings, deliberate as well as unconscious, often irrational when apparently they are most objective and scientific: but even when they are uncontrollable they are not external. Choice manifests itself in society in small increments and moment-to-moment decisions as well as in loud dramatic struggles; and he who does not see choice in the development of the machine merely betrays his incapacity to observe cumulative effects until they are bunched together so closely that they seem completely external and impersonal.
>
> *(Mumford, 1934: 6–7)*

> The ideology that underlies and unites the ancient and the modern megamachine is one that ignores the needs and purposes of life in order to fortify the power complex and extend its dominion. Both megamachines are oriented toward death (…) As with all modern technical performances, the mass infliction of death has been both expanded and speeded up (…) Yes: the priests and warriors of the megamachine can exterminate mankind: so, once again, if Von Neumann is all right, they will.
>
> *(Mumford, 1971: 260–262)*

> But for those of us who have thrown off the myth of the machine, the next move is ours: for the gates of the technocratic prison will open automatically, despite their rusty ancient hinges, as soon as we choose to walk out.
>
> *(Mumford, 1971: 434–435)*

References

Mumford, L. (1924) *Sticks and Stones*, New York: Boni and Liveright.

Mumford, L. (1931) *The Brown Decades*, New York: Harcourt Brace.

Mumford, L. (1934) *Technics and Civilization*, New York: Harcourt Brace.

Mumford, L. (1961) *The City in History*, New York: A Harvest Book.

Mumford, L. (1967) *The Myth of the Machine, I. Technics and Human Development*, New York: Harcourt Brace.

Mumford, L. (1971) *The Myth of the Machine, II. The Pentagon of Power*, New York: Harcourt Brace.

4

THE ARTEFACT [I]

Diversity and coherence

Summary

In this chapter we focus on the many aspects, dimensions or facets of technology using analysis of reality as developed by the philosophers Dooyeweerd and Vollenhoven. As the 'vehicle' for our analysis we use an industrial robot. A technological artefact like a robot proves to be exceptionally multi-faceted. In total as many as 15 different aspects can be distinguished, each with its own nature and character. Yet it is one and the same robot. In this chapter it transpires that the different dimensions all refer to one another and are inextricably connected. From this analysis it also emerges that there are many laws and norms connected with the design of a technological artefact which an engineer should consider. We also show that technology is not something that 'happens' to us, but that we as human beings are actively engaged in the process of designing and so can give direction to technology.

In the previous chapter we explored the challenging complex world of the engineer. Engineers occupy themselves with various aspects of technology, not only the spatial and physical dimensions of a design but also with, for instance, the social, economic, juridical and moral sides of it. Moreover, in the whole process of designing engineers have to consider the interests of different groups. They not only have to do justice to the interests of the client but also to other groups, like customers, co-workers, suppliers, authorities, governing bodies, etc. And finally engineers also have to cope with the complexity of the design itself. Modern technological artefacts not only require knowledge of the different classical and modern technologies but also of the integration of these different technologies into one single design.

This chapter is based on the ontology (theory of reality) of the philosophers Dooyeweerd (1894–1977) and Vollenhoven (1892–1978). For the analysis we make use of a technological artefact: an industrial robot. There are 15 different dimensions to this robot, all of which play an important part in the designing process.

4.1 An industrial robot

The Czech writer Karel Capek introduced the word 'robot' in his play *Rossum's Universal Robots* (1920). The word is derived from a Czech expression for boring and monotonous work. Capek's robots were androids made by means of a secret process. The reason for their existence was to provide cheap and flexible labour. *Rossum's Universal Robots* is a science fiction story about the unintentional consequences of technological overconfidence. In the course of the story the robots acquire human traits. They no longer accept being subordinates and revolt against the people. In a bitter struggle just about all of humankind is exterminated by the robots. The science fiction writer Isaac Asimov also wrote a great number of stories in which robots play the leading part. These stories were published in a volume entitled *Robot Dreams* (1986).

In industry robots are used on a large scale.[1] They are – as in Capek's story – extremely well suited for rendering cheap and flexible labour. The industrial robot is a typical example of a *general purpose* or universal machine. According to the ISO-norm 8373 a robot is: 'An automatically controlled, reprogrammable, multipurpose manipulator programmable in three or more axes, which may be either fixed in place or mobile for use in industrial automation applications.'

So robots are automatically controlled, the programmed movements or functions can be altered without physical adaptations to the robot itself and the robot can be used for different purposes without having to be fully converted. Figure 4.1 gives a photo of an industrial robot, including power supply and control. Figure 4.2 gives a schematic drawing of a six-axised robot arm.

Robots are often used in the automobile, metal and synthetic fibre industry, and also for assembling household appliances and other consumer products. They are used for moving materials, transporting parts, loading and unloading machines, assembling parts, changing tools, joining parts and applying coats of paint to products. First of all there are economic arguments for using robots. A robot replaces one or more workers, works at a consistent speed and is available 24 hours a day, seven days a week. This mostly results in a substantial decrease in the cost price for a specific operation. Subsequently there are considerations regarding the quality of the operation. In some cases robots are used to eliminate variations in human acts and to ensure that all operations are carried out with the same quality. In other cases robots are essential for successfully carrying out specific operations, as in the *high-tech* industry. Robots are also used for repetitive operations, strenuous work and dangerous or unhealthy operations. In these situations robots are essential to improve the quality of the labour. Finally scarcity in the labour market sometimes compels manufacturers to use robots in the production process.

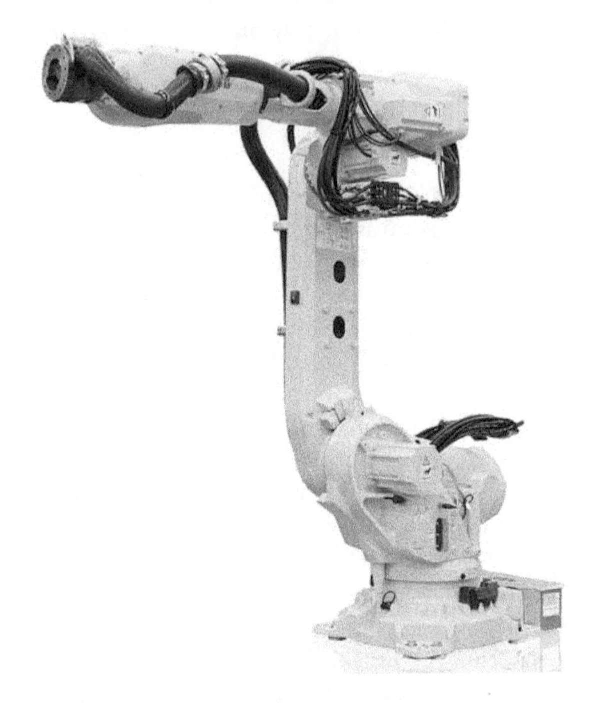

FIGURE 4.1A An industrial robot: mechanical manipulator

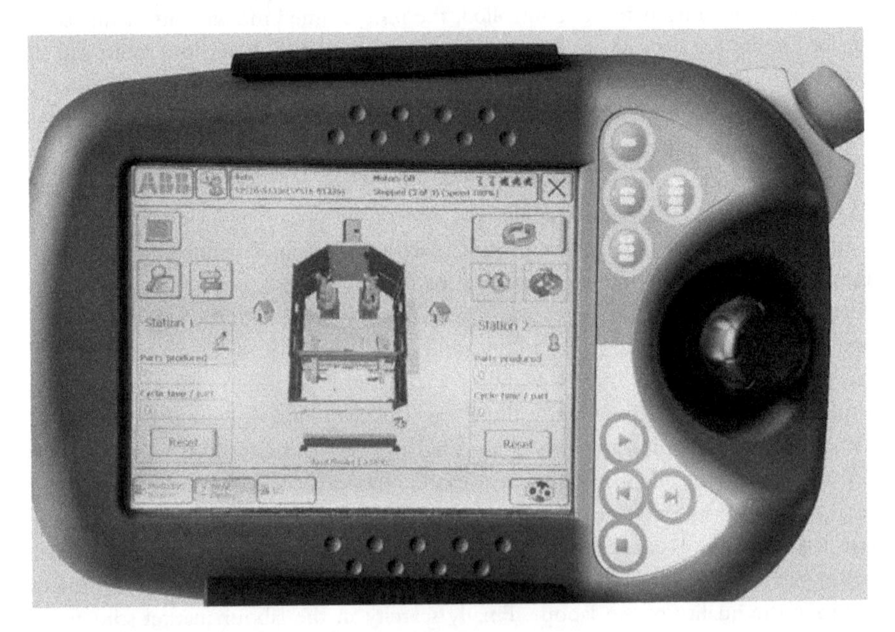

FIGURE 4.1B An industrial robot: power supply

FIGURE 4.1C An industrial robot: control system

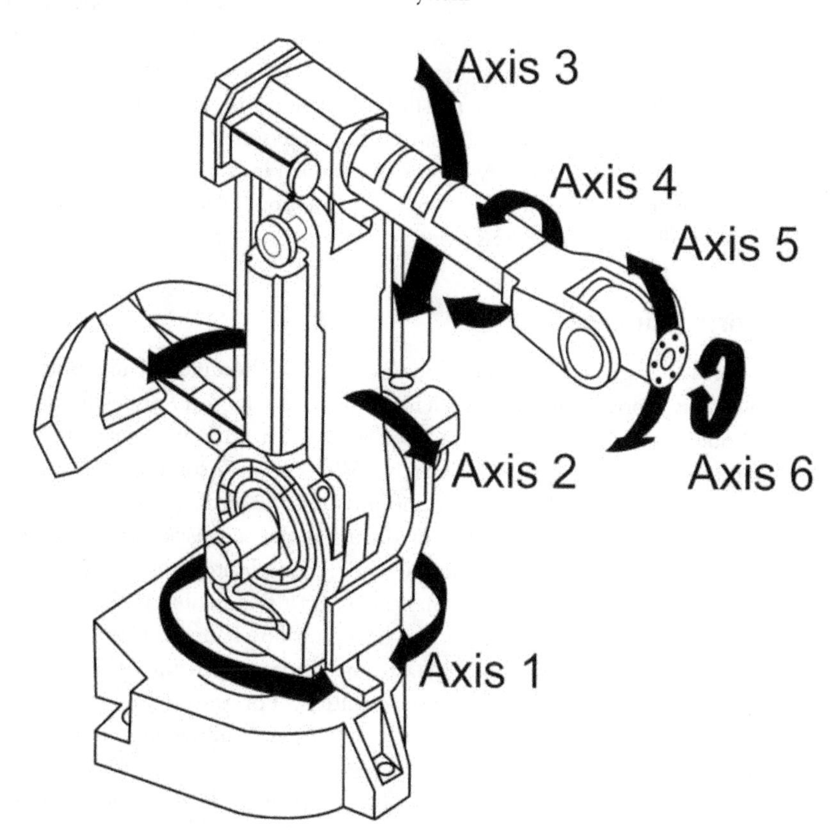

FIGURE 4.2 A schematic drawing of a six-axised robot arm

A robot consists of three basic components: the mechanical manipulator, the electronic control system and the electric power supply (see Figure 4.1). The mechanical structure of the robot is mainly determined by the number of axes and the nature of the axes. Thus for a three-axised robot we can distinguish four types: the *Cartesian robot* with three translation axes, the *cylindrical robot* that has two translation axes and one rotation axis, the *spherical robot* with one translation axis and two rotation axes and finally the *articulated robot* compounded of three rotation axes. In the control system we can also distinguish four types: *sequential control* in which the manipulator goes from one position to the other, *course control* in which control of the course of the robot arm takes central position, *adaptive control* in which the movement is adapted by means of a sensor, and *tele-operated control* in which the robot can be manipulated with remote control by a human operator. In the current generations of robots the quality of the operations is no longer determined by the mechanical systems but by the quality of the control. Of great importance is the accuracy in repetition. In sequential control this relates to the accuracy of the position when the robot arm goes from position 1 to position 2 and in course control to the borders of the real course compared to the theoretical course. An ideal robot has infinitely stiff arms, carries out the movements without hysteresis and consists of parts all made with *zero* tolerance. But in practice robots change under influence of the load, aberrations develop when a certain position is approached from different directions and the composition of the parts exhibit size differences in comparison to other robots from the same series. An attempt is made to correct this kind of aberration by means of various kinds of modern technologies – like automatic calibration, laser measuring, sensors, electronic stabilisation, clash detection and an active brake system.

4.2 The first four aspects

In this section we begin with a systematic discussion of the different dimensions.[2] The philosophers Dooyeweerd and Vollenhoven present these in a specific order: they talk about the so-called 'earlier' and 'later' aspects. They use these terms to prevent the order from being seen as a hierarchy. We start with some of the 'earlier' aspects: the arithmetic, spatial, kinematic and physical-chemical.

The first aspect that we would like to discuss is the *arithmetic.* Simple robots consist of one, two or three axes and the more complicated robots of five or more axes. Every technological artefact has the aspect of countability. A motor car has four wheels, a shaver three heads and a television 36 channels. Countability is therefore one aspect of all things or artefacts and distinct from the other aspects. Dooyeweerd characterises this as 'discreet quantity'. For the arithmetic facet a number of specific laws are valid, for instance, adding up, subtracting, multiplying and dividing.

The second, *spatial* aspect is encountered in various ways in the robot: the different positions that can be taken up by a robot arm, the course made by the robot arm, the range of the robot and, of course, the form itself of the robot. Spatial extent also is inherent to all things or artefacts, and has its own particular nature and character. The particular nature of this facet we can characterise as 'extent' or 'unbroken extent'. The total range of a Cartesian robot, for instance, is determined by the space formed by the range of the different translation axes. If a cylindrical robot moves only around its rotation axis, the course covered has the form of a circle. For a circle the law applies that the distance from each point to the centre is equal. So here we encounter laws that apply to spatial figures and the relations between them as categorised in geometry and stereometry.

Third is the *kinematic* aspect. In all applications movements play a great role: transport of parts (movement from position 1 to position 2), assembling one part on another (for instance by means of a screwing movement), the execution of a joining operation (movement over the length of the total load seam) and the application of a coat of paint on products (movement to reach the total surface). The particular nature of this aspect is 'movement' or 'continuous movement'. We know a great number of movements like the uniform rectilinear movement, the oscillating movement, the circular movement, vibration and waves. One of the fundamental laws of classical mechanics is that an object on which no force is exerted, moves rectilinearly and uniformly. In robotics in particular we come across accelerated and retarded movements. For a uniformly accelerated movement the distance covered is directly proportional to the acceleration, and quadratically proportional to the time. And for a retarded movement a comparable relation holds. If a loaded robot arm moving at a specific speed is abruptly stopped it will not come to a halt immediately but the movement made can be described as a cushioned vibration. Not only robots but all things and artefacts can be subject to the laws that apply to the kinematic aspect. We are referring here to the laws of kinematics and wave theory.

The *physical* or *physical-chemical* aspect plays a great part in the designing and use of robots. The materials used for the manipulator must have a certain strength, hardness and firmness. To set in motion the manipulator a certain amount of electric energy is needed. The sensors and measuring apparatus used are all based on physical principles and physical laws. Dooyeweerd describes the particular nature of this aspect as 'energy'. The physical aspect comprises a number of different phenomena and areas. Think, for instance, of the expansion of solid bodies, fluids and gases. Thus the expansion of the arms of a robot is directly proportional to the length of the arm and to the difference in temperature to which the robot is exposed. To accelerate a loaded robot arm uniformly from a stationary position to a maximal speed demands a specific amount of energy that is directly proportional to the mass of the arm and the weight hanging from it. In the control box of the robot there are numerous electronics to control the robot arm. In this box heat is generated as a consequence of the resistance that the electrical current encounters in the circuits and in the wiring. The heat generation is amongst other things directly proportional to the resistance of the system and to the square of the current intensity.

Dooyeweerd and Vollenhoven gave account of the versatility of reality in their theory of modal aspects. Roger Henderson described the history of the development of this theory in his doctoral thesis *Illuminating Law* (1994). 'Modal' comes from the Latin *modus* which means 'way' or 'manner'. The theory of modal aspects deals with the different ways or manners in which we experience reality, or better still: we experience reality in different ways because it is so multifaceted. It relates to the different *aspects, modalities, modes of existence, facets* or *dimensions* of reality.

Each modality is distinguished by its 'particular core', the arithmetic aspect by 'quantity', the spatial by 'extent', the kinematic by 'movement' and the physical by 'energy'. This 'particular core' is mostly denoted by Dooyeweerd as 'meaning nucleus'. The prefix 'meaning' is used by him to make it clear that these aspects are not detached from one another but are mutually connected in a meaningful coherence (see Chapter 2). The different dimensions cannot be traced back to one another or reduced to one another. In other words, the arithmetic facet does not merge with the spatial facet, nor the spatial with the kinematic, nor the kinematic with the physical. Here are two examples.

The first is the intriguing paradox of Zeno of Elea. Achilles and the tortoise decide to have a race. Let us say the total distance is 100 metres. Achilles begins at the starting point. The tortoise gets a head start and starts at the halfway mark. When Achilles has done 50 metres – having reached the point where the tortoise started – the tortoise is 5 metres further. When Achilles has covered these 5 metres, the tortoise is another 50 centimetres further. When Achilles has covered these 50 centimetres, the tortoise is another 5 centimetres further. If we continue this argumentation it will transpire that Achilles will never overtake the tortoise. But this conclusion conflicts with our experience. The origin of the paradox is that Zeno does not recognise the particular character of 'motion' but reduces it to the spatial aspect (distance). This paradox disappears when we realise that the speed represents the covered distance per time unit. If Achilles runs ten times as fast as the tortoise it means that he covers ten times as great a distance as the tortoise in the same time. So after some time Achilles will simply overtake the tortoise.

A second example is the belief of the instrument maker Franz Reuleaux (1829–1905) about the essence of technology. In his book *Theoretische Kinetik: Grundzüge einer Theorie des Maschinenwesens* (Theoretical kinetics: outlines of the theory of the essence of machines) from 1875 he states that the essence of the science of machines is the theory of motion. In itself this was not such a mad idea: at the time machines consisted of moving parts. What Reuleaux did not fully realise, however, is that in instrument manufacturing the issue is not merely movement but *controlled* movement. Controlled movement is brought about in technology only by supplying energy. And this supply should take place in a controlled manner. So we see that Reuleaux did not sufficiently distinguish between movement (the kinematic aspect), energy (the physical aspect) and control (the formative or forming aspect; see Section 4.4).

The irreducibility of the different aspects is also denoted as 'sphere sovereignty'. This expression emphasises the particularity of a facet or sphere (physical laws apply to the physical facet or the physical sphere) as opposed to the particularity of other facets (physical laws, for instance, are not dominant in the biological facet or the biological sphere).

In total Dooyeweerd and Vollenhoven distinguish 15 aspects: the arithmetic, spatial, kinematic, physical, biotic, psychic, analytical, formative, lingual, social, economic, aesthetic, juridical, ethical and the pistic aspect (aspect of faith). All these have a universal character, all 'things' in our reality function in 15 aspects. Not only utensils, clothing and technological products, but also plants, living beings, events (weddings, concerts, revolutions) and social structures (marriage, family, church, state, business).

4.3 Object functions and subject functions

There is a huge difference between the way in which technological artefacts like robots function in the first four aspects and in the later 11 aspects.[3] This difference emerges in the so-called object functions and subject functions.

In philosophy the words 'object' and 'subject' have a specific meaning. In the seventeenth century there occurred in philosophy the so-called 'reversal to the subject'. The great philosophers left the track of philosophy inspired by the church and sought their certainty in *ratio* or human reason. This development led to a division between subject and object. The human being as a reasoning subject placed itself in relation to the cosmos as an object that can be known. The slogan 'knowledge is power' of the English scholar Francis Bacon became the compass of Western humankind which attempted to escape the grip of natural reality by means of knowledge. In modern philosophy the subject–object relation becomes a polarisation that denies the coherence between both 'poles' and that is characterised by hierarchy, power and subordination. From a philosophical point of view there are serious objections to be brought against an essential separation between the human subject and natural object and the accompanying hierarchical relation between the two. The separation and the hierarchy do not do justice to the fact that the human being is a part of natural reality. Human beings are connected with every 'fibre' to the natural environment in which they live and to societal relationships in which they function. Moreover, the human faculty of cognition is an integral part of reality (Chapter 2).

In this chapter we would like to introduce a different meaning to the words 'subject' and 'object'. We start with a simple example. There is a huge difference in the way a rock functions in the spatial and in the aesthetic aspect. A rock takes up space (is a subject in the spatial aspect) and we may find it beautiful (is an object in the aesthetic aspect). There is a comparable difference with reference to the physical and moral dimension. A rock rolls down the mountain (is subject in the physical dimension) and can be used to injure somebody (is object in the moral dimension). In these meanings of the words 'subject' and 'object' a rock has a subject function in the spatial and physical modality and an object function in the aesthetic and moral modality.

A systematic investigation of an industrial robot yields the following. A robot functions as a subject in the first four dimensions and as an object in the later 11 dimensions. For instance a robot has a number of axes, takes up space, has a moving arm and conducts the current through electric wiring. However, a robot cannot perform any economic or juridical acts. On the other hand, however, it is the *object* of economic or juridical acts by human beings. To elaborate on this: a manufacturer who plans automatisation will calculate beforehand how much can be saved by using a robot. But a robot in itself has no inkling of the economic aspects of automatisation. A manufacturer knows that for a more sophisticated robot – definitely if it is a new and innovative model – s/he would have to pay a higher price, but the robot itself knows nothing about this. At the end of the negotiations the manufacturer and the seller sign the contract, not the robot, in any case it does not have a 'right' to do so. Therefore a robot has an *object function* in the economic and juridical aspect. These object functions naturally only emerge in the actions of human beings. Thus the economic and juridical functions are *latent* in the robot and are *unfolded* by the acting human being. This example also shows (although this will not be elaborated on) that the manufacturer and the seller do function as subjects in the economic and juridical dimension.

The above distinction helps us to describe the different relations more finely. If a manufacturer buys a robot we can say there is an economic *subject–subject* relation between the manufacturer and the seller, for both have a subject function in the economic facet. But we can say there is an economic *subject–object* relation between the manufacturer and the robot, for the manufacturer has a subject function and a robot an object function in the economic facet. When a robot joins one part to another we can speak about a physical *subject–subject* relation between the two parts. They both have a subject function in the physical aspect.

4.4 The 'later' aspects

We now focus on the 11 'later' aspects from the perspective of an acting human being who functions as a subject in all aspects and who also makes use of technological artefacts in his/her actions by means of which their object functions are unfolded.

The fifth aspect is the *biotic* that relates to biotic processes taking place in living organisms. Here we are thinking of processes like reproduction, growing, metabolism, adaptation to circumstances, ageing and dying. The meaning nucleus of the biotic facet is usually described as 'life', 'organic', or 'vital'. It is clear that this aspect relates to all processes concerning 'impregnating, growing, blooming and dying'. People are living beings: they function as subjects in the biotic aspect. In the biotic life of human beings technological artefacts function as objects.

By way of example: pacemakers have no subject function in the biological dimension, yet they play an important part in watching over the heart rhythm. By emitting an electrical impulse they support the biological working of the heart. Medical robots do not have any subject function in the biotic dimension either, but they have been developed in such a way that they can assist the surgeon when operating on a patient.

The sixth, *psychic* aspect, relates to the perceptive and sensitive dimensions of experiences and the conduct of human beings. The meaning nucleus of this aspect is described by most authors as 'feeling', 'sensitive' and 'sensory'. In the perceptive dimensions we are concerned with the various processes by which human beings 'are in contact' with reality around them. This concerns sensory observations like seeing, hearing, smelling and feeling, but it also concerns signals like temperature, balance and pain. In the sensitive dimensions we are concerned with feelings like longing, desire, passion, affection, joy, uncertainty, anger, fear and irritation. We would like to emphasise that the psychic aspect may not be identified with an experience or the event itself. When one is glad because one has solved a difficult technological problem, the psychic aspect (joy) comes prominently to the fore in the situation and the other aspects recede. In the psychic life of human beings technological artefacts function as objects. A robot has no subject function in the psychic dimension: it cannot observe and has no feelings. But the robot can be observed by human beings and can evoke feelings in human beings. So it happens that in a factory the operators regularly get cross with a robot because it does not 'do' what it is 'supposed to do'.

The seventh, *analytical* aspect relates to the ability of human beings to analyse reality and make various logical distinctions. The meaning nucleus of this aspect is indeed described as '(consciously) discerning', the 'logical', 'intellectual' or 'rational'.

In the first section of this chapter we gave a general introduction to the world of industrial robots with various logical differentiations. We defined a robot as an automatic manipulator with three or more axes that can be reprogrammed and that can be used for various purposes, an analytical definition of the identity of a robot. The robot itself does not function in the analytical dimension but is an object of human beings' knowing and thinking. In our view to think logically is therefore a *function* of human beings and not *the* characteristic; this is in opposition to, for instance, the philosophy of the Enlightenment that absolutises logical thought. The analytical facet, finally, has a *normative* character. A human being cannot violate a physical or biological law, but s/he can indeed infringe a logical law by, for instance, using a specious argument or giving an inconsistent report. Apart from the analytical, all following facets discussed also have a normative character.

The eighth aspect is called the *historical, cultural* or *formative* aspect. The word 'historical' can easily conjure up misconceptions since it refers to events that have had an influence on history. The word 'cultural' can also be misleading. We therefore prefer to speak about the formative or forming aspect. This aspect is concerned with the ability of human beings to form reality around them. This forming does not only concern material reality as it emerges par excellence in technology but also the forming of social relations, economic structures and artistic creations. It is concerned with the influence, directing, control or power of individual persons, societal relationships or nations. Dooyeweerd describes the meaning nucleus of this aspect as the 'controlling formation of a *particular capacity, structure or condition*'. One can also describe the meaning nucleus of this aspect as 'power of freedom', or simply 'power' or 'dominion'. In technological activities – from development, manufacture, up to and including use – the controlling aspect is expressed in *optima forma*, and therefore it is understandable that Henk Hart (1984) speaks about the technological-formative aspect. Culture always has to do with giving form to reality. People give form to their social activities, economic actions and artistic expression. It hardly needs to be explained that technology functions not as subject but as object in this formative modality. Thus a robot itself exercises no power but is a powerful instrument in the hands of human beings. The formative aspect implies that the technological operation of the engineer is subject to norms. And in the designing process the 'power' of a robot is regulated by the activities of the engineer.

The ninth aspect is the *lingual* one and it relates to the meaning attached by human beings to lingual expressions, symbols or acts. The meaning nucleus is described as 'interpretation' or 'symbolic meaning'. The lingual dimension of reality eminently comes to the fore in, for instance, the American English language or the Midland American English dialect, but cannot be identified with these. Languages and dialects have their own structure and function in all dimensions. Technological artefacts also have a lingual dimension namely as object function. They can be discussed and denoted by human beings. Thus we speak about a Cartesian and a cylindrical robot. But the meaning can also be wider. Robots used for heavy and dangerous labour will more readily get a positive meaning while robots that lead to higher unemployment are often denoted negatively. The lingual facet is also found in all kinds of encoding, symbols and signs used in technology. So, for instance, on traditional electric resistances we find a colour coding from which the magnitude of the resistance can be inferred. The capacity of a condenser is expressed with the aid of the symbol 'F' (from Farad). In technological drawings resistances and condensers are often denoted with the sign of a 'saw tooth' and 'two parallel lines' respectively.

The *social* aspect concerns the connection between people and groups, the customs in mutual association, the rules of human intercourse, and solidarity and communion. The meaning nucleus is described as 'intercourse and association' and 'connectivity and coherence'. The social aspect is found in every societal form: from marriage and family to society and church, and from friendship and relationship to trade and commerce. It will be clear that technological artefacts only have a social object function. People use technological artefacts, give them a certain role in their lives and society. The introduction of technological innovations is therefore often coupled with a different form being given to social relations. Think of the influence of computer games on family life. Sometimes technology also has a compelling influence on the way in which people interact mutually. A moving belt, for instance, at which co-workers stand in a fixed position and where a new product comes along every so many minutes, gives limited opportunity for mutual interaction. If the distance between the colleagues is small and the noise is restricted, social contact is easily possible, but if the distance is greater and the noise dominates, then it is hardly possible. Changes in the technological structure of a factory like the introduction of industrial robots therefore always lead to a change in social relations.

The *economic* aspect we see in business and sales, money transfers and banking. The meaning nucleus of this aspect is described by Dooyeweerd as the 'control of rare goods', by Goudzwaard (1979) as 'stewardship, bearing fruit' or 'careful management', or simply as 'productivity' or 'effectivity'. In our view acting in the economic sphere is to act in a normative way, for the fulfilling of human beings' needs is also regulated by norms. The image of the steward probably best portrays this. In the past it was the steward's task to control the estate entrusted to him as best as he could so that it was maintained, that all that lived and worked there were provided with a livelihood, and that goods were handed over to the owner. The steward had a responsibility for all who were involved in the estate. Currently we see a similar idea of 'responsibility for everybody' recurring amongst other things in the various approaches of *corporate social responsibility* based on the 3P-model (*profit, people, planet*) and the *stakeholder*-model (Freeman, 2001). The idea of 'responsibility for everybody' is exactly what is missing where the economic aspect is narrowed down to financial parameters and where the interests of society have to yield to power and self-enrichment as described by, amongst others, Noreena Hertz in *The Silent Takeover. Global Capitalism and the Death of Democracy* (2002). In the introduction to this section we have already remarked that an industrial robot itself has no inkling of the economic aspects of automatisation. A robot is an object in the economic activity of human beings, first when such an artefact is purchased and subsequently when it is used in a production line. This means that the economic responsibility for the automatisation rests with human beings.

The 12th aspect relates to the experience of beauty and harmony. This *aesthetic* aspect emerges in particular in the arts: creating and enjoying paintings, statues, music and song. We can also intensely enjoy the beauty of nature. The aesthetic facet also plays a significant role in all kinds of utensils made by human beings. When we buy a table and chairs, we want to use it in the first instance for taking meals and for other activities like meetings. But it is also important that the table and chairs should look good and should harmonise with the interior of the house or office. The aesthetic facet is characterised by 'harmony' or 'beauty'. Technological artefacts themselves do not experience beauty, but they are the object of the experience of human beings. Thus an industrial robot can be designed in a harmonious way so that it embodies a certain beauty. In designing apparatus, increasing consideration is given to the meaning of the design.[4] However, technology also has a great influence on our aesthetic experience. Think, for instance, of the influence of medical technology on our opinions of beauty and the (female) body, an influence aptly voiced in an advertising slogan '*Beauty is Natural. Perfection is Surgical*' by the firm *Eurosilicone*.

The 13th aspect is the *juridical*. In Western countries every citizen is bound by the law, and impartial justice is guaranteed. The juridical aspect is not only concerned with the law and the juridical system, but also with the underlying concept of justice and righteousness. In this aspect the justified interests of the various persons or parties take central position. Injustice arises when the interest of one person holds sway over that of another or when one party dominates the other. Each individual and each societal relationship is entitled to its rightful place in our society. So the juridical aspect does not restrict itself to legislation and jurisprudence. Of course engineers have to comply with national and international directives and laws when designing a new product. But in the final instance the issue is to what extent technology in the hands of human beings can contribute to the development of a just society. In Part III of this book this kind of question will be dealt with further. In the application of industrial robots juridical questions are at stake especially in the case of safety and the intellectual ownership of the design.

In the *moral* aspect, the 14th, the love and care for the reality in which we live are expressed. This love or care can take the form of the love between partners, parents and children, and brothers and sisters in a church community. We also think of care between neighbours, friends, colleagues at work and the care given by doctors and nurses to patients in hospitals, nursing homes and care institutions. Furthermore we should not forget the care for our natural environment. The meaning nucleus of this is described as 'love', 'fidelity in friendship and marriage' and 'willingness to serve'. In everyday speech the word 'moral' is used for the entire constellation of values and norms regarded as important in social life and for the duties and virtues to which individual citizens should adhere. In our approach the moral aspect pertains to norms concerning love, care and willingness to serve. The word 'ethics' is reserved for the whole constellation of values, norms, virtues and duties regarded as important to society. In the design of technological artefacts moral norms clearly play an important part. The moral dimension of industrial robots can be seen, for instance, concerning the performance of dangerous operations. The engineer faces the challenge of designing an apparatus in such a way that the care of and serviceability to the client take central position. But s/he also has to pay attention to the care of the environment.

The 15th and last aspect is the *pistic* aspect or aspect of *faith* (*pistis* is the Greek word for faith). This facet of reality may not be equated with a church or religion. The aspect of faith denotes the ability of human beings to believe and is expressed specifically in religion. It cannot be reduced, as many theologians and philosophers have done, to the emotional life or the intellectual capacity of human beings but has its own particular nature and character. Believing in God means that one knows for certain that He exists, that one accepts something on His authority and that one relies on Him. These different elements of 'faith' we clearly find in the Catechism of Heidelberg, a creed from the sixteenth century which describes true faith as a 'sure knowledge' and 'firm confidence'. Dooyeweerd describes the essential core of this aspect of faith as 'transcendental certainty'. The human being's ability to believe is manifested in the first instance in a fundamental religious sense. People put their trust in God (Jewish–Christian faith), Allah (Islamic faith) or Buddha (Buddhist faith). But even the humanist 'faith' in rational man is an example of this. It is not easy to explain the relationship between the aspect of faith and religion adequately. Sometimes the image of the branch and roots of a tree is used. Just as a branch is nourished by the roots, the faith of human beings is nourished by religion. In the second instance faith plays an important part in various relationships within which human beings function. Believing means that one relies on the other (person) and that one trusts him/her. Without trust life and living together in relationships is well-nigh impossible. Not for nothing is trust sometimes called the cement or lubricant of human relationships. Finally faith plays an important role in relation to various objects. In this framework the word 'trust' is mostly used. People trust in technology. Put otherwise, technological artefacts have an object function in the aspect of faith. Thus every driver of a car trusts that when he steps on the accelerator the speed of the car will increase. Likewise a worker in a factory trusts that a robot will perform the programmed actions. It might be noted that trust in technology can take on religious traits (Noble, 1997). This can be seen among futurologists who expect that technology holds the promise of realising a dignified and just society. In this way of thinking the religiously laden expectance of delivery has been replaced by a technologically laden expectance of delivery (see further Chapter 14).

Table 4.1 gives an overview of all aspects and their meaning nucleus as described by Dooyeweerd, Vollenhoven and others.

TABLE 4.1 Overview of the modal aspects and their meaning nucleus

Aspect	*Meaning nucleus*
arithmetic	discreet quantity, number
spatial	extent, unbroken extent
kinematic	movement, continuous movement
physical or physical-chemical	energy, interaction
biotic	life, organic, vegetative, vital
psychic	feeling, sensitive, sensorial
analytical	logic, rational, analytical distinction, conscious distinction
formative	controlled forming, power of freedom, power, domination
lingual	denotation, meaning, symbolic meaning
social	intercourse, communion, interconnectedness, coherence
economic	control of rare goods, stewardship, fertility, productivity
aesthetic	harmony, beauty, allusion, full diversity of shades
juridical	retribution, justice, law
moral	love, care, fidelity, willingness to serve
faith	transcendental certainty, reliability, credibility

We have emphasised several times that 'things' – utensils, clothing, plants, living beings, events and social structures – function in all dimensions. This statement evokes questions. Is there really no special relation between, for instance, numbers and arithmetic, biological organisms and the biotic, languages and the lingual, laws and the juridical, and religion and the aspect of faith? And do not technological artefacts like industrial robots function in a special way in the formative modality? Formulated differently, is there no particular relation between the 'things' mentioned and the modality in question? The reply to the last question has to be in the affirmative. In Chapter 5 we will investigate further the structure of concrete things and see that they have a particular quality that we can analyse further with the aid of the theory of modalities. It transpires that numbers are *qualified* by the arithmetic aspect, biological organisms by the biotic aspect, languages by the lingual aspect, laws by the juridical aspect and religion by the aspect of faith. It will also become evident that industrial robots are *qualified* by the formative aspect and pacemakers by the moral aspect. In order to make a proper design for a technological artefact it is of the utmost importance to know by which modal aspect such an artefact is qualified (see Chapter 5).

In the hands of the engineer the theory of modal aspects can be a good 'instrument' for getting a better view of the many aspects of technology. This theory shows that technology cannot be reduced to kinematics or physics but functions in all aspects. This approach also clarifies – amongst other things by differentiating between subject functions and object functions – in which way technology exerts power over human beings and nature. A last remark: this analysis gives insight into the different kinds of normativity which an engineer should keep in mind. In other words, there is a pronounced interaction between the analytical, critical and directional function of philosophy (Chapter 1).

4.5 Coherence in diversity

Our analysis evokes a great number of questions. What is the relation between a technological artefact and the 15 dimensions? And how do the different dimensions relate to one another?

The polished diamond is a well-known metaphor to explain the nature of the different dimensions. Such a diamond has many facets. If one looks from one specific angle, one facet catches the light, and if one looks from another angle, another facet catches the light. Only one diamond is referred to and this one diamond shows many facets. The image of the polished diamond and its facets is appropriate for illustrating the relation between a technological artefact and its dimensions. If one looks at a technological artefact from a physical perspective diverse physical phenomena are highlighted. If one looks from the formative point of view one catches sight of control and power. And when one looks from an economic perspective the various economic mechanisms catch the attention. So only one technological artefact is looked at and that one artefact exhibits no less than 15 different aspects, like the one diamond having many facets that can catch the light.

The coherence between different aspects emerges in the so-called analogies. An analogy is a phenomenon in which one modal aspect refers to another. For instance, when we say that an engineer has designed an efficient control mechanism for a robot we mean that with a minimum of components s/he has obtained a maximal result. In this example we could speak of reference from the formative to the economic aspect. There are two kinds of reference: a reference to an 'earlier' dimension and a reference to a 'later' dimension (the terms 'earlier' and 'later' were introduced in Section 4.2).

We start with references to earlier aspects. In the first section of this chapter we said that the spatial range of a Cartesian robot (a robot with three translation axes) is determined by the space formed by the range of the different translation axes. If the range of the x and y axis is 1.5 metres and the range of the z axis is 0.5 metres then the spatial range of the robot is $1.5 \times 1.5 \times 0.5$ m^3. First of all we need numbers to determine the dimension of the range: in this case we speak of a three-dimensional range. To express the size of this range in the different dimensions we also need numbers: in this case the numbers 0.5 and 1.5. Hence, we can only describe spatial phenomena properly by making use of numbers. In the terminology of the Dooyeweerdian reformational philosophy: the spatial aspect refers to the arithmetic aspect. Or, the numerical characteristics of spatial figures (length, width, height, surface, volume, dimension) are retrocipations in the spatial aspect to the arithmetic aspect (Latin: *retro* is back or backwards and *capere* is take or seize). We find comparable back references in the kinematic aspect. In robotics we come across accelerated movements amongst other things. These movements may take place in a straight line (e.g. for a Cartesian robot), but also in a circle (e.g. for a cylindrical robot). For a uniformly accelerated movement the distance covered is directly proportional to the acceleration and quadratically proportional to the time. In the case of a rectilinear and a circular movement we are dealing with a retrocipation to the spatial aspect. For a straight line and a circle are spatial figures. But when we speak about the acceleration and distance covered we make use once more of numbers and this has to be denoted as retrocipation to the arithmetic aspect. Thus we see that the kinematic refers to the spatial and the arithmetic. In other words, we are here dealing with retrocipations in the kinematic to the spatial and arithmetic aspects.

In the relation between the arithmetic, spatial and kinematic aspects on the one hand and the physical aspect on the other we also come across various references. A moving robot arm itself has an amount of energy, also called kinetic energy. 'Kinetic energy' is a retrocipation within the physical to the kinematic facet. A loaded robot arm transports a part or product of a certain weight. This part or product is at a certain height above the ground. So this part or product has an amount of potential energy that is directly proportional to the weight and the height. This potential energy depends on the place: so here we are dealing with retrocipation within the physical to the spatial facet. In all these retrocipations a certain amount of energy is at stake. 'Energy' denotes an amount of interaction and thereby refers from the physical to the arithmetic facet. So here, too, we can speak of retrocipation.

Retrocipations are also found in the 'later' aspects discussed above. For instance, if we say that a businessman makes use of his economic power we are dealing with an economic phenomenon from which there is a reference to the formative aspect. Thus, speaking of a feeling for language, we can say there is a reference to the psychic dimension, in juridical causality to the physical dimension and in the religious expression 'communion of the saints' ('saint' in this connection pointing to the relation of the believers to God) to the social dimension. From all these examples it becomes evident that retrocipations are no simple repetitions of earlier in later aspects but *qualified resumptions*. In a retrocipation the earlier aspect does not return to its own qualification but takes on the qualification of the later aspect. The particular characteristic of a retrocipation is that within the qualification of the later aspect the earlier one is still to be recognised. When we say that a businessman makes use of his economic power – for instance by negotiating for higher discounts with his suppliers when he orders in bulk – we are dealing with a typical economic phenomenon in which the formative aspect can still be recognised but is no longer present in its own particular qualification.

At the same time earlier aspects also refer to later ones. These references are called anticipations (Latin: *ante* means before and *capere* means to take or seize). In the design of an industrial robot all possible kinds of anticipations can be found. Much consideration is given, for instance, to the ergonomic aspects of a robot. For instance in the spatial construction the control panel must be at working height and the critical parts have to be easily accessible. Here we are dealing with anticipation within the spatial modality to the biotic modality. The robot has to be designed in such a way that social contact between the workers during work time remains possible and is even stimulated. The noise level during the technological operation is therefore limited to a certain maximum. All movements and all physical processes have to meet specific noise norms to enable cooperation between colleagues. These kinds of aspects of a design are anticipations within the kinematic and physical to the social modality. In designing a robot an engineer has to take into consideration various safety regulations. These regulations refer, for instance, to the position of an emergency button with which the robot can be switched off and the spatial screening of the turning range of the robot; here we are dealing with anticipation within the spatial to the juridical modality. The design of the electric power supply also has to comply with various regulations and hence we can speak of anticipation within the physical to the juridical modality. The spatial construction, the various movements, the whole constellation of physical processes and the electronic control all have to be reliable. The reliability of an apparatus or part is sometimes expressed in technology by the concept '*mean time between failures*'. In this case we are speaking of anticipation within the spatial, kinematic and physical modality to the modality of trust. These examples explain that in anticipation a later aspect does not 'come back' in its own qualification but in that of an earlier aspect.

It proves that references are a fundamental phenomenon, for in the meaning nucleus of every aspect we find references that resemble the meaning nucleus of other aspects. This phenomenon is called *sphere universality* because these references are characterised by their own meaning nucleus, not the one of the aspect to which reference is made. Let us give another example. When we say of a business concern that its annual statement looks reliable then we are saying that the different ratios as, for instance, the relation between own/particular and foreign capacity meet certain requirements. Then it is an economic phenomenon that resembles or refers to (anticipation) the aspect of faith (trust).

'Sphere universality' shows that the reality in which we live is multifaceted. To use an image: we see different colours (modal aspects) and diverse shades (analogies). Every attempt to reduce reality to only one or a few dimensions (reductionism) is doomed to fail. Reality is not more of 'the same' but more of 'the other' and this versatility is also found in technology. In our analysis of an industrial robot we have named all the different modal aspects and identified some analogies, but even this limited analysis shows how complex technology is. Technology is more than controlling materials, in it we find the versatility of the whole of reality. This conclusion is of great significance for the ethics of designing. For it means that we as engineers are not only dealing with technological but also with various other norms.

The phenomenon 'sphere universality' at the same time explains why many scientists and engineers have fallen into the trap of reductionism. Over the years attempts have often been made to reduce all phenomena in reality to one denominator. This is where we encounter the so-called 'isms' like materialism, biologism and economism. Because every aspect reflects all other aspects it is a temptation to reduce the complete versatility of reality to that one aspect. This explains why a physicist sees physics as the foundation of reality, why a biologist points out the fundamental character of biotic processes and an economist explains everything from the angle of the idea of shortage. However tempting, this kind of reduction does no justice to the versatility of our reality. It is not strange that an engineer identifies the formative power of technology as the driving force of our society. Within the formative aspect all other aspects are reflected and conversely in all other aspects we find references to the formative.

In philosophy of technology the autonomous character of the development of science and technology is often pointed out. Amongst other things we find this opinion in the work of the French philosopher Jacques Ellul (1912–1994). The idea is that science and technology have a particular dynamics that human beings cannot influence. But in this opinion no recognition is given to the fact that technology functions as an object in the activities of human beings. It is man's responsibility to 'unfold' technology. This unfolding not only takes place under the influence of technological norms but also under the influence of social, economic and moral norms (see Chapter 10).

We have shown that the coherence between the various aspects can be analysed from two 'directions': from the later to the earlier aspects and from the earlier to the later aspects. These different directions emerged clearly in the different analogies: retrocipations and anticipations. The direction from the earlier to the later aspects is called the *basic* direction. This means that the earlier aspects form the basis or foundation for the later ones. Thus the arithmetic, spatial and kinematic dimensions formed the basis to the physical functioning of a robot. The direction from the later to the earlier aspects is called the *transcendental* direction. This implies that the earlier dimensions are unfolded or deepened by the later dimensions. The different American and European laws, for instance, have a great influence on the spatial design of a robot. So here the spatial dimension is unfolded by the juridical dimension. In the analysis of technology we will regularly return in direct and indirect ways to the basic and the transcendental direction. The basic direction emerges in particular in the arithmetic, spatial, kinematic and physical grounding of technology and the transcendental direction in the normative unfolding of technology.

The question now is how we should interpret this analysis of reality. Is the theory of Dooyeweerd and Vollenhoven a *model* projected on reality or is it a model grafted onto reality? Do we here have a *systematic philosophy* in fixed categories or a systematic philosophy that helps us to analyse technology and design technological artefacts? These are important questions that we would like to consider here. First, Dooyeweerd and Vollenhoven very consciously attempt to link up with everyday experience. They try to express this non-scientific experience in a scientific manner and that gives their model a basic openness. Second, the theory of modalities is not yet fully rounded according to Dooyeweerd and Vollenhoven. They expected that scientific research would lead to adaptations and corrections. During the course of time various modifications have been suggested. For example, it has been proposed to split up the psychic aspect into a perceptive and a sensitive aspect or to insert an aspect of purposiveness between the social and the economic. By this we want to emphasise the open character of this approach, for in this way a specific approach can stay inventive. Recent studies have shown that this approach is still exceptionally fertile (Strijbos and Basden, 2006).

4.6 Other approaches

What is the relationship between the analysis recounted above and the studies of other philosophers? The most topical is the analysis developed in the framework of the research programme *The Dual Nature of Technical Artefacts* carried out at the Delft University of Technology. But first let us consider the insights of Gilbert Simondon, a French philosopher (1924–1989). Gilbert Simondon is one of the few people who, even when philosophy of technology was just beginning, called attention to the nature of technological artefacts (Vries, 2008). He wrote the monograph *Du mode d'existence des objets techniques* (The mode of existence of technological objects) (1958). Until now Simondon's book has not been completely translated into English, and perhaps this is why this very original work remains so little known. A part of his ontology of artefacts is devoted to the relation between human beings and technology (see Chapter 12). Here we are primarily concerned with what he wrote about the relation between the functions of an artefact and its physical features. An important concept in the work of Simondon is *concretisation* by which he means that the origin of an artefact is a matter of the intended functions in a physical object being crystallised further and further. In this process an ever more sophisticated way is developed of having multiple functions performed by one part. A designer always starts with a collection of functions that have to be performed by the artefact. Often one part per function is invented and so an artefact comes into being as a compilation of parts. In the subsequent development it happens more and more that multiple functions are combined in one and the same part. They then increase in what Simondon calls *technicity*. It is as if they are more and more formed according to technological requirements of functionality. To explain this Simondon uses some detailed examples, namely the four-stroke engine and the electronic amplifier tubes. In the four-stroke engines one sees that in the course of their development the cooling fins served not only to cool but also to make the engines more stable. Simondon says the parts become *over-determined*. They had already been determined by their original function (cooling for cooling fins) but become determined further by an extra function (affording stability). Simondon calls this a *convergence* of functions in one part. In the electron tubes separate parts for isolation (rubber parts) gradually disappear and the active components are redesigned in such a way that in certain places they themselves isolate instead of conduct. But despite the process of concretisation, according to Simondon, technological artefacts will always be less determined than natural objects where one finds an ideal 'fit' between functions and parts. The other extreme is the (technological-)scientific, abstract representations of objects, for instance the mechanical representation of any falling object as a point mass that propels itself according to a certain formula. These are still entirely underdetermined according to Simondon. Such a falling object can still become anything in terms of possible functions and its form is still in no way adapted to functional requirements. Therefore, for Simondon, as far as their mode of existence is concerned technological artefacts are an intermediate form between abstract scientific representations of objects and natural objects.

Subsequently Simondon shows how one should distinguish different levels: parts, complete artefacts and 'ensembles' (greater entities of artefacts having the character of a system). So one can take a needle as a part, the sewing machine itself as an artefact and the textile factory in which clothes are made as the 'ensemble'. According to Simondon, a kind of evolutionary process is taking place not only in the concretisation but also in the relationship between elements, artefacts and levels. Elements grow into artefacts which in turn grow into ensembles. When an ensemble at any particular time no longer meets the requirements of the time a new element can come into being in it that will lead a separate life and eventually grow into a new ensemble. According to Simondon this process, which he metaphorically denotes with the term 'saw tooth mechanism', explains the rise of information technology. Unfortunately he does not elaborate any further on this. This part of his analysis remains rather speculative. Still, it is an interesting thought.

One can, for instance, think here of the idea of feedback that in a steam engine is responsible for keeping the pressure in the kettle constant. When the steam engines were already obsolete, completely new technologies were invented entailing that idea of feedback (like an automatized production line). The element of feedback belonging to the old 'ensemble' (the steam engine) grew into a new ensemble (automatized production line) when the old ensemble had become obsolete.

We see that Simondon drew attention to the influence of the functional demands on the evolution of (the different parts of) a technological artefact. What he emphasised was the insight into its technological complexity. In our analysis the emphasis is on the interaction between human being and artefact and thereby insight is gained into the multidimensional complexity. Simondon's analysis of technological artefacts demands closer study. Due to their limited empirical basis the concepts he developed and the mechanisms he proposed have a predominantly speculative character. Yet it is important to point out that his view of the relation between technological artefact and functional requirements was in fact a forerunner of the Delft *Dual Nature* programme – although he is not mentioned in a single publication by them.

The engineers and philosophers involved in the Dutch research programme *Dual Nature of Technical Artefacts* have focused in particular on the nature and character of technological artefacts. They distinguish between two kinds of characteristics: structural and functional characteristics. The structural characteristics refer to the physical aspects of an artefact like measurements and materials. The functional characteristics refer to the purpose for which the artefact has been made. To characterise a technological artefact properly one needs both characteristics. The relation between structure, function and artefact gives rise to a number of questions. What connection is there between the structure and the different functions of an artefact? In which way do engineers establish a relationship between the different functions and the purpose of an artefact? Take, for instance, a light bulb. Its purpose is to light up a dark space. In order to realise this purpose a bulb has to contain an element that creates light when a current flows through it. This function is performed by the incandescent filament. But it is not enough to describe only the function. One also has to know which material the spiral has to be made of, the thickness of the filament and how it should be wound. From the angle of the purpose of a bulb one can describe the function of the various parts of the bulb. Thus the fitting has two functions: conducting the current and keeping the glass bulb in its holder. The difference between structural and functional characteristics is that structural characteristics give a good characterisation of the materials used without saying anything about the function of the artefact, and the functional characteristics give a good characterisation of the characteristics of an artefact without saying anything about the physical-chemical characteristics. Those who support this approach point out that the technological function contains a normative element: the technological artefact should work in a certain way. If this does not happen we can call it a bad product. Thus a bulb should convert electricity into light. If it does not do so the bulb is defective. This approach implies that an engineer has to design materials and structures by means of which a desired function can be realised. Secondly, s/he should establish a direct relationship between these materials and structures on the one hand and the desired functions on the other hand. The engineer not only wants to know *that* the artefact 'works' but also *why* it 'works'. Finally, s/he has to develop instructions so that the users can reach their goal with the help of the technological artefact. In other words, during the process of designing the engineer already has to think of the artefact's use. For a more detailed description of this approach see *Philosophy and Design* (2008) by Pieter Vermaas et al.

It is not difficult to find some parallels between our approach (the theory of modal aspects) and that of the Delft research group (*dual nature).* First, both approaches distinguish between different aspects or characteristics. The theory of modal aspects distinguishes 15 different dimensions and the *dual nature* approach distinguishes between structural and functional characteristics. The subject functions of the artefact (that are only present in the lower aspects) relate especially to the physical nature of an artefact and the functional nature is particularly determined by the object functions of the artefact in the higher aspects. One could almost say that the aspects are a further division of the two natures into 15 natures (we do sometimes call the aspects 'modes of existence' or 'modalities' which correspond with the term 'nature' in the *dual nature* programme). Second, both approaches depart from the unity of the technological artefact. This unity is deeply rooted in the theory of modal aspects (for it is concerned with the fact that the one artefact – the entity – functions in all aspects) and in the *dual nature* approach in the realisation that structures and functions are closely related. Third, both approaches use the word 'normative'. The theory of modal aspects does so in connection with the different aspects (laws for the earlier aspects and norms for the later aspects) and the *dual nature* approach with a view to the function the artefact has to perform. Finally, both approaches do justice to the fact that technology is embedded in society. In the theory of modal aspects it emerges in the different object functions of technological artefacts and in the *dual nature* approach in the relationship between the function and the (social) objective. There also are striking differences between the two. The theory of modal aspects gives a more detailed analysis of what is called structural characteristics by the *dual nature* approach. Thus a difference is made between arithmetic, spatial, kinematic and physical-chemical characteristics. Further we see that the *dual nature* approach focuses primarily on the functional characteristics with a technological character, the modal theory also looks at other functional features. Finally the modal theory gives more insight into the complex relationship between the different aspects or characteristics, amongst other things by means of the concepts anticipation and retrocipation.

4.7 Conclusion

In this chapter we have given a philosophical analysis of technological artefacts. Such a detailed analysis is vital to enable us to (a) shed light on various questions about technology, (b) analyse critically the part of technology in society, and (c) give direction to the designing of technological artefacts.

Our analysis proves that technology is exceptionally complex. No less than 15 different aspects could be discerned to one technological artefact. Each one of these aspects has its own particular nature and character that cannot be reduced to one of the others. Moreover we have analysed the relationships between these aspects. We have pointed out the phenomenon of 'referring back' (retrocipation) and 'pointing forward' (anticipation) by means of which many different 'shades' come into being within a certain aspect. Since each aspect is characterised by laws (earlier aspects) and norms (later aspects) this analysis also affords insight into the laws which an engineer has to bear in mind when designing, and the norms which a technological artefact should meet.

We have also seen that there is a difference between the first four aspects within which a technological artefact functions as a subject, and the later aspects within which a technological artefact functions as an object. This analysis gives deeper insight into the relationship between technology and society. We can therefore say that the statement that technology is autonomous, as the French philosopher Ellul claims, is too absolute. On the contrary our analysis makes it clear that humans as acting beings develop technology. If technology gets an autonomous character it is primarily due not to the nature of technology but to the acting human being who in designing does not sufficiently consider the normative aspects of technology.

Notes

1 In this section we give grateful recognition to the presentation *Introductie robotica* (Introduction to robotics) (2004) by Cafmeyer of the firm ABB.
2 The 'theory of modal aspects' or 'theory of modalities' was set out by H. Dooyeweerd in his *A New Critique of Theoretical Thought*, Volumes I, II, and III (1969) and by D.H.Th. Vollenhoven in his *Isagoogè Philosophiae* (2005). H. Hart gives an introduction to the theory of the modal aspects in *Understanding our World* (1984).
3 For products of biotechnology the difference lies between the first five and the later ten aspects.
4 A good example is that Philips' section for Small Household Appliances discovered at the beginning of the 1990s that their market no longer had to be defined as technological but as *fashion*. In other words, in the design of the product the aesthetic appearance was central and not the various nifty technological devices.

References

Asimov, I. (1986) *Robot Dreams*, New York: Berkley Publishing Group.

Cafmeyer, O. (2004) *Introductie robotica* (Introduction to robotics), Powerpoint presentation, Rotterdam: ABB.

Dooyeweerd, H. (1969) *A New Critique of Theoretical Thought*, Phillipsburg, NJ: The Presbyterian and Reformed Publishing Company.

Freeman, R.E. (2001) 'The stakeholder theory of the modern corporation', in T.L. Beauchamp and N.E. Bowie (eds), *Ethical Theory and Business*, 6th edn, Upper Sabble River, NJ: Prentice Hall.

Goudzwaard, B. (1979) *Capitalism and Progress*, Grand Rapids, MI: Eerdmans. Translation of *Kapitalisme en voortuitgang* (1976).
Hart, H. (1984) *Understanding our World*, New York: University Press of America.
Henderson, R.D. (1994) *Illuminating Law*, thesis VU University, Amsterdam.
Hertz, N. (2002) *The Silent Takeover. Global Capitalism and the Death of Democracy*, London: Arrow.
Noble, D.F. (1997) *The Religion of Technology. The Divinity of Man and the Spirit of Invention*, New York: Knopf.
Reuleaux, F. (1875) *Theoretische Kinetik: Grundzüge einer Theorie des Maschinenwesens* (Theoretical kinetics: outlines of the theory of the essence of machines), Braunschweig: F. Vieweg.
Simondon, G. (1958) *Du mode d'existence des objets techniques* (The mode of existence of technological objects), Paris: Aubier.
Strijbos, S. and Basden, A. (eds) (2006) *In Search of an Integrative Vision for Technology. Interdisciplinairy Studies in Information Systems*, New York: Springer.
Vermaas, P.E., Kroes, P., Light, A. and Moore, S.A. (2008) *Philosophy and Design*, Heidelberg: Springer.
Vollenhoven, D.H.T. (2005) *Isagoogè Philosophiae (1967)*, Sioux Center: Dordt College Press.
Vries, M.J. de (2008) 'Gilbert Simondon and the dual nature of technical artifacts', *Techné*, 12(1): 23–35.

ALASDAIR MACINTYRE

(1929)

PORTRAIT 4 Alasdair MacIntyre

Alasdair Chalmers MacIntyre is a moral philosopher. He was born in Glasgow and taught at various universities in the United Kingdom. In 1969 he moved to the United States of America, where he taught, among others, at Yale University and the University of Notre Dame. In 1981 he wrote *After Virtue. A Study in Moral Theory*. In this book he criticizes modern moral philosophy. Inspired by Aristotle and Thomas Aquinas, he defends a virtue ethics in which concepts like virtue, social practice and moral tradition are key elements. Other important books are *Whose Justice? Which Rationality?* (1988) and *Dependent Rational Animals* (1999). He started as a Marxist analytical philosopher, but became more and more interested in the philosophy of Aristotle. In the early 1980s, MacIntyre converted to Roman Catholicism.

The work of MacIntyre is not widely known among philosophers of technology and management. However, that does not mean that MacIntyre's virtue ethics is not relevant for this field. On the contrary, his study of social practices and the accompanying distinction between internal and external goods may be very useful for a philosophical understanding of technology and management sciences. In Section 9.2 of this book MacIntyre's concept of social practices is introduced in order to get a clear understanding of the engineering profession. Due to MacIntyre, not only utilitarianism and deontology, but also virtue ethics has its place in contemporary moral philosophy. Utilitarianism focuses on the consequences of a moral act and deontology on moral duty. Virtue ethics, however, focuses on the moral character of a person.

In *After Virtue* MacIntyre hypothesises that we have lost a comprehensive framework to understand our moral world. What we possess are just the fragments of a coherent conceptual scheme that has fallen apart. By reintroducing the concept of a virtue, he wants to bring the scattered fragments together again into a comprehensive moral framework. Therefore, he first of all introduces the idea of a social practice, followed by a narrative concept of selfhood and the concept of a moral tradition.

One of the dominant figures of modern society is, according to MacIntyre, the manager. Management is often seen as a morally neutral activity. A manager's task is to organize social reality in the most effective way. MacIntyre, however, holds that managerial effectiveness is a moral fiction. Managerial effectiveness wrongly presupposes a domain of morally neutral facts, as well as the existence of law-like generalizations of human behaviour. Such a neutral domain, however, does not exist in the social world. The authority and power of managers therefore rest on an illusion. Managerial effectiveness is a believed-in reality that is keeping us captivated.

MacIntyre is quite critical about the role of management. At the same time, his analysis of social practices also provides space for a more generous approach. According to MacIntyre, social practices cannot exist without institutions that distribute money, power and status. These so-called 'external goods' are necessary to sustain a practice. They need to be carefully distinguished from the 'internal goods' of a social practice, but that does not mean that they are not important (Section

9.5). It can be argued that it is especially this concern with the external goods of a practice that legitimates the role of managers. One may even go a step further and analyse management as a social practice itself, with its own intrinsic normativity.

Some quotes that present MacIntyre's view on management and social practices:

> Managers themselves and most writers about management conceive of themselves as morally neutral characters whose skills enable them to devise the most efficient means of achieving whatever end is proposed. Whether a given manager is effective or not is on the dominant view a quite different question from that of the morality of the ends which his effectiveness serves or fails to serve. Nonetheless there are strong grounds for rejecting the claim that effectiveness is a morally neutral value.
>
> *(MacIntyre, 1984: 74)*

> By a 'practice' I am going to mean any coherent and complex form of socially established cooperative human activity through which goods internal to that form of activity are realised in the course of trying to achieve those standards of excellence which are appropriate to, and partially definitive of, that form of activity, with the result that human powers to achieve excellence and human conceptions of the ends and goods involved, are systematically extended.
>
> *(MacIntyre, 1984: 187)*

> And if the tradition of virtues was able to survive the horrors of the last dark ages, we are not entirely without grounds of hope. This time however the barbarians are not waiting beyond the frontiers; they have already been governing us for quite some time. And it is our lack of consciousness of this that constitutes part of our predicament. We are waiting not for a Godot, but for another – doubtless very different – St Benedict.
>
> *(MacIntyre, 1984: 263)*

References

MacIntyre, A.C. (1984) *After Virtue. A Study in Moral Theory*, 2nd edn, London: Duckworth.

MacIntyre, A.C. (1988) *Whose Justice? Which Rationality?*, Notre Dame, IN: University of Notre Dame Press.

MacIntyre, A.C. (1999) *Dependent Rational Animals*, Chicago, IL: Open Court.

5

THE ARTEFACT [II]

Identity, function and structure

Summary

This chapter offers an analysis of the identity of technological artefacts. This is not only important for getting a better view of the character of our technological society but also for deeper insight into the various aspects of the process of designing. The analysis shows that a technological artefact is the product of the forming activity of human beings (foundational function), that it performs a particular duty or has an operation (operational function), and that it functions within a certain context (qualifying function). The qualifying function in particular is of great importance since it guides the process of designing. This chapter also describes the relationship between technological artefacts, modules, parts and materials. The term 'interlaced structures' is introduced, since we are dealing with different structures incorporated into one greater entity. Three different relationships can be identified: between technological artefacts, between artefacts and modules and parts, and between parts and materials. Each one of these relationships has its own particular features.

It is important to have an understanding of the many aspects of technology. On the basis of the 'universal' analysis in Chapter 4 we have already made some critical remarks about the role of technology in our society (critical function of philosophy) and about the way in which technology should be developed (directional function of philosophy). We have pointed out that the various (latent) functions of technology should be disclosed in a normative way. The idea of disclosure fits in with the thought that human beings are *responsible* for the development of technology, that technology has to be embedded in society as a whole, and that several groups are involved in its development.

So far we have investigated different phenomena encountered in technological artefacts without paying attention to their specific structure. That is to say, we did not probe into the identity of technological artefacts. We all know the difference between, for instance, a drilling machine, a robot, a television, a corkscrew and a pacemaker. We can also easily identify a motor car although they come in various different kinds, sizes and forms. Even when essential parts are missing, like the engine, the steering wheel, the tyres and the seats, we can still see that it is a car. In other words, a technological artefact has a certain *identity*. An identity that is distinct from that of other artefacts (identity-in-diversity) that is recognisable within a series of products (unity-in-diversity) and that can be denoted in spite of changes (constancy-in-variability). But how can one characterise this identity? We know there is a difference between a seam of iron ore and raw steel. We also sense that there is a difference between an industrial robot and a medical robot. Further we are convinced that a church hall and a courtroom differ substantially from each other. The question is whether we can somehow describe these differences. Can we express the identity of these various artefacts in a satisfactory manner and will that lead to a deeper insight into the requirements we have to set for the process of designing technological artefacts? In this chapter we offer a philosophical analysis of the structure of technological artefacts.[1]

5.1 Foundational function

How is a technological artefact made up? What is characteristic of the whole? How can we characterise the parts? One of the ways of answering these questions is to follow the whole manufacturing process from the extraction of the different materials to the point where a technological product gets its form. By way of example, let us look at one of the parts of an industrial robot: the mechanical manipulator. The 'birth' of the arm of an industrial robot begins with the different materials which are in the soil, like iron ore, nickel ore and coal. These materials are extracted by means of the various mining techniques and reduced to a manageable form. In a blast-furnace the raw metals are extracted from these ores. For the extraction of iron and nickel completely different furnace processes are used. Then the raw metals are converted into steel in a subsequent process where small quantities of other materials are added. The steel is poured into certain forms which are easy to process. The different parts of the arm of the robot are made from the semi-processed steel products by means of a mechanical process. Finally these parts are assembled into an arm.

How exactly should we characterise the different stages? Let us start with the difference between iron ore lying as a deposit in the earth and pre-treated iron ore ready to be processed in a blast furnace. A seam of ore in the earth exists independently of the activity of human beings. The seam originated long ago as a result of physical-chemical processes under the influence of high pressures and high temperatures. As a consequence of this process the ore got its own characteristic features. In this regard one can think of the physical structure and the chemical composition. A seam of ore is the product *of a natural* process. Expressed philosophically, a seam of ore is based in the physical-chemical processes *of nature*. But what about the pre-treated ore that lies ready in the yard of the blast furnace to be converted to metal? This ore has come through several stages. First the ore is dug out and transported. Thereafter pre-treatment takes place with the removal of undesirable materials, the crushing into particles of a specific size and the washing of these particles. In some cases these particles also undergo thermal treatment. Only after these steps is the ore suitable to be processed in a blast-furnace. This pre-treated ore no longer is the result of natural processes but is the consequence of *technological activity by human beings*. Expressed philosophically, pre-treated ore is based in the technological, form-giving or formative work *of human beings*.

From the above analysis it emerges that the difference between the seam or iron ore and raw steel lies in their origin. A layer of iron ore originates in physical-chemical processes of the earth and raw steel in the technological or form-giving activity of human beings. In other words, their foundational functions are different. The concept 'foundational function' is important, for all technological artefacts are based in the form-giving or formative aspect. Products of our culture like literature, films and art are also based in the formative aspect. But natural 'things' like seams of ore and rock formations are based in the physical-chemical aspect, and shrubs and trees in the biological aspect. The marble found in the mountains of Carrara (Italy), is a 'natural thing' that is based in the physical-chemical modality, but a church built with marlstone is a 'product of culture' based in the formative modality. A tree in the wood is based in the biological dimension but a bench in the same wood made from the trunk of a tree is based in the formative dimension.

The difference between a seam of iron ore and a heap of pre-treated ore can also be expressed with the help of the terminology developed in the previous chapter. We distinguished between subject functions and object functions. A seam of iron ore and a heap of pre-treated ore both function as *subjects* in the physical-chemical dimension and in the formative dimension as *objects*. In this last dimension we see a difference emerging, for in the seam of iron ore the formative object function has not yet been disclosed (or actualised) but in the heap of pre-treated ore it actually is the case because human beings have processed it in various ways. The fact that iron ore can be mined and pre-treated is an essential feature of that ore. Formulated differently, the formative object function is latent in the iron ore and this object function is disclosed or actualised by the activity of human beings. The formative object function therefore is a feature of the ore! In the disclosure or actualisation human beings are bound by the specific characteristics of that ore. So, for instance, there is a huge difference between mining iron ore for the steel industry and mining clay for the brick industry. The technologies used are determined by the location and the stratification (deep down in the soil, near the surface) and by the physical-chemical characteristics (hardness, structure). We can close by saying that technological artefacts are always the result of human activity.

We can give a comparable analysis of biotechnological products as in the case of 'genetically modified soya beans' in Chapter 3. The original material – the wild soya bean – is the result of various biological processes that took place in the course of time in nature and is therefore biologically based. The modified soya bean is the result of human activity and is therefore – although it may at first sight look strange – formatively based. We want to emphasise that the possibilities of modification are limited: the engineer is bound by the specific characteristics of the biological structures s/he takes as a starting point.

5.2 Qualifying function

What exactly is the difference between an industrial robot and a medical robot? What is the difference between a church hall and a courtroom? And between a passenger car and a lorry? For an engineer it is important to be able to answer this kind of question accurately. The concept of foundational function is of no help to us in distinguishing these technological artefacts. For they are all the result of human activity. So how can we differentiate between these artefacts?

In Chapter 2 we pointed out that there are different contexts within which human beings act. One of these contexts is the world of science and technology. There are others like marriage and family, a commercial organisation, and an institution for care-giving or a religious community. The way in which people use technology depends considerably on the context. A technological artefact is well-designed when it is completely tailored to and fully embedded in a specific context. The specification of the context compels the engineer not to see the artefact to be designed as a *stand-alone* thing. Thus the context greatly influences the process of designing and the final design. This concept now offers us a second handle for analysing the identity of a technological artefact. Let us illustrate this with an analysis of the difference between an industrial robot and a medical robot. Both have the same subject and object functions and are based in the formative aspect. But there is a huge difference between the contexts in which they are applied and which determine their identity. The industrial robot is used in particular in production lines and factories, and here the manufacture of technological artefacts takes a central position. We have to do with a context characterised by control and form-giving which determines the way in which the robot has to function. So, for instance, the features of a robot used for applying a coat of paint to the bodywork of a motor car are to a great extent determined by this specific application. The robot has to be developed in such a way that the necessary movements and actions to apply a coat of paint in a controlled manner can be realised. An industrial robot is therefore qualified by the formative aspect. But what is it like for a medical robot used in a hospital, for instance, for performing eye operations? In this context the focus is on the performance of a medical intervention, on care for a sick human being. Such a robot has to be developed in such a way that the necessary movements and actions needed for performing a sensitive eye operation will indeed be possible. Here, too, the context therefore determines the specifications for a robot. Thus we can say that a medical robot is qualified by care-giving (moral dimension).

Let us look next at the difference between a church and a courtroom. Both are based in human activity and therefore in the form-giving or formative aspect. But here, too, we see a difference in context. A church hall takes a central place in the functioning of a religious community, while a courtroom has a great part in upholding the law. An architect will therefore design a church hall in a different way to a courtroom. Designing a church hall is determined by the religious activities that take place in it, such as listening, praying and singing. The architect will give much consideration to the form of and the atmosphere in the hall, the positioning of the pulpit and the place for an organ or other instruments. Designing a courtroom is determined by the activities related to jurisprudence, like trials, defence and passing judgment. In this design safety will play an important part. In both a church hall and a courtroom the specifications for the building will be determined to a great extent by the specific context. The pistic aspect qualifies the church and the juridical aspect qualifies the courtroom.[2]

The meaning of the qualifying function for the design of technological artefacts can be seen particularly well in Gothic cathedrals (Norberg-Schulz, 1975). These buildings were designed in such a way that they 'automatically' direct our thoughts to God. We recognise this in, amongst other things, the height of the hall that invites the believers to look up, and in the form of the windows that point upwards like a finger. In their design the builders also expressed the idea that God is great and human beings are small. This emanates from the immense space of the nave: one is just a wisp of a person in the great cathedral. The Christian notion that God is not only the Exalted One on high but also is near human beings has been given form in various ways. A Gothic cathedral has many and huge windows through which the light from the heavens can shine on earth in profusion. It is as if heaven and earth are joined directly. The location of the cathedral is also meant to impress on the people the presence of God. The cathedral preferably stands in the centre of the city on the market square. So we see that the builders used various instruments of design (form, situation, space) to express the religious function of this building.

The above examples show that the qualifying function characterises the identity of a technological artefact. First of all we see that this function categorises the whole design: all components are concentrated on the performance of the one activity. If, for instance, one takes apart an industrial robot to analyse the different modules and components it would be proved that they are all necessary for performing the particular operation at the conveyor belt. Expressed in a philosophical way, this function characterises the *internal structure* of a technological artefact. The qualifying function also plays an important part in the process of designing. Thus, the builders of cathedrals were always considering how to contribute to the functioning of the religious community and in which way certain biblical notions could find a place in their design. We see that this function exhibits a *leading role* in the process of designing. Finally the qualifying function characterises the specific role that the technological artefact plays in everyday life. Every citizen knows that in a courtroom trials are conducted, verdicts are given and those pronounced guilty are sentenced. And everyone knows that the function of a building like a prison is for locking up convicts. Thereby this function characterises the *specific destination* of a technological artefact in our society.

The qualifying function thus typifies the identity of a technological artefact. By means of this we can denote the difference in identity of, for instance, a passenger car and a lorry, or a games computer and a CAD system (*computer aided design*). The social dimension qualifies a passenger car (bringing people into contact with one another) and the economic dimension qualifies a lorry (transport of goods). The social facet characterises a games computer (recreation, joining in a game) and the technological or formative aspect a CAD system by means of the technological aspect (designing new products).

In practice it is not always easy to identify the qualifying function. Take, for instance, the difference between the Swinton Football Club football field and the Manchester United stadium Old Trafford.[3] It might seem as if there is a considerable difference in context. All technological artefacts (field, lighting, stand and canteen) of the field belonging to Swinton Football Club are concentrated on the sports meeting. All interest centres on the game and the show. It is an entirely social activity. The stadium of Manchester United makes a different impression. Most of the technological artefacts (grandstand, business clubs, recording facilities, catering facilities) are focused on maximising income. The design of the Old Trafford seems to have been inspired by 'big money'. This impression is confirmed by the fact that many top clubs are companies floated on the stock exchange. At first sight one is inclined to qualify the stadium of Manchester United by the economic dimension. Yet this conclusion is incorrect. In spite of the fact that economic considerations play an important part in the process of designing, football is and will always be a social activity. Both the playing of the game and watching it. *First and foremost*, therefore, the stadium should be designed in such a way that the football players can play and the spectators can watch. For this reason the Old Trafford stadium is also qualified by the social aspect.

We conclude that the qualifying function is of great importance to enable us to characterise the identity of a technological artefact. Under the direction of the qualifying function all preceding (earlier) functions are disclosed and an artefact gets its coherence and identity. This means therefore that the identity cannot be comprehended directly by departing from the theory of modal aspects. A technological artefact does express itself in all aspects, but the structural unity of this thing can only be comprehended by looking at the qualifying function.

5.3 Operational function

Let us now focus our attention on the technological artefact itself. What does a technological artefact actually 'do'? What kind of 'operation' does it have? What 'task' does it perform? In order to describe this element of the identity of a technological artefact we introduce the concept 'operational function'.

Keeping the robot as an example: one might wonder what the difference is between a robot that transfers a component from one belt to another and a robot that fixes one component on another by means of spot welding. And what is the difference between this robot and one that covers components or products with a coat of paint? This kind of difference can be described with the concept 'operational function'. Transferring a component from one belt to another concerns a spatial operation, fixing one component on another by means of spot welding concerns a physical operation, and applying a coat of paint is a different kind of physical operation. In all these cases we are dealing with the execution of controlled operations. This can be seen very well in the application of a thin coat of paint by means of the spraying process. This presupposes that both the spatial positioning and the kinematic movements take place in a controlled manner: the paint should end up in the right place (positioning) and it has to be applied in a homogeneous layer (movement). The operational function therefore gives a modal characterisation of the task or function of a technological artefact in a specific context. It describes what a machine 'does'.

Here are further examples. The operational function of a walking frame is to support the movement of a human being (kinematic facet) and that of a pacemaker to emit an electrical pulse (physical facet). The operational function of a building is to create a sheltered space (spatial aspect) and of a parking lot to mark out parking space for motor cars on a plot of ground (spatial aspect).

The concept 'operational function' does justice to the similarities between technological artefacts functioning in different contexts. Take, for instance, an industrial robot that 'cuts through' material by means of a laser and a medical robot that 'cuts off' a thin layer from the lens of the eye by means of a laser. The operational function of both robots is the physical dimension: the controlled removal of material by means of a laser. This function is, however, carried out in different contexts, a fact that has a huge influence on the design of the robots.

For the practice of designing it is of the utmost importance to differentiate between the operational function and the qualifying function. In designing a new artefact the first concern is to realise the operational function. Unless this basic condition can be met, further development of that artefact makes no sense. Second, the engineer has to realise that the new artefact will be used in a specific context. This context will have to be directional in the design of the relevant artefact. It goes without saying that there are two pitfalls here. The first is that too little consideration may be given to the design of the operational function: in that case the artefact will not function properly. The second is that too little consideration may be given to the qualifying function: then the artefact will not be very practical or user-friendly.

To provide good insight into the different concepts that characterise the identity of an artefact there is an overview of these concepts for the different stages of things in Table 5.1.

5.4 Modules, compound components and components

So far we have given no consideration to the components, compound components and modules of which an artefact is made up, nor to the relationship of these to the whole. Technological artefacts like pacemakers, robots, televisions and buildings are

TABLE 5.1 Overview of the diverse functions for the different stages of technological artefacts

Thing	*Foundational function*	*'Highest' subject function*	*Qualifying function*	*Operational function*
Tree in the wood	Biotic	Biotic	Biotic	not applicable
Tree sawn up into planks	Form-giving	Physical (dead tree)	Form-giving	not applicable (mostly)
Hut made of planks	Form-giving	Physical	Social	Spatial
Pieces of wood for open hearth	Form-giving	Physical	Social	Physical (warmth)

Thing	*Foundational function*	*'Highest' subject function*	*Qualifying function*	*Operational function*
Ore in a seam	Physical	Physical	Physical	not applicable
Sheet metal made from ore	Form-giving	Physical	Form-giving	not applicable (mostly)
Hut made from sheet metal	Form-giving	Physical	Social	Spatial
Electric wire made from ore	Form-giving	Physical	Form-giving	Physical (conduction)

'wholes'. These are 'things' which are the result of human activity and are used in a particular context. Wholes are characterised by a 'complete internal unity' (Dooyeweerd, 1969). This internal unity is 'guaranteed' by the qualifying function. Wholes have a certain autonomy. Within the context of a factory a robot can perform a certain operation, for instance placing a component on a conveyor belt. This robot performs this function in an 'autonomous' way within this technological context. But the electricity supply of the robot knows no autonomy. It only functions within the whole of the robot. Neither does the control unit of the robot have any autonomy.

Therefore the whole has features that the components, the compound components and the modules of the artefact do not possess. A robot can place a component on the conveyor belt, but the manipulator, control unit and power supply cannot do that. It is therefore important to investigate the relationship between wholes on the one hand, and modules, compound components and components on the other hand.

When, for instance, we look closer at the control unit of a robot we encounter various different modules like a processing unit, a display and a keyboard. The processing unit has a number of submodules consisting of compound components and simple components. Likewise the display and the keyboard are made up of submodules and compound components. We can analyse the manipulator and the power supply in the same way.

By the term 'module' we understand a number of (compound) components forming a 'unity' and characterised by a specific 'technological' function. So, for instance, the engine forms an important module of a motor car. This module consists of several submodules like ignition, combustion unit, transmission and cooling, each with its own specific 'technological' function. Other important modules of a car are the steering-mechanism and the brake system. By a compound component we mean different components brought together but not performing a specific 'technological' function. In industry compound components are much used. One of the reasons is that it can significantly simplify the assembling process in a factory if a supplier does not supply all components separately but as a compound component. Thus in the motor car industry the electric wiring is supplied not in the form of loose cables and connections but in the form of a 'cable harness' with the different cables including the connections as a 'unit'. The use of this kind of compound component often also has a positive effect on the quality, because fewer mistakes are made in the electric connection when the cables are fitted.

How should we characterise such a module in a philosophical discussion? How can we describe the identity of such a module? Would it be possible to use the concepts 'foundational function', 'qualifying function' and 'operational function' for this? '*Foundational function*' refers to the origin of the particular thing. Of every module it can be said that it is the result of formative activity by human beings. In technological form-giving each module gets its own 'identity', But we have to admit frankly that the concept 'foundational function' is of little help in distinguishing between the different modules for the reason that they are *all* based in the formative activity of human beings.

Above we introduced the concept 'qualifying function' for 'wholes' that function as autonomous artefacts in a context. We saw that the *qualifying function* characterises the internal structure of an artefact and is directional in the process of designing. Every module of a technological artefact is therefore characterised by the same qualifying function as that of the specific artefact itself. In fact this means that the different modules of an industrial robot are qualified by the formative function and the different units of a court building by the juridical function. The concept 'qualifying' function helps us to distinguish the different modules of one technological artefact from those of another artefact, but not to distinguish the different modules within one technological artefact.

Finally let us look again at the different modules of an industrial robot which are all qualified by the formative aspect. They exhibit a distinct difference in *operational function*. Thus the operational function of the manipulator is the kinematic aspect (it is concerned with movement), of the control unit it is the physical aspect (concerned with physical signals) and of the power supply it also is the physical aspect (concerned with energy). 'Operational function' therefore leads to a closer differentiation between the modules of a technological artefact. The operational function of the manipulator has a kinematic dimension and that of the control unit and power supply a physical dimension. The operational function of the control unit and power supply is indeed characterised by the same modality (physical) but has different mechanisms (generation of a signal, generation of energy). A comparable analysis of a medical robot leads to the same operational functions, namely the kinematic dimension for the manipulator and the physical for the control unit and power supply. The concept 'operational function' therefore offers a better insight into the nature of the various modules.

This is complicated by the fact that in the industry many modules have been standardised, as in the industrial robot technology with standardised arms, control units and power supplies. Standardisation means that the same module can be used in different types and applications. The same robot arm can place one component on top of another when in one position, and can weld a component on a product when in another position. In that case only the 'head' of the robot is different. The same applies to control units and power supplies. Standardisation means a certain loss of individuality because a module to some extent is 'neutral' towards the individual destination of the manipulator.

The standardisation of modules can go a step further when the same robot arm, control unit or power supply has been applied in two differently qualified contexts, for instance in a technological context (precision-mechanical factory) and in a medical context (hospital). It cannot be determined when seeing the module lying 'on the shelf' of the warehouse for what application it will be used. This 'neutrality' is therefore only 'cancelled' at the moment it is used in building a technological artefact for a particular context.

We can also find the same modules in completely different products. We encounter similar control units and power supplies in industrial, medical and domestic apparatuses. We see this particularly in 'peripheral equipment' like monitors and key boards. Here the neutralisation has gone even further, for it cannot be seen from the module in what kind of apparatus and in what context it will be applied. Here, too, it is true that the 'neutrality' of these components is cancelled the moment they are used for building a technological artefact.

In modules designed for *one specific* technological artefact and for *one specific* context no neutralisation takes place. These can be characterised unambiguously by a qualifying and an operational function. In the case of modules designed for *different* technological artefacts and for *different* contexts a considerable degree of neutralisation has taken place and the module can no longer be characterised by the qualifying function. The most important remaining qualification for describing the identity is the operational function.

Almost all components used in technology are standardised and are characterised by the formative aspect as their foundational function. Since neutralisation leads to loss of identity, in most cases speaking about the qualifying function of a component is not useful. Only in the case of specific components used for *one* artefact in *one* context may it be useful. In many cases we can indeed pinpoint the operational function of a component unambiguously. An axle, a gear, chain and wheel are used to enable movement (kinematic aspect). A resistance is applied to weaken an electric current (physical aspect) and a capacitor for storing energy (physical aspect). Wooden shelves are used for creating space (spatial aspect) and windows for letting in light (physical aspect). A woodscrew is used to fix in place two components in relation to one another (spatial aspect) and a spring to modulate movements (kinematic aspect). In the concrete application many components often fulfil several operational functions. So, for instance, in many electrical apparatuses nuts and bolts are used not only for fixing two components in relation to one another (spatial aspect) but also to conduct the electric current (physical aspect). Moreover, designers often attempt to realise more than one function in *one* component.[4]

Standardisation plays an important part in all fields of technology. Neutralisation takes place to a greater or lesser degree. Architects use standards in the field of materials (stone, wood, synthetics) and dimensions (windows, doors, surfaces), chemical engineers in the field of raw materials (specifications) and installations (processes), mechanical engineers in the field of working (moulding, tools) and systems (transport, working), and agricultural engineers in the field of materials (substrates, nutrients) and biological processes (improvement, breeding). Standardisation always leads to neutralisation. The result is a certain loss of the 'individuality' of intermediary products, semi-manufactured articles, modules and end products. Standardisation on the one hand facilitates the process of designing for the engineer (s/he can make use of the standards and therefore needs not design 'everything') but on the other hand it restricts the process of designing (s/he may not deviate from the prescribed standards).

5.5 Interlaced structures: the relationship between wholes

So far we have focused our attention on the characterisation of technological artefacts, modules and components. We did this by means of the foundational function, the qualifying function and the operational function. The next step is an analysis of the relationship between technological artefacts, artefacts and modules, modules and components, and components and materials. For these relationships we use the general term 'interlaced structures'.

The term already indicates some important aspects of our analysis. First we will look at the identity of interlaced structures. 'Structure' in this context is used for wholes, modules, components and materials. To what extent do the structures have their own or autonomous identity? Subsequently we analyse to what extent the particularity or independence of these structures remains intact or is lost when they are interlaced with others into a greater whole.

Individual artefacts are increasingly combined with one another in a greater system. One of the most striking examples is the *World Wide Web*, where a great number of computers and servers are connected with one another by means of a network. Another example concerns the measuring apparatus in laboratories, hospitals and factories that are linked to the network so that data can be processed directly. We also see that in shops the cash registers send the data directly to a central computer so that new stock can be brought in on the basis of the sales. For this phenomenon Simondon uses the term 'ensemble'. Technology has – to use an expression of Ellul's – become a system. Technological artefacts are at present almost invariably included in a greater system. Motor cars are dependent on a particularly complex infrastructure of roads, sign systems, petrol stations, etc. Each element of this infrastructure in turn is a component of another network. Roads are linked to asphalt machines, asphalt factories, cranes for digging away soil and suction dredgers for mining sand.

What is the relationship between these different artefacts within such a system? Let us have a look at an automobile factory in which industrial robots are installed. The central point in such a factory is the conveyor belt on which the motor cars are. Next to the conveyor belt there are many robots and measuring apparatus. Without further discussion we can call such a production line a whole. It performs a function that the various 'components' cannot do independently: assembling motor cars. The elements of this line, like the speed of the conveyor belt and the work done by the different robots, are carefully coordinated. The production line as a whole therefore also consists of wholes. Its core is the conveyor belt which transports the products. This mechanism likewise has the features of a whole. It has its own particular foundational, qualifying and operational function, a certain autonomy. In principle it can function without robots provided that the necessary operations are carried out by people. The robots that carry out the various operations also have all the features of a whole. Each one of them has an independent identity. In principle they can likewise function independently of the conveyor belt. So, a conveyor belt is a whole consisting of different wholes each of which has its own identity.

For characterising wholes consisting of other wholes we use the term 'interlaced structure'. The different wholes of which a production line consists each has its own identity. We can distinguish between them in particular by means of the operational function. In the case of the conveyor belt it is the kinematic facet. A robot that places products on the belt is also characterised by the kinematic facet but there is a difference in the mechanism of motion. The operational function of a robot that welds one component onto another is the physical dimension. In the case of a robot that measures certain physical characteristics of a product the operational function is also in the physical aspect, but then a different physical mechanism is used. So we see that a production line consists of structures that are different in operational function and in the underlying mechanism. These different structures have not been designed at random nor have they been placed indiscriminately in the production hall. They have been carefully coordinated and in this process their own autonomous character has remained intact. We could speak of a mutual influence from one whole on the other. Thus the speed of the conveyor belt determines the speed of the different robots, but it can never move faster than the 'slowest' robot. We also see that both the conveyor belt and all robots are provided with standard ports so that information can be sent to one central computer. We increasingly find that technological artefacts are designed in such a way that they can function in a greater whole. This mutual influencing is described with the term 'disclosure'.

In this example we are dealing with an interweaving of wholes that are mutually coordinated during which process the identity of these wholes remains intact. For this Dooyeweerd introduced the term 'encaptic interlacement of structures'. He borrowed this term from the philosopher T. Haering who in turn took it over from the anatomist M. Heidenhain. With this term Heidenhain expressed the idea that the whole is more than its (relatively autonomous) components (Dooyeweerd, 1969: III, 627 et seq.), to describe the relationship between wholes which each has its own identity (qualifying and operational function). We can say that there is a so-called 'correlative encapsis' when there is mutual dependence of the different wholes.

There are other examples of interlaced structures of wholes. Take, for instance, the installation of sound equipment in an old church. In such a case the identity of the church is maintained. Nor does the installation change anything of the identity of the sound equipment. But it is true that the choice of the different elements of the design, like microphones, amplifier and loudspeaker is coordinated with the size of the church building and with the exact function it has to perform (amplifying the spoken word, amplifying solo singing, etc.). It would be a different situation were a new church to be built in which the technological amplifying of the sound was taken into consideration from the start. Then there would be contact between the architect and the designer of sound equipment from the start in order to have the building and the sound attuned as much as possible. Here, too, the identities of the different wholes (the building, the installation) remain intact, but they do influence one another. We can say there is a specific disclosure of both the building and the installation.

5.6 Interlaced structures: the relationship between wholes, modules and components

Simple technological artefacts consist of just a few components. Complex technological artefacts consist of modules, modules often consist of submodules, and the submodules consist of components. So what exactly is the relationship between a whole and its parts, between a whole and its modules and between a module and its components, or what is the identity of the different elements and the nature of the interweaving? In the case of modules and components we cannot speak of wholes because they cannot function autonomously in a certain context.

Let us analyse a simple technological artefact like a hammer. A hammer is foundationally based in the formative aspect, since it has its origin in human activity, and it is also qualified by the formative aspect since it functions in a form-giving operation. A hammer consists of two components: the head and the handle. A hammer can function autonomously as an artefact, but the components cannot. What is the relationship between the hammer and these components? In Sections 5.2 and 5.4 we saw that the qualifying function (a) characterises the internal structure of an artefact, and (b) should be directional in the process of designing. In other words, the components 'head' and 'handle' have no autonomous identity but derive their identity from that of the hammer. The head and the handle have been formed with a view to the formative function of the hammer. In this simple technological artefact the identity of the whole (the hammer) therefore determines the identity of the components (head, handle). Here we see a huge difference with an interweaving in which the different wholes each have an autonomous identity. The design of the hammer – and therefore of the head and the handle – is also determined by the context of its use. A carpenter sets different requirements for a hammer than a cobbler or a road worker. The qualifying function of the whole is directional in the design of the components. In this example, therefore, we are dealing with an interweaving of parts in a whole in which the identity of the parts is determined by the whole. For this Dooyeweerd (1969: III, 638) introduced the term 'whole-component relationship'.

A somewhat more complex technological artefact like a paper guillotine consists of 10–30 components. The most important components are the table on which the paper is laid, strips to clamp it down, rulers to determine the measurements, the shield of the cutting part, the blade and the handle. Besides these components there are a number of standard parts like springs, screws, nuts and bolts. The relationship between the guillotine and the table, strips, rulers, shield, blade and handle is a whole–component relationship. The design of these components – the internal structure – can only be understood from the formative function of the whole. The clamps are made in such a way that the paper is firmly clamped down and the shield is developed so that no fingers can come under the blade during the cutting process. The blade has a specific form so that by one movement of the hand the paper can be properly cut off. The handle is designed in such a way that it fits well into the hand to carry out the desired movement. None of these parts has its own identity; it is derived from the identity of the whole. The qualifying function of the paper guillotine determines the design of the components. One complication presents itself here: a number of standard components were used. These components were developed without considering the specific destination of the paper guillotine and they can be applied in many different artefacts. Their identity is not or only slightly determined by the whole. It is a matter of 'levelling of identity' which one could call a 'neutralisation regarding the destinational function of technological things'. The moment standard components like springs, screws, nuts and bolts are built into a paper guillotine the specific destination of these components are determined. But when the guillotine is taken apart these components can once more get another specific destination in another artefact.

Complex technological artefacts consist of modules which in turn are made up of submodules, and these submodules again sometimes consist of subsubmodules, etc. We restrict ourselves to the relationship between the complex artefact (the whole) and the different modules. An artefact can function autonomously but a module cannot. In Section 5.4 we give an analysis of the modules of an industrial robot. We conclude that the qualifying function characterises the internal structure of an artefact and with it the internal structure of every module in it. We also see that the qualifying function is directional in the process of designing and therefore for the design of all modules. Further it is apparent that the different modules are distinguished by the qualification of the operational function: for the manipulator it is the kinematic aspect, for the control unit the physical (generating a steering signal) and for the power supply the physical (generating energy). Manipulator, control unit and power supply have no autonomous identity but derive their identity to a great extent from that of the robot. The different modules were formed with a view to the formative function of the industrial robot. Therefore, in the relationship whole–module we are dealing with modules that are not autonomous and that only reach their destination in the interlaced situation. In this relationship we are also dealing with the complication of standardisation. Technology often uses standard modules for different products. One of the best examples are displays and keyboards. As in the case of components we can say there is loss of identity and some neutralisation regarding the destination. Only when installed in a concrete artefact is the specific destination determined. This complication does not alter the conclusion that the relationship between a whole and its modules can best be described as a whole–component relationship.

The relationship between a module and a submodule also has the character of a whole–component relationship in which the identity of the submodule (just like that of the module) is determined (or fixed) by the specific destination of the technological artefact. The same applies to the relationship between a module and its components.

The relationship between wholes, modules and components can be described as an interlaced structure. The most important feature is that the identity of the different parts is determined by the specific destination of the whole. As a result of the use of standard modules and standard components a certain loss of identity ensues. And the destination is determined only when concrete use is made of this kind of module or component in an artefact.

5.7 Interlaced structures: the relationship between components and materials

One can also speak of an interlaced structure between the various components and their materials. In the process of designing the selection of material plays an important part. When a hotel is built some architects choose peaceful and dark materials to create a respectable and cosy atmosphere. Other architects choose light colours to create a cheerful and sportive impression. Mechanical engineers, on the other hand, hardly pay any attention to the aesthetic aspects of the materials used when designing but focus on the physical characteristics like hardness, strength and elasticity.

The relationship between components and materials has some special features. For instance, in many countries, like England, Germany and the Netherlands, a lot of sandstone is found in the soil. In the past stone was cut from these layers for building houses. In this application a cut stone has a social destination. However, what is the relationship between the physically-chemically qualified material and the socially qualified stone?

Sandstone is a sedimentary rock consisting mostly of grains of sand. At sedimentation silt (clay particles), lime (alkaline salts of calcium), gravel, glimmer (mica-like minerals), feldspar (aluminium tectosilicates) and other stone fragments are also deposited. In the material there are many pores. In the beginning the material is a golden brown. As a result of the presence of certain oxides and minerals the more reddish brown shades emerge. What happens when stone is cut from this material and a house is built? The material is retained as 'material'. The composition and the characteristics of the material are still exactly the same as in the layer in the soil. The stone is stacked in such a way that a house comes into being. In this whole process the physical-chemical characteristics of the material (*inter alia* strength and colour) are harnessed or disclosed in the service of the social function of the house. So we see that in the stone cut for building a house an interweaving has taken place of the physically-chemically qualified material and the socially qualified stone. This stone can only be 'released' from its structural unity with the socially qualified house if the latter is demolished.

A second example is the relationship between the screw and the steel. The steel is physically-chemically qualified and the screw is formatively qualified. Even when the steel has been formed into a screw all characteristics like hardness, elasticity and conductivity are preserved. In the formation into a screw these characteristics are disclosed in such a way that it can fulfil its technological or formative function. It is understandable that engineers in the past chose steel as the material for screws and not silver. Steel can be successfully fashioned and is very strong. Silver can also be easily fashioned but is not strong. The relationship between component and material is asymmetrical. Without steel there would be no screw but without the screw there still is steel. We can say there is interweaving between the physically-chemically qualified steel and the formatively qualified screws.

Information and communications technology offers a third example where the physical-chemical characteristics of different materials (silicon, silicon oxide, *dopants,* aluminium, copper, gold) are interlaced into a formatively qualified chip.

In the above examples we can say there is encaptic interlacement of a particular kind, namely the 'irreversible foundational relationship'. A typical feature of this is that during the interweaving the material and the component retain their own qualification. For the stone this is the physical-chemical and social modality respectively, and for the screw and the chip the physical-chemical and formative modality respectively. The next feature is that the components are based in the materials and not the other way round. The house cannot exist without the sandstone but the sandstone can exist without the house, and this also holds for the screw and the chip.

5.8 Conclusions

From our analysis it appears that we can characterise the identity of a technological artefact by means of three concepts: the foundational, the qualifying, and the operational function. The foundational function denotes the origin, the qualifying the context and the operational function the working. These concepts can in principle also be used for the characterisation of modules and components. A complication in this analysis is that standardisation leads to 'loss of identity'. We conducted a detailed investigation into the relations between technological artefacts, modules, components and materials. We can speak of 'interlaced structures': structures with a more or less pronounced own identity having varying degrees of autonomy and taken up into one greater whole. We found three different relationships: (a) a whole–whole relationship between technological artefacts (correlative relationship), (b) a whole–component relationship between artefacts and modules and between modules and components, and (c) the component–material relationship between components and materials (irreversible foundational relationship). Each one of these relationships has its own particular features and characteristics.

Notes

1 The 'theory of individuality structures' has been set out by H. Dooyeweerd in his *A New Critique of Theoretical Thought*, Volume III (1969). H. Hart gives an introduction to this theory in *Understanding our World* (1984).
2 Of course a courtroom can be used for religious activities and a church hall for jurisprudence. But we all feel that these spaces are not suitable for this. In order to render them suitable one would like to make all kinds of minor and major adjustments. From this it clearly emerges that the internal structure of a courtroom and that of a church hall differ distinctly.
3 Swinton is a small town north-east of Manchester.
4 See the analysis of Simondon in Section 4.6.

References

Dooyeweerd, H. (1969) *A New Critique of Theoretical Thought*, Phillipsburg, NJ: The Presbyterian and Reformed Publishing Company.
Hart, H. (1984) *Understanding our World*, New York: University Press of America.
Norberg-Schulz, C. (1975) *Meaning in Western Architecture*, Londen: Rizzoli.

Further reading

Kroes, P.A., and Meijers, A. (eds) (2000) *The Empirical Turn in the Philosophy of Technology*, Amsterdam: Elsevier.
Mitcham, C. (1994) *Thinking Through Technology. The Path between Engineering and Philosophy*, Chicago, IL: Chicago University Press.

GILBERT SIMONDON

(1924–1989)

PORTRAIT 5 Gilbert Simondon

Gilbert Simondon was born in 1924. He was a student of the French phenomenologist Maurice Merleau-Ponty. He studied at the Ecole Normale Supérieure and at the Sorbonne in Paris. Some philosophers of technology have remained unknown because they published only in their mother tongue. Such a philosopher is the French philosopher of technology Gilbert Simondon. He was one of the first philosophers of technology who was interested in an analysis of technological artefacts in a time when most philosophers were busy with views on the relationship between technology and society. However, in contemporary philosophy of technology, the philosophical analysis of technological artefacts receives a lot of attention. A 'revival' of Simondon might be in order.

Simondon's dissertation *L'individuation à la lumière des notions de forme et d'information* (The individual in the light of ideas about form and information) was published in two parts. The first part was published in 1964 under the title *L'individu et sa genèse psychico-biologique* and the second part in 1989 under the title *L'individuation psychique et collective*. He also wrote *Du mode d'existence des objets techniques* in 1958. The latter was meant as a complement to his thesis. His oeuvre was very small but exceptionally original. In philosophy of technology we are particularly concerned with his last book. His first book deals with the question what it is that distinguishes something individual from the rest of the world. This distinction is, according to Simondon, especially contained in a process of genesis (the so-called 'individuation'). In his last book he applies this to technological artefacts.

When Simondon speaks about a technological object he does not mean the concrete thing but the object in the history of its development. A motor car as a technological object, for instance, is not one particular motor car, or a specific type of car, but the car as it was gradually developed in an evolutionary process. In this process the design became increasingly sophisticated because the functions of the object were combined more and more. In the past a number of components were needed to realise different functions, but nowadays the designer can make do with fewer components because each component realises more than one function. This Simondon calls the process of concretisation. What is interesting about Simondon's book is that he gives a number of detailed examples to illustrate this. He deals, for instance, with amplifier tubes, motor car engines and electric locomotives. Simondon also shows how technological objects can differ by functioning more or less dependent on their context. A steam train has to take in coal, but during the journey it is not dependent on the environment, while an electric train cannot move a metre without overhead electric wiring. But they both need the rails. There also are vehicles which are still less dependent on their environment as for instance jeeps or 4x4s. Finally, Simondon distinguishes three levels of objects: elements (the components of which an autonomously functioning object is made up), individuals (the autonomously functioning object) and ensembles (objects in the context in which they function). These levels coupled with an analysis of concrete examples have to give insight into the nature of different technological objects.

Simondon's ideas are represented in the following quotations from *Du mode d'existence des objets techniques:*

Culture is focused on a defence system against technology; this defence pretends to be a defence of human beings and supposes that the technological objects have no human reality. We would like to demonstrate that culture denies that behind the technological reality there is a human reality, and that culture can only fulfil its part fully if it includes the technological matters in the form of knowledge and in the sense of values. The awakening to the mode of existence of technological objects has to be effected by philosophical reflection. In this process it will prove that there is a duty to be taken up that resembles the events accompanying the abolition of slavery and the affirmation of human dignity.

(Simondon, 1958: 9)

In this way culture carries with it two contradictory attitudes towards technological objects: on the one hand it treats these like material compounds, devoid of true meaning, and only entailing some usefulness. On the other hand it supposes that the objects are also robots and inspired by hostile intentions towards human beings, that they hold a constant threat of aggression and revolt to them. He who wants to conserve the first feature, wants to prevent the manifestation of the second and then speaks of placing the machines in the service of human beings, convinced that in subjecting them to slavery one is sure to find a means to prevent any revolt.

(Simondon, 1958: 10–11)

Far from being an overseer to a band of slaves, a human being is a constant organiser, a constant interpreter to a community of technological objects that needs him in the way an orchestra needs a conductor. The conductor cannot lead the orchestra because he is playing with them in the piece with equal intensity; he restrains them or encourages them, but he himself is restrained and encouraged by them; … he interprets from the one to the other. In this way a human being has the function of being a constant coordinator and inventor of machines around him. He is standing in the midst of the machines that work together with him.

(Simondon, 1958: 11–12)

References

Simondon, G. (1958) *Du mode d'existence des objets techniques*, Paris: Aubier.
Simondon, G. (1964) *L'individu et sa genèse physico-biologique*, Paris: PUF.
Simondon, G. (1989) *L'individuation psychique et collective*, Paris: Aubier.

Further reading

Simondon, G. (2005) *L'invention dans les techniques: Cours et conférences*, Paris: Seuil.

CASE STUDY I

Nanotechnology

This case study is about a field of research that has stirred up much interest in recent years: nanotechnology. We begin with a technological-philosophical analysis of this technology and then focus on the meaning of nanotechnology for human beings and society. In this case study it becomes clear how the aspect analysis from Chapter 4 and the structural analysis from Chapter 5 can offer insight into the nature and functioning of technological artefacts.

1 A rising new technology

The term 'nanotechnology' literally means 'dwarf technology' ('nanos' is the Greek for 'dwarf'). The term refers to the extremely small dimensions of the objects with which the work is done: atoms and molecules.[1] However, we have been working with atoms and molecules for ages, without calling it nanotechnology. Chemistry, for instance, is nothing other than the manipulation of atoms and molecules. In physics, too, it has been happening for a long time. According to some experts 'nanotechnology' is a new term for something we have been doing for a long time. In recent years a hype has come about in connection with this concept. Nanotechnology creates enormous expectations in society. Especially in the fields of electronics and medical technology the possibilities appear to be infinite. The question is where these expectations come from.

The 'holy grail' of nanotechnology is the act of moving atoms and molecules one by one so that they form new structures with the properties that we desire. Formerly engineers, chemists and physicists had to work with great quantities of atoms and molecules together. In chemistry, for instance, one could never have one molecule reacting with one other molecule. One used a whole test tube full or an Erlenmeyer flask full. Physicists, too, were always dealing with whole beams of electrons or vessels full of gas molecules. The ultimate claim made by nanotechnology is that it will enable engineers to stick together atoms and molecules like Lego bricks and thus work 'bottom-up' (building greater structures from smaller ones upwards). Since it can be applied in various fields 'nanotechnology' actually is more of a collective term for a family of technologies. Applying a layer merely a few atoms thick, for instance, in sun tan cream or as a dirt-repellent layer on clothing, is called nanotechnology. Building molecules for medicines that can react with the organ cells for which they are intended without affecting other cells also falls under this concept. It is quite possible that in future there will be different names for the various technologies and that the term 'nanotechnology' as a collective term may become obsolete.

What exactly is envisaged for applications of nanotechnology? In the first instance this technology could render all kinds of new materials with interesting applications. We have already mentioned special coatings for clothing. Part of this application already is in a commercial stage at the moment. The so-called '*buckyballs*' are known. These are molecules consisting of sixty carbon atoms bonded to one another in a grid having the form of a football. *Buckyballs* can be used for storing smaller atoms like hydrogen atoms. The latter application which is still in the laboratory phase, is of major concern to a 'hydrogen economy' in which hydrogen becomes the most important raw material. Another important area of application is electronics, because the possibilities of miniaturising by means of current technologies are running out. By means of nanotechnology it would be possible to build extremely small circuits existing of only a few atoms or molecules. With such a circuit one can, for instance, measure the composition of blood with very little material, the so-called '*lab-on-chip*'. There also are high expectations concerning medical nanotechnology. We have mentioned the special medicines that only attack sick cells, but one could also think of the possibilities of intervening in the body by means of extremely small robots ('nanobots'). People who have seen the film *Fantastic Voyage* will be able to picture for themselves something along this line.[2] Growing sophisticated tissue could be possible. Even the word 'immortality' is heard now and again. By constantly renewing dying tissue a human life could be prolonged indefinitely. By means of this not only would we be able to overcome every illness but also death itself. However, the question whether all of this can be realised is answered in the negative by many experts. Yet we have to discuss these claims as well. Often things which formerly seemed impossible are realised later on. But even more importantly, even the non-viable claims give an impression of the direction in which people want to expand their technological possibilities. And the ambition itself is, philosophically seen, perhaps even more relevant than what actually eventually emerges.

2 Aspects and functions

We have seen that an analysis of aspects gives insight into the nature and functioning of technological artefacts (see Chapter 4). So how about nano-artefacts, are they different from 'ordinary' technological artefacts? If one looks at the end product there are great similarities between for instance an *integrated circuit* (IC) produced with the help of a classic technology and one produced with nanotechnology. Both products consist of a great number of particles. It is 'easy' to produce this product by means of classic technology but with nanotechnology it has to be realised by taking hold of all these particles one by one and placing them. We therefore should not analyse a nano-artefact as a product in itself, but look at the way in which it comes into being, namely via nanotechnology.

Analysing nanotechnology from the viewpoint of different aspects, we could begin with the arithmetic or 'numerical' aspect. Numbers are an awkward problem when we are dealing with the ambition to build up bigger structures from individual atoms and molecules. For one then has to manipulate an inconceivable number of things. Nanotechnology would be able to avoid this problem by means of '*general assemblers*': small machines which in turn build small machines to take care that the atoms and molecules are constructed in the correct way. According to Drexler (1986) this is a practicable route: the mechanism is known from nature where the ribosomes perform this function. Criticism of this view maintains that ribosomes are not at all as 'general' as Drexler supposes. The next aspect is the spatial dimension. It is awkward manipulating the atoms and molecules because they are so small. For this, too, 'tricks' have been devised. By means of a so-called *Scanning Tunnelling Microscope* one can observe atoms, take hold of them and move them. All of this happens indirectly. One cannot see directly what one is doing. An essential consideration for nanotechnology is the quantum behaviour of the particles. Therefore we here take together the kinematic and the physical aspects. The behaviour of nanoparticles deviates, for instance, from that of billiard balls which can be described by means of classic mechanics. Quantum behaviour exhibits all kinds of idiosyncrasies of which we do not yet know and understand all when it comes to nano-artefacts.

The problems we have observed up to now in nanotechnology all relate to their behaviour as subject. In the later aspects they can only act as object.[3] When we come to the biotic aspect of nano-artefacts we have to keep in mind that these are so small that they can penetrate living beings unseen. This recalls the asbestos problem. Asbestos consists of such small particles that the users thereof initially were unaware that the particles penetrated into their lungs. It is already envisaged that nanoparticles could cause similar problems.[4] The psychic (perceptive) aspect shows which problems accompany the small dimensions: they are hard to detect. We have seen that in principle for this there are only indirect methods. The essence of the analytic dimension is making a distinction. In this respect nano-artefacts also confront us with problems. One of the ambitions of nanotechnology is that eventually one would be able to grow tissue. This means that one starts with non-living nanoparticles which one builds up more and more until at a given moment features of life appear. Does this mean that the border between living and non-living becomes blurred? Since nanoparticles practically cannot be distinguished this is a difficult question to answer. Regarding the lingual or symbolic aspect we think of the magic effect that the very name 'nano' carries with it. It is confusing enough that the term is a collective noun but still more annoying that the magic connotation of the name contaminates various discussions and decisions. The form-giving or formative aspect amongst other things points out that the development of an artefact is an on-going process. Technological development relies on earlier designs. Nanotechnology is at the very beginning of a development in which the old technology offers little on which to build.[5]

Many of the nano-artefacts will have a social destination as a function. It will, for instance, lead to problems if one part of society has access to this technology and another part does not. Nano-artefacts could mean an important economic stimulus to companies. But will it really bring in as much as the nano-enthusiasts claim? We discover a problem in the juridical aspect. Does anyone have a view of what can be regulated by legislation about nano-artefacts? This is something into which one should have insight as early as possible. However, the actual effects of nano-artefacts are still difficult to surmise. In the aesthetic aspect harmony is the keyword. In dealing with nano-artefacts under this aspect, the question arises whether there is the possibility of a harmonious combination of nanotechnology and the more classic technologies. The ethical aspect already arouses much interest when dealing with nanotechnology. Thus, one can wonder whether it is ethically justified to develop something when one does not understand completely how it will behave, or when one does not know whether it will be possible to guide it with adequate legislation. The aspect of faith plays a great role in dealing with nanotechnology. The belief in this technology for some gurus surpasses their sense of reality. They make the most fantastic promises to the public.

In short, each aspect proves to raise questions regarding the development and use of nano-artefacts. But in this analysis we have not yet expressed our ideas on what would be a desirable development and what would not.

3 Social and ethical discussions

Much discussion goes on about the future of nanotechnology. Maybe it is the not so realistic stories of nano-gurus like Eric Drexler which create the discussions in society. It is precisely because these predictions mention far-reaching consequences that they give rise to all kinds of questions. Do we really want it? Can the consequences be foreseen? From a number of events in the USA it seems that policy makers are not willing to let this new technology just run its course to development.

The first event is the evidence given by philosopher of technology Langdon Winner in the American Senate (see Portrait following Chapter 9). Winner is known for his critical attitude towards many technological developments and the Senate therefore knew not to expect a glowing speech in favour of nanotechnology from him. Apparently it was an intentional choice to hear the other side as well. Winner posed three fundamental questions: (1) Do we have to keep on developing new technologies to overcome nature or should we rather promote harmony with natural structures and processes? (2) Do we have to start a development which in the long run will lead to technological means determining social goals? And (3) Is it wise to experiment with technological applications which most probably will have irreversible consequences? There is much to be said against these questions. Yet, the Senate could not simply dismiss them. In his publications Winner has shown that various technological developments in the past have been accompanied by the problems on which he touches in his questions. Seen from the various viewpoints we will later discuss (see Part III) these questions also are more relevant than they appear at first sight.

A second indication of policy makers' interest in nanotechnology in the USA was the conference on nanotechnology organised in 2000 by the *National Science Foundation* (NSF). This is an important American organisation of natural scientists that has great influence on the development of natural science and technology. The conference was devoted almost entirely to the social and ethical questions surrounding nanotechnology. It is significant that the conference was organised by the NSF: the scientists wanted it to be known that they are conscious of their role in determining policy on nanoscience and nanotechnology. The extensive report was made available to everybody on the Internet. Evidently the NSF did not want any secret consultations but an open contribution to the social debate.

One could however wonder how much effect such reports have. The flow of money channelled by government (both the USA and the European) to nanotechnology is enormous and is on the increase rather than the decline. So questions like those posed by Winner do not in practice lead to a check on the pace. Moreover much of the research done under the name of nanotechnology is little more than a continuation of research that already has been fully accepted by society but formerly did not bear the name of nanotechnology. Of course it would be strange if such research were suddenly broken off only because the research was given a new label. Still, even in this research one should more often raise the question about which motives lie behind it and in which direction these developments are driving us.

Ethical questions also play a part in the social discussions on nanotechnology. Just as in various other technologies these questions are partly concerned with the possibility that they could be used for designing new weapons. It could be clear that nanotechnology offers unprecedented possibilities for this. However, the ethical discussion in this case goes much further than the question whether weapons might emerge from it. Sadly they are mostly reduced to a discussion on the possible consequences and risks. In politics it is easier to reach consensus on possible consequences than on fundamental motives or general obligations. However, this means a huge reduction of the ethical discussion. It often is aggravated by the fact that people also express the risks in numbers. Such a number creates the impression that one has a grip on and a view of the possible risks. We then soon forget that behind the numbers there often are various assumptions and uncertainties. Thus the discussion is poisoned by a pseudo-accuracy that creates a false feeling of calm. The actual questions, for instance as Winner put them, then remain in the background.

In what follows we will see that technology is fully bound up with our being human (see Chapter 12). Technology can be seen as an extension of human limbs and organs. However, it remains a question whether typically human characteristics like the power of thought and emotions can actually be realised in technological artefacts. Such questions are also evoked by the far-reaching ambitions of nanotechnology. There are nano-gurus who claim that with nanotechnology one can grow human tissue and in this way *ad infinitum* keep on replacing dying tissue in humans. In this way one would be able constantly to delay one's dying day. Would not conquering death be the ultimate victory of human reason? One can brush aside this question by saying that nanotechnology will never make this possible. But is it not more important to ask whether one really wants to go in this direction, irrespective of how far one can get? The consequences would in any case be very radical because our way of living is also determined by the knowledge that our life is finite.

The question surrounding nanotechnology and being human, however, lies even nearer. Nanotechnology already enables us to deal with disabilities in the human body. More and more members and organs can be replaced by prostheses. This will increasingly draw forth the question of how far one can stuff a human being with technology. When can one no longer speak of a human being but rather of a cyborg? What will it be like to be a human being with physical and psychic sides while one's body has become technology altogether? Here, too, we have to put the question whether we really want to go this way. The answer to this demands much more than a factual assessment of possible consequences. It concerns essential questions not to be suppressed by speaking only about consequences.

Notes

1 One nanometre is one millionth of a millimetre.
2 In the film the members of a medical team in a kind of submarine are made so much smaller that they can be injected into the blood stream of an atomic expert who has had a cerebral infarction. They travel through the arteries in their boat and remove the blood clot.
3 For the moment we set aside the possibility of nano-artefacts becoming so complex that they start showing signs of life.
4 For a detailed analysis of the influence of nanoparticles on the human body it is important to point out that the physical structures of the human body are 'encaptically' interlaced with the biological structures (see Chapter 5).
5 This is put into perspective by the fact that not everything called nanotechnology is altogether new. But certain developments like building up tissue 'from scratch' on the other hand certainly are not merely 'more of the same'.

References

Drexler, E. (1986) *Engines of Creation*, New York: Anchor Press.

Further reading

Roco, M.C. and Bainbridge, S. (2002) *Societal Implications of Nanoscience and Nanotechnology*, Dordrecht: Kluwer Academic Publishers.

6

KNOWLEDGE OF DESIGNING

The role of the engineer

Summary

Technological knowledge can be categorised in various ways. Walther Vincenti uses the following categories: fundamental design concepts, criteria and specifications, quantitative data, theoretical tools, practical considerations and design instrumentalities. Michael Polanyi differentiates between 'tacit' and 'codified' knowledge. Yet another approach is: knowledge of the physical nature of artefacts, the functional nature of artefacts, the relation between the physical and functional nature of artefacts, and knowledge of processes. Finally one can also relate different aspects of reality to different types of knowledge. An important feature of technological knowledge is its integrated character: a combination of knowledge from different kinds of sciences. One feature that distinguishes technological knowledge from some forms of scientific knowledge is its normative character: the content of the knowledge refers not merely to what is, but also to what should be. However useful technological knowledge has proved to be, it has its limitations. This is partly due to the fact that reality sometimes presents itself in a different way from what it is (think of mirages, for instance), that we sometimes think things are true since we would like them to be true, and that technological knowledge, no matter how broad it is, always contains only a part of reality.

One of the most important inventions in the field of Integrated Circuits (ICs) is the process LOCal Oxidation of Silicon (LOCOS). For many years this process was *the* industrial standard which more or less compelled every producer of ICs to work with this process. LOCOS is a technology of making ICs from slices of silicon. The essence of this technique is that areas in the silicon are provided with another material that causes an excess or a shortage of electrons in that place. Applying materials of this kind is called 'doping'. By doping the current can be sent from one place to another. It is important that the materials only enter into the silicon in the predetermined places. To achieve this the rest of the silicon surface is screened off. This screening off is effected by letting that surface oxidise. The oxidation therefore has to take place only in parts which later may not contain any doping. In the LOCOS process a thin layer of silicon nitride is first applied locally by means of a lithographic process. This is used as a 'mask' during the oxidisation of silicon. In places where silicon nitride is present the underlying silicon is protected against oxidation, and where no silicon nitride is present, oxidation does take place. During the oxidation the oxide layer partly sinks down into the silicon (see Figure 6.1).

The LOCOS process was developed in Philips' Physical Laboratory (Nat. Lab.) in Eindhoven (De Vries 2005a). By referring to the documents in the archives one discovers that in the development of this process various types of knowledge were applied. One part of the knowledge was of a very practical nature. The 'assistants' (the name for the employees who conducted the experiments) had knowledge of experimenting with silicon. They knew exactly how to set the temperature of an oven to get a good IC. The researchers had knowledge of the mechanisms that take place when silicon oxidises. They knew exactly how they could influence the properties of the oxide layer. Another part of the knowledge was very theoretical. Scientists in another research group did quantum mechanical calculations on this kind of layer. But these scientists had no contact with the group working on LOCOS. In this way it was only later discovered that the work of the quantum theoretical group could have been particularly useful for the LOCOS group which was more experimentally oriented. In short, a chance was lost to bring together different kinds of knowledge.

The example of LOCOS shows that in the development of technology different kinds of knowledge can be distinguished. In this chapter we will be going into the kinds of knowledge and their properties. The important theme of the integration of different kinds of knowledge will also be dealt with.

6.1 Kinds of technological knowledge

6.1.1 Vincenti

Another example of a technological field that employs a variety of kinds of knowledge, is the designing of aeroplanes. Walther Vincenti, an American professor in aeronautics, wrote *What Engineers Know and How They Know It* (1990) about a number of historical case studies in this technological field. Vincenti is a historian, not a philosopher. Yet in later literature on philosophy of technology his attempts to find a categorisation of types of technological knowledge are often quoted. This is reason enough to discuss his categorisation here. However, we will also see that it lacks the systematics usually pursued in a more philosophically oriented book. But for a start his categorisation is certainly interesting. We will try to distil from it some features of technological knowledge.

From his series of case studies Vincenti therefore inferred a categorisation, or *taxonomy* for technological knowledge. In this he differentiated between the following types:

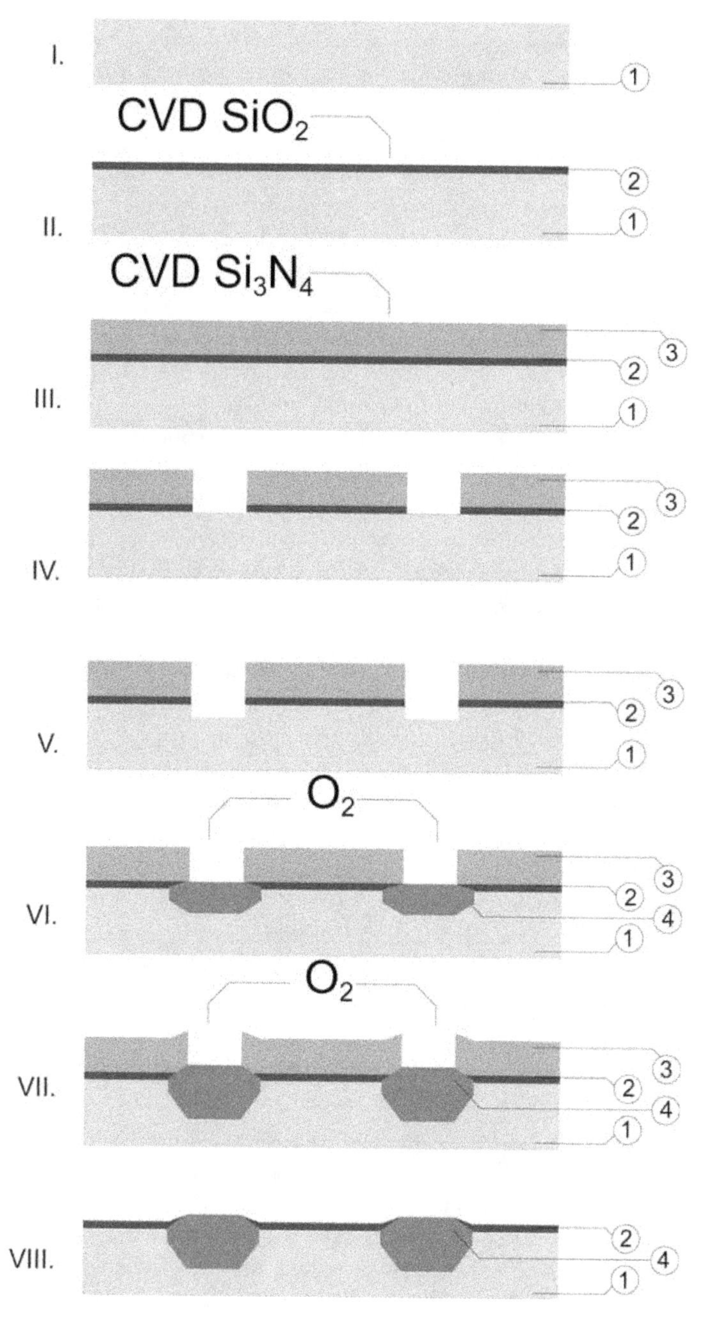

FIGURE 6.1 The LOCOS process

a *Knowledge of fundamental design concepts.* By this Vincenti meant that an engineer has fundamental knowledge of the way designs are made up. A trivial example: an architect knows that when designing a house, s/he should consider a foundation, walls and floors, and a roof.

b *Knowledge of criteria and specifications.* Engineers have knowledge of all kinds of demands and criteria which the design will have to meet.

c *Knowledge of quantitative data.* Amongst other things this has to do with physical constants and properties of materials. An engineer will for instance find data of this kind in a Technological Handbook.

d *Knowledge of theoretical tools.* This means knowledge of the different formulae and arithmetic models with which engineers work. Many of these formulae and models are derived from the natural sciences.

e *Knowledge of practical considerations.* These are considerations that do not emerge from scientific research, but are 'plain' common sense. So, for instance, an engineer would know that it is wise to choose a certain material because it is available locally and furthermore is not expensive.

f *Knowledge of design instrumentalities.* This is the knowledge of how to approach problems of design. In other literature the expression 'procedural knowledge' is used.

The second part of the title of the book ('*how they know it*') denotes Vincenti's research into how engineers acquire the different kinds of knowledge. He distinguishes the following kinds of knowledge-generating activities:

1. *Learning from natural science.* Although we would probably intuitively see this as one of the most important sources of knowledge for engineers, research proves that only a relatively small portion of the knowledge is derived directly from the natural sciences. For this knowledge often proves to be too abstract and idealised to be applied directly to design by engineers. It is almost as if such knowledge has to be 'translated' first to the practice of designing.
2. *Inventing.* The invention of a new method, new process or product often renders new knowledge as well.
3. *Theoretical technological research.* This kind of research usually consists of calculations for theoretical models and usually renders quantitative knowledge.
4. *Experimental technological research.* This kind of research consists, for instance, of experiments with scale models.
5. *Experience in designing.* The completion of a design process often renders knowledge that can be used in a subsequent design process.
6. *Experience in production.* Making products can lead to new technological knowledge.
7. *Trying out directly.* Often directly trying out ideas also renders some concrete technological knowledge.

Finally Vincenti gives an overview of which activities lead to which kind of knowledge. The most striking fact is that the natural sciences only contribute to two kinds of knowledge, namely to theoretical aids and to quantitative data. So it seems that engineers have a lot of knowledge that is not derived from the natural sciences. This is a significant conclusion for it has often been claimed that technology is nothing more than applied natural sciences. Vincenti's studies prove that this belief is incorrect. It seems that for designing an aeroplane much more knowledge is needed than only knowledge from the natural sciences. And this does not apply to aeroplanes only.

Another conclusion from Vincenti's categorisation is that some of the knowledge of engineers is indeed derived from scientific research but that another important part of it comes from the direct experience of technological reality. It is only when we have got to know reality from experience that we are able to view certain aspects separately. We can then detach that one specific aspect from the rest – this we call abstraction – and go about researching it systematically (see Chapter 3). The knowledge that flows from it relates exclusively to that separate aspect, and no longer describes the full reality. An engineer has to be aware of this.

As we have said earlier the contents of Vincenti's categorisation lack systematics. Therefore we also look at alternatives that do not originate from empirical case studies but from a more theoretical reflection on technological knowledge.

6.1.2 Polanyi

In the literature in the field of business administration there is a division into '*tacit knowledge*' and '*codified knowledge*'. Technological knowledge can also be either tacit or codified. What is the meaning of these terms? The word '*tacit*', used for the first time in this context by Polanyi in *The Tacit Dimension* (1966), means that knowledge is implicitly present without being articulated or written down. It is present in the minds of people only. This can be inconvenient for a company, for when people go to another company the knowledge they have leaves with them, since it is in their minds. Therefore Polanyi states that such knowledge should be brought 'out in the open'. Stated in a different way: the knowledge should be transposed from *tacit* to *codified* knowledge. *Codified* thus means that the knowledge is made accessible to others. Making knowledge explicit falls under what is called 'knowledge management'. *Codifying* is very important, for example, when one wants to develop an expert system. Such an expert system comprises the knowledge of various experts, for example medical expert systems bringing together the knowledge of diseases that could match specific symptoms. However, such an expert system can only be established after knowledge has been made explicit.

The categorisation by Polanyi links up very well with our own experience. Companies know from experience how much knowledge suddenly disappears when an employee leaves the company. By failing and trying again everyone gains knowledge that is never really written down. In every company there are 'unwritten' rules that exist only in the minds of the employees and are nowhere on paper. One important aspect of knowledge therefore is that it is '*embodied*'. There is a separate speciality concerned with that aspect of knowledge, so-called '*social epistemology*'. But according to Polanyi the categorisation of knowledge also has its disadvantages: it seems that nothing is added to the knowledge by it. The same applies to the case of knowledge management: it seems that the content of the knowledge has no effect on managing it. This representation of matters is not satisfactory. It cannot be that in transposing knowledge from *tacit* to *codified* it is of no concern what the knowledge relates to. One can of course imagine that one body of knowledge is easier to make explicit than another. So Polanyi's categorisation is not adequate either for reaching a satisfactory categorisation of technological knowledge.

6.1.3 Knowledge of the two-fold nature of artefacts

A third attempt restricts itself to knowledge relating to technological artefacts. Such artefacts can be described in two ways (De Vries, 2005b). A pair of pincers, for instance, can be described in terms of its form, weight, colour, and possibly smell, roughness of the surface, size, kind of material, etc. These are all geometric, physical and chemical properties. One could call it the physical nature of the pincers. But one can also describe the same pincers as an object that can be used for something. Such pincers are meant to pull out nails somewhere. But with a little imagination one can also allot various other functions to them. For instance, if the drawings in the workplace are blown away by the draught, one could use the pincers as a paperweight. When describing the function one looks at what can be called the functional nature of the artefact. What kind of technological knowledge can now be inferred from this dual mode of description? First an engineer can have knowledge of the physical nature of the artefact. S/he can, for instance, have knowledge of properties of forms and materials. Second, an engineer can know the functional nature of an artefact. S/he can, for example, know exactly what is needed so that a pair of pincers can be effective when removing nails from planks. A third kind of knowledge concerns the relation between the physical and the functional nature of the artefact. An engineer can know that a specific kind of material is more suitable for making pincers than some other material. By this s/he finds a connection between the physical nature (the properties of certain materials) and the functional (what one should be able to do with a pair of pincers). In the last instance an engineer can have knowledge of processes linked to the artefact. For instance, s/he can know what movements one should carry out with a pair of pincers to remove nails from a plank. But s/he can also have knowledge of the processes necessary for making a pair of pincers from the materials chosen for it.

6.1.4 Knowledge of the various aspects of technological artefacts

By the categorisation of knowledge of the physical nature, of the functional nature, of the relation between physical and functional nature, and of processes, we reach – better than by Vincenti's categorisation – a certain systematics in typifying technological knowledge. But we have differentiated relatively few types, while every student experiences that technological knowledge is extremely complex. How can we do better justice to the complexity of technology? In Chapters 3 and 4 we distinguished 15 different aspects or dimensions of technological artefacts. In this section we would like to demonstrate that these different aspects can be connected to different kinds of knowledge. Let us consider this by referring to the example of the development of the material GLARE (GLAss REinforced laminates).

GLARE is used in the Airbus 380, a huge aeroplane with two storeys. The development of this material took place mainly at the Delft University of Technology in cooperation with some industries (Vlot, 2001). By referring to this development one can see very well how many different kinds of knowledge were necessary not only to produce a material with the right properties but also to apply that material in the design of a plane.

GLARE is a material consisting of alternate layers of aluminium and fibreglass. How many layers there are depends on the place in the plane where the material is used. Sometimes three layers are needed and sometimes five. So it is a matter of knowledge of the numerical aspect of GLARE when designing it. Moreover, the direction in which the fibres of the fibreglass run is important. This knowledge in turn concerns the spatial structure of the composite. When this material is incorporated into an aeroplane, all kinds of forces are exerted on it during flying. Not only does the material bend somewhat as a result of its weight, but when taking off the fuselage is somewhat inflated like a balloon, and shrinks back again when descending. This, too, causes forces on the material that one should know. In other words, what is at stake here, is knowledge of the physical aspects of GLARE.

The development of this material has a long history. GLARE succeeded Arall. All knowledge gained in the development of Arall was used when developing GLARE. So here we are dealing with knowledge of material development and material forming under the aspect of form-giving. For the development of a successful material for aeroplanes even language plays a role. This became evident when the Delft researchers found out that their French partners with whom the material was being developed, did not like the chosen term. For the word 'GLARE' to French ears sounds like 'glaire' which in French means 'slime'. And who would want to make the walls of a plane out of slime?

The development of GLARE took place in an international consortium of partners. The researchers also had to have some knowledge about the way in which the different partners interacted socially. Not only did they have different nationalities but they also represented different interests. Constantly keeping the whole consortium together in the development process was a delicate game. Naturally GLARE also has an economic dimension of which the developers had to have knowledge. For GLARE had to compete with aluminium, not only as far as functionality was concerned, but also in cost. And, just as in the case of many technological developments, the juridical aspect also came into play, namely the issue of patent rights. This, too, required specific knowledge and expertise. In this case it was especially important since the application for patent rights was a much higher priority to the industrial partners than to the university community. The scientific researchers were more interested in publications and therefore the patent rights had to be arranged in such a way that publishing would also be possible. To some researchers there was a moral dimension to the development of GLARE. This material makes a plane lighter and can therefore contribute to better environmental properties (less use of energy and fewer emissions). To many researchers GLARE was the object of an almost unshakeable faith and trust in its potential. It is important to know which properties of the material fostered this faith. Such an aspect of faith is often found in technological developments. Faith in the potential of an invention often counterbalances all arguments against further development on the basis of other aspects like the economical.

One can have knowledge of each of the 15 different aspects or dimensions that we differentiated. In the work of Vincenti engineers have not only knowledge resulting from experience of reality but also scientific knowledge of the separated, abstracted aspects. The example of GLARE demonstrates that engineers in some way or another come up against all aspects. Apart from the knowledge gained from experience that as a whole touches on all aspects, they can therefore often make good use of scientific knowledge of the different aspects. They then have recourse to different scientific disciplines like engineering sciences, natural sciences or social sciences. This can also be seen in the example of GLARE. A technological artefact consists of materials of which one can gain knowledge. Before long one has recourse to physics and chemistry. But such an artefact also has a juridical existence. New inventions are often patented and in that case one would have recourse to jurisprudence. Further, sociological knowledge can help to understand the area of tension between social actors.

The conclusion that one can have knowledge of all 15 aspects of a technological artefact automatically leads us to question whether we have hereby altogether left the track of the four kinds of knowledge mentioned above in Section 6.1.3. The answer is 'no', since out of the 15 a number are connected with the physical nature and a number with the functional nature. Knowledge of the form of a coffee cup, which is a part of its physical nature, is connected with the spatial aspect of the artefact. Knowledge of being regarded as beautiful or ugly, which is part of its functional nature, is connected with the aesthetic aspect. Figure 6.2 depicts which aspects are connected with the physical and which with the functional nature. The biological dimension is found under the physical nature because there are technological products not only from dead material but also from living material as in biotechnology. With the aid of the terminology from Chapter 4 one can also formulate it as follows: in the dimensions connected with the physical nature the technological artefact functions as a subject and in those with a functional nature as an object of the acting human being.

6.2 Integration of technological knowledge and the nature of the technological sciences

It is evident that engineers, apart from direct knowledge gained from experience of reality also use knowledge from the different scientific disciplines. But engineers do

Aspect	Nature
arithmetic spatial kinematic physical biotic	non-intentional (physical)
psychic analytical historic linguistic social economic aesthetic juridical ethical faith	intentional (functional)

FIGURE 6.2 Modal aspects and the dual nature of artefacts

not gather knowledge merely for the sake of knowledge. They want to use the knowledge to design new products or technologies. Abstract scientific knowledge cannot be used for this right away. So now we will investigate how abstract knowledge can still be used in real engineering work. We will see that knowledge of different aspects of reality can be combined in two ways. In the first instance the engineer can enhance his direct knowledge gained from experience of reality with knowledge from the disciplines. S/he then more or less returns to the level of 'non-abstracted reality' from which s/he departed, that is before dividing reality into dimensions. Second, within the level of abstract scientific knowledge s/he can integrate different kinds of knowledge. Then s/he returns in a manner of speaking to the level of 'abstracted reality'. Let us take a closer look at the latter.

In order to be able to design a technological artefact it is not sufficient to have different types of knowledge. One way or another the various elements of knowledge have to be integrated to reach a good result. Since technological knowledge nowadays is very specialised, integrating knowledge often begins with bringing together different disciplines in a multidisciplinary design team. The term 'multidisciplinary' is currently very often heard in technology. A mechatronic design like a cd player requires knowledge from at least three disciplines: electrical engineering, mechanical engineering and information science. The word 'mechatronics' is a contraction of 'mechanics' and 'electronics'. And then we are only speaking of the strictly technological disciplines and not even about knowledge of, for instance, the economical and juridical aspects of designing. It is said about the new field of nanotechnology (isolating and manipulating separate atoms) that it brings together at least the disciplines of physics, chemistry, material science and biology, not to mention knowledge of the ethical aspects. There are companies who pursue the ideal that not only engineers determine what the design will look like but that at an early stage other specialists are also introduced. People with marketing expertise or from the patents department join the discussions in the design process. An example of a method of designing that stimulates this is Quality Function Deployment (QFD). With the help of this method the wishes of clients are translated into technological parameters. Carrying out this method is mostly in the hands of a team of staff from different departments of the company. In such a team the core disciplines of marketing, design and manufacturing are always represented. Often such a team is enlarged with specialists in the fields of quality, application or finances. In this way different kinds of knowledge of technological products are combined. Here we once more find what we have already seen in the work of Polanyi: knowledge lodges in people. Stated in a different way: knowledge lodges in actors. Combining knowledge therefore to some extent consists of bringing together people with specific knowledge.

Experience has taught that the integration of different kinds of knowledge, for instance in a QFD team, is not always plain sailing. Sometimes serious communication problems arise between experts from different core disciplines. Marketers sometimes do not understand what designers are talking about and designers sometimes have little understanding of the contribution of manufacturers. It seems that different disciplines communicate about their knowledge in different languages. Each discipline has its own vocabulary or jargon. And even when people use the same words there often are (subtle) differences in meaning. Chapter 7 will deal with how different kinds of knowledge are integrated in the design process.

Apart from distinguishing between kinds of technological knowledge, one can also look for a typology of different (scientific) disciplines. For this there are different options. So, for instance, a distinction is made between sciences that do research on cause and effect relations, and sciences that do research on intentions. Physics focuses on regularities within cause and effect relations. Elementary particles, for instance, do not have intentions. Their behaviour is entirely determined by cause and effect relations without any intentions. They collide, they attract or repulse one another but they have no consciousness. However, in psychology one does look at intentions. Cause and effect relations, as such, do not offer a good basis for the description of what people do and why they do it. Although in the past psychologists did search for explanations for human conduct in terms of cause and effect relations. Behaviourism attempted to explain all psychic aspects of people in terms of responses (consequences) to stimuli (causes). But most psychologists nowadays are of the opinion that this approach falls short of an understanding of the specific nature of people.

Another distinction sometimes made is that between sciences looking for universal laws and sciences wanting to describe the unique. The first kind is called nomothetic (literally, posing laws) sciences and the second kind ideographic (literally, describing the particular) sciences. Physics is an example of a science looking for universal regularities. Newton's law is valid always and everywhere. History is an ideographic science. It is more concerned with describing as accurately as possible how and why a battle was fought in 1600 at Nieuwpoort in Belgium than with formulating universal laws that explain why the battle had to take place in that year and in that place. For that matter there has also been a kind of conflict between methods just as in psychology. There have been historians who thought that in the historical sciences one actually had to look for universal patterns and 'laws'.

The two ways of categorising sciences are not entirely detached. The fact that electrons can be described in terms of laws that are applicable always and everywhere is closely connected to the fact that they know no intentions. The fact that people in many respects cannot be adequately described by universal laws results from the fact that their reactions vary so much. Whenever one is dealing with people it is difficult to practice nomothetic science. This can be seen very clearly in a science like economics. Very often attempts are made to draw up models based on cause and effect relations but often it appears that the forecasting value of these models is disappointing.

A third way of categorising the sciences is related to the distinction we introduced between laws and norms. We have pointed out that the earlier aspects (namely the arithmetic, spatial, kinematic, physical-chemical, biotic and psychic) have a character of regularity. Every 'thing' from this reality is subject to these laws and nothing can withdraw from them. The later aspects, on the other hand, from the logical-analytic up to and including the aspect of faith, have a normative character. The human being who functions as the subject in all these aspects has the task of giving form to these norms. In the technological sciences this distinction can be used with good effect. Thus an engineer designing a new product cannot withdraw from the laws of physics: an object always falls downwards and a metal will always conduct a current. But s/he can transgress juridical laws, for instance by infringing the patent of a rival and s/he can ignore moral laws by not considering ergonomic aspects.

6.3 Technological sciences

How should we see the technological sciences in the light of these categorisations? This cannot easily be stated. As early as the work of Vincenti we noticed that they cannot adequately be described as 'applied natural sciences'. Herbert Simon also was convinced of this when he wrote his well-known book *The Sciences of the Artificial* (1969). The title in itself expresses the contrast between the 'sciences of the natural', the natural sciences, and the 'sciences of the artificial', the technological sciences (by which Simon meant information technology in particular). Technological sciences are not so easily categorised according to the different typologies we have discussed. All of this is connected to the fact that the knowledge of engineers not only comprises the abstract knowledge of separate aspects, but also the systematically hoarded direct knowledge gained from experience of non-abstracted reality. In order to see this more exactly we first have to look at the differences between science and technology.

The philosopher Egbert Schuurman (2009) has pointed out some differences between science and technology that are still useful. The aim of science in the first place is knowledge that is abstract and universal. Abstract, since it refers to only *one* aspect that has been detached from the whole of reality in order to be studied separately. Thus a physicist only focuses on the physical aspects and an economist in principle only on economic aspects. Science also is universal: it is valid always and everywhere. The third part of Newton's main work of 1687, the *Mathematical Principles of Natural Philosophy*, has as its title: *The System of the World*. By that he meant that his mechanical laws are applicable to all mechanical phenomena, irrespective of their place and time. Technology, on the other hand, is always concrete and specific. An engineer cannot permit him/herself to involve only the physical aspect of reality in their designs, or to consider only the interests of one group, but must take into consideration all forms of complexity. Schuurman pointed out that scientification of technology (applying features of science to technology) in certain cases has led to undesirable results. The abstract and the universal is then expressed in the uniformity of technological products: from terraced houses in architecture to all kinds of identical consumer products. Ready-made clothes, the same for everyone, replaced suits made to measure. The table or chair bought from the furniture giant replaced the unique and authentic chair made by a craftsman. In our postmodern times with greater emphasis on individuality we are conscious that technological products should be characterised less by the abstract and universal. Technology and science are not the same and should not be treated equally. We will return to this issue in Chapter 9.

With the aid of the difference between laws and norms from the previous section we can pinpoint better the difference between natural science and technology. In physics and chemistry we do research on the regularities in nature: in technology we do research on technological artefacts. That means a closer investigation into the physical-chemical regularities of materials and constructions used in the particular artefact, like the research on material done in the framework of the GLARE project. It also means a closer investigation into the normative object functions of the materials and construction, as found within the economic and moral dimensions of the material GLARE.

As a *researcher* an engineer consciously detaches him/herself from the broad complexity in which a technological artefact functions. S/he focuses on *one* or a few aspects of the artefact, the interests of *one* or a few groups, or the features of *one* technology. This restriction makes sense for it is the only way to collect detailed knowledge of a certain element of the artefact or of certain phenomena while using the artefact. But as a *designer* the engineer may not detach him/herself from the broad complexity in which an artefact functions. S/he has to take into consideration all aspects, all groups and all related technologies. In this respect the technological sciences are actually positioned somewhere between science and technology. On the one hand they are sciences in the sense that systematic knowledge is collected on technological artefacts but on the other hand in the design process they are closely linked to technological practice. Formulated in a different way, in the technological sciences one not only wants to acquire technological knowledge that exceeds an individual problem of design but also to do justice to the uniqueness of each problem of design. Technological sciences therefore have a hybrid character.

Handling the complexity of technological developments is therefore not quite so simple. Due to the complexity knowledge of very diverse aspects is needed which then has to be integrated too. This complexity is sometimes experienced as taxing. Yet many engineers find it inviting exactly because the technological designs are so challenging and complex. They have appreciation of the complexity of reality and certainly of technology. This makes technological designing no easier but it does make it much more interesting.

6.4 Norms in technological knowledge

In technology we are not only concerned with reality as it is but also with the way we can change it by technological means. Technological knowledge therefore relates to the existing reality and also to how we would like it to be, or not to be. We can speak in a normative way about this situation that has not come into existence yet: What do we consider good and what do we consider as not good? This results in all kinds of normativity in technological knowledge. We will here be dealing with three examples of normativity: knowledge of requirements, knowledge of functions and knowledge of 'good practice'.

Norms for what we call technologically good or not good are in practice formulated in the list of requirements that a designer uses as his/her guideline. Such a list of requirements comprises the wishes of the client, the needs of the future users and the interests of other groups. If it is good it is written from the perspective of and in the language of the client, user and other groups. The normative knowledge that relates to it then revolves around knowledge of requirements and criteria. The same knowledge can afterwards be used to check whether the design meets the expectations.

A second form of normative knowledge concerns the functions of a technological artefact. When an engineer says 'I know that this is a drilling machine' what he really means to say is: 'I know this is an apparatus with which one should be able to drill holes'. Please note: 'should be able to'. The words 'should be able to' likewise denote a form of normative knowledge. Let us go one step further. An engineer could also say: 'I know this is a good drilling machine'. Here the normativity in the knowledge of functions becomes even clearer. The word 'good' means we are dealing with a normative statement that says something about *one* specific drilling machine. We call this normativity at the level of the individual apparatus. The engineer could also have said: 'I know that drilling machines of this kind are good drilling machines'. This is normativity at the level of the type. A third statement could be: 'I know that drilling machines are good for drilling holes'. This means a normative statement on drilling machines as a kind of tool. Normativity can therefore relate to different 'levels' and this also becomes evident when the apparatus is broken and cannot be used for drilling holes. In that case an engineer will still say it should be a good apparatus with which one should be able to drill holes.

However, an engineer can also have normative knowledge of technological activities that s/he has to perform during the process of designing. This relates to knowledge of methods and technologies. So for instance, there is much knowledge in the ceramic industry about the way in which new materials should be developed. Part of this knowledge has been explicated in procedures, *codified*, and part of this knowledge is applied unvoiced, therefore *tacitly*, according to Polanyi. In the components and assembling industry certain methods and technologies are used to analyse systematically the problems that could occur in a design, the so-called *Failure Mode and Effect Analysis*. An engineer should have knowledge of various environmental and safety regulations which have to be complied with by design processes. But we can also mention what is called 'good practice'. Hereby we understand various rules of thumb that tell how best to handle something, mostly based on experience.

In technology we therefore come up against different kinds of normativity. By making a modal analysis of technology we are able to define these more accurately. The first we deal with is what is sometimes called 'functional normativity' that relates to the functioning of the apparatus, otherwise called the formative object function of a technological artefact. In a good drilling machine the formative aspect is adequately developed. In the second instance we are dealing with 'non-functional' normativity. This form relates to the later object functions of a technological artefact. A drilling machine is good only when moral object functions like safety and energy conservation have been properly disclosed. Under this form of normativity the juridical object function is also reckoned, which emerges amongst others in infringing or respecting patents of rivals. In our opinion functional and non-functional normativity should be integrated. So a design is only good when an engineer has responsibly considered the justified interests of the various groups involved in the design. In philosophical terms, a design is good only when all object functions have been disclosed in a normative way.

6.5 The limited nature of engineering knowledge

The application of technological knowledge has demonstrated impressively what the power of knowledge is. Yet technological knowledge has its limitations and it is important to keep this in mind. In studying reality we should have knowledge of the way in which it presents itself to us. There are philosophers of science who infer that our theories can do no more than succinctly summarise our observations. They would say no more about the reality behind the observations. It is true that the way in which we interpret our observations is not always correct, for example, a mirage in which travellers in the desert see an oasis that in reality is not there. But experience has taught us that we can trust that our interpretation of our observations generally is correct (see Chapter 2). For the rest most scientists work while departing from a kind of intuitive confidence in the regularity of reality and its accessibility to human understanding.

Another limitation on our knowledge relates to what we think is true. A good example from technology is the realisation of the transatlantic cable (Coates and Finn, 1979). The magisterial physical scientist William Thomson (Lord Kelvin) had initially calculated meticulously that an electric signal would never be able to bridge the enormous distance over the transatlantic ocean. But the entrepreneur Cyrus Field had set his mind on getting the transatlantic cable on his name and he persevered. All the effort it took to lay the cable was amply rewarded when finally, in spite of Kelvin's calculations, it was proved that the cable was indeed able to transmit a signal. If Field had depended on the calculations done by Lord Kelvin he would never have laid the cable and then he would have missed the success of this enterprise. This is an example that makes it seem favourable to not take as true the forecasts of science. But in general this will not be the case and engineers have to be aware of this.

In the third instance there is the mistake often made of absolutising the abstract knowledge of separate aspects of reality to such an extent that one loses sight of the fact that reality is more than the sum of the aspects. In the final instance the engineer has to work with the entire complex reality while knowledge of separate aspects makes only a small contribution to the knowledge of reality.

It would be good if these limitations on our technological knowledge brought us to some degree of modesty in our technological activity. Unfortunately this is not always the case. This is shown in the way risk factors are ignored in technological society (in Chapter 11 we will return to the society of risks). Uncertainties about risks are camouflaged in impressive calculations of chances of accidents and chances of certain consequences. Instinctive objections by 'ordinary people' are swept aside by experts as being inaccurate, while in all their calculations they should be taking into account all kinds of reservations. Philosophical reflections on technological knowledge can make us aware of the limitations of this knowledge. Being conscious of the limitations should save us from overconfidence when we use this limited knowledge for technological intervention in reality.

References

Coates, V.T. and Finn, B.S. (1979) *A Retrospective Technology Assessment: Submarine Telegraphy, the Transatlantic Cable of 1866*, San Francisco, CA: San Francisco Press.

Polanyi, M. (1966) *The Tacit Dimension*, Garden City, NY: Doubleday & Co.

Schuurman, E. (2009) *Technology and the Future. A Philosophical Challenge*, Grand Rapids, MI: Paideia Press. Translation of: *Techniek en Toekomst. Confrontatie met wijsgerige beschouwingen* (1972).

Simon, H. (1969) *The Sciences of the Artificial*, Cambridge, MA: MIT Press.

Vincenti, W. (1990) *What Engineers Know and How They Know It*, Baltimore, MD: Johns Hopkins Press.

Vlot, A. (2001) *Glare. History of the Development of a New Aircraft Material*, Dordrecht: Kluwer Academic Publishers.

Vries, M.J. de (2005) *80 Years of Research at the Philips Natuurkundig Laboratorium (1914–1994): the Role of the Nat.Lab. at Philips*, Amsterdam: Pallas Publications.
Vries, M.J. de (2005) *Teaching About Technology*, Dordrecht: Springer.

Further reading

Chalmers, A. (1999) *What is this Thing called Science?*, 3rd edn, Queensland: University of Queensland Press.
Strijbos, S. and Basden, A. (eds) (2006) *In Search of an Integrative Vision for Technology. Interdisciplinary Studies in Information Systems*, New York: Springer.

HERBERT SIMON

(1916–2001)

PORTRAIT 6 Herbert Simon

Herbert Simon studied political science and econometrics at the University of Chicago. He first worked at the University of California in Berkeley, later at the Illinois Institute of Technology and subsequently at the Carnegie Mellon Institute of Technology. In the 1950s he worked with Allen Newell in the field of artificial intelligence. They were especially interested in the way in which people and computers solve problems. Their book *Human Problem Solving* (1972) for a long time was a standard work in psychological research on the way in which beginners and experts solve problems (for instance chess problems and the solving of puzzles). For his work in economy, in particular his theory on decision-making in *profit* organisations, Simon received the Nobel Prize in 1978.

Although Simon is often seen as a psychologist and a sociologist he also is a philosopher of technology. In particular in his book *The Sciences of the Artificial* which was originally published in 1969 and was reprinted many times. In this Simon was one of the first to attempt to describe what was specific to the engineering sciences. In Chapter 5 of his book Simon describes what he calls the science of designing. The engineering sciences fall under this but also other sciences in which new things originate, for instance new laws in the juridical sciences or new medicine in medical science. According to Simon designing is incorrectly regarded as something which has shut itself out from the possibility of scientific research. Nor does he agree with the statement that engineers do not practise science but only apply it ('technology as applied science'). According to Simon a 'science of design' is definitely possible. Particularly at the interface between design and information technology there is an urgent need for such a science according to him. His book was to be a first initiative in this. Given his earlier work it is not surprising that he seeks innovation particularly in the direction of research on the role of decision-making and problem solving when designing. Simon especially sees possibilities in the structuring of decision-making and therefore also the possibility of simulating and supporting with computers the making of decisions and solving of problems. Critics of his work see a significant limitation in it, since it seems as if he does not do justice to certain ways of designing, as for instance architecture. In this creativity and fantasy play such an important part that designers see the structuring of the design process more as an impediment than as a support. However, in other areas, for instance mechanical engineering, such an approach was well received. Simon further points out that design is situated on the interface between an 'inner world' (the structure of the artefact itself) and an 'outer world' (the context in which it has to function). It reminds one very much of what the French philosopher Simondon claimed in his book *Du mode d'existence des objets techniques* (About the mode of existence of technological objects) (see the portrait of Simondon). In the design process the artefact has to be well tailored to both sides of this interface.

We close this portrait with a few quotations from *The Sciences of the Artificial*:

> In view of the key role of design in professional activity, it is ironic that in this century the natural sciences have almost driven the sciences of the artificial

from professional school curricula. Engineering schools have become schools of physics and mathematics; medical schools have become schools of biological sciences; business schools have become schools of finite mathematics. The use of adjectives like 'applied' conceals, but does not change, the fact. It simply means that in the professional schools those topics are selected from mathematics and the natural sciences for emphasis which are thought to be the most nearly relevant to professional practice. It does not mean that design is taught, as distinguished from analysis.

(Simon, 1969: 129–130)

My main goal in this chapter has been to show that there already exist today a number of components of a theory of design and a substantial body of knowledge, theoretical and empirical, relating to each. As we draw up our curriculum in design – in the science of the artificial – to take its place by the side of natural science in the whole curriculum, it includes at least the following topics:

The evaluation of designs

1. Theory of evaluation theory, statistical decision theory
2. Computational methods
    a Algorithms for choosing optimal alternatives such as linear programming computations, control theory, dynamic programming
    b Algorithms and heuristics for choosing satisfactory alternatives
3. The formal logic of design: imperative and declarative logic

The search for alternatives

4. Heuristic search: factorization and means–ends analysis;
5. Allocation of resources for search
6. A theory of structure and design organization: hierarchic systems
7. Representation of design problems.

(Simon, 1969: 155–156)

References

Newell, A. and Simon, H. (1972) *Human Problem Solving*, Englewood-Cliffs, NJ: Prentice Hall.
Simon, H. (1969) *The Sciences of the Artificial*, Cambridge, MA: MIT Press.

7

DESIGN AND REALITY

Methodological obstinacy

Summary

Design methodology studies the way in which designs come into being, in particular the methods used for this. The design process is often based on the steps analysis-synthesis-evaluation. Reality is much more complicated and therefore such ideal schemes should always be treated with caution. Towards the end of the previous century people doing designs started giving more attention to the consumer and the market. The realisation also dawned that in the design process different products may require different approaches. For market-oriented design all kinds of new methods have been developed in which the concept 'quality' takes central position. There is a trend to do designing not purely individually and independently, but in an open cooperative context. All these innovations do greater justice to the intricacy of reality but they remain ideal models.

In his book *What Engineers Know and How They Know It* Walther Vincenti describes how the designers of planes over a process of some decades gradually learned to make use of the experience of pilots. This mainly concerned the controllability of the plane. Pilots initially wanted a machine that was not too stable, since that was easier to handle. They expressed it concisely but forcefully by saying that the plane had to be 'stable, but not too stable', in other words that it had to react swiftly to the controls but not so fast that it would no longer be stable. The engineers found that this was too vague to be useful for their work of designing. They were therefore inclined to ignore these statements by the pilots. The pilots were moreover not consistent in their wishes either. For piloting a plane that is not very stable takes constant attention and intervention. Initially, when all flights were comparatively short, this was not much of an obstacle. But later on when pilots were flying long hours on end, continuously adjusting the steering proved to be exhausting. The research done by Vincenti showed that the pilots only learned as time went by what their actual wishes were. Moreover they had the impression that their experience was hardly taken into account in the designing of planes, which one can understand seeing the attitude of the engineers.

Engineers in the meantime were completely occupied by the question which variables of the plane played a role in the relation between stability and manoeuvrability. By experiments in wind tunnels one got insight into the variables. And the designers learned to relate the variables to the different theories from mechanics and aerodynamics. Only at that stage did they once more catch sight of the wishes of the pilots which had up to then been more or less ignored. Vincenti describes how the work of a 'mediator' between the pilots and developers played an important part in establishing a connection between the wishes of the pilots and the technological properties of the plane. It seems that the 'mediators' were able to make a connection between the vague and qualitative statements by the pilots on the one hand and the carefully defined and quantitative approach of the engineers on the other hand. In this way the designers eventually reached guidelines for the design of planes that were explicitly synchronised with the wishes of the pilots.

This case study demonstrates various aspects of the design process. It becomes clear how difficult it is to take into consideration multiple factors like the possibilities and limitations of technology as well as the wishes of the pilots. Not to mention the legal regulations for planes, which are currently much more elaborate than they were in the period described by Vincenti. We also see that translating client wishes into technological properties – nowadays a precondition for design – was definitely not taken for granted. Furthermore it appears that both designers and consumers had to go through a long learning process. Pilots only learn in the course of using the planes what their wishes are, and designers only gradually learn what the relevant design parameters are. The design process is complicated; not only is knowledge applied, knowledge is also gained. Understanding exactly what happens in such a design process therefore requires careful study. Vincenti's book is a good example of this. The branch of science that is oriented in particular towards processes as well as designs is called methodology.

Why do we have a separate chapter on design? To the (future) engineer the answer to this question stands to reason. Designing is thinking out new products and processes. Without new designs there can be no progress. The design process forms the heart of technology. Still, it is not altogether obvious why we devote a separate chapter to the design of technology. For designing has not always been a separate activity, at least not as independent as at present. In traditional technology designing and making often happened simultaneously. It was mostly one person who designed a new piece of furniture and then also made it. In the nineteenth and twentieth centuries there began a gradual separation between designing and making. A specialisation came about between the various core processes in technology. And science gained substantial influence on the way in which technological designs came into being. All this justifies giving specific attention to designing.

In this chapter we will concentrate on the methodological side of designing. We look at how designers devise ways of giving materials a meaningful destination. We will pay ample attention to development 'for the market'. We close with a critical reflection on the developments as sketched. However, we will not be looking in detail at the manner in which groups like government, universities, action groups and multinational concerns exert power and influence on the different design processes. Neither will various ethical aspects be dealt with explicitly. In Part III of the book we will return to these points. The line of approach of this chapter is mainly methodological. We will mostly be using the word 'design', except when using examples from the electronic industry, we will adapt to the language used in this sector and use the word 'develop'.

7.1 What is methodology?

The term 'methodology' can lead to misunderstanding. The first misunderstanding is that methodology is seen as a subdivision of the science describing which methods can be used to do research. The word 'methodology' is then used as a synonym for the word 'method'. 'I use this and that methodology to research this or that subject.' But the suffix '-logy' in the word shows what the real issue is, for it points out that we are dealing with a field of knowledge, as in bio-logy and socio-logy (and actually techno-logy, too). Methodology therefore is the field of knowledge concerned with studying methods. But this definition can lead to a second misunderstanding. Methodology is mostly taken to mean the systematic of the research or the orderly structure in which designing activity takes place. Such a definition has a solid normative character: engineers should do their work like this and this. Methodology, however, not only looks at the way in which researchers or designers should be working, but also at the way in which they work in practice. Literally method means the way ('hodos') along which ('meta'). This can be a straight highway which many people have in mind when they think of a 'genuine method'. But it can also be a tortuous path full of bends and unexpected horizons that in the end does lead the researchers and designers to the desired goal.

In traditional philosophy, methodology is mostly regarded as a subdivision of philosophy of science. Philosophy of science in general deals with the question of what science is and is not; methodology with the question along which way (meta-hodos) scientific knowledge is and is not generated. In technology we also have methodology of design which focuses on the question along which way technological designs are generated. We now focus on the designing of new products, processes and technologies.

7.2 Proper and non-proper functions

In Chapter 4 we gave an analysis of the complexity of technology and we saw that technological artefacts in the so-called 'earlier' aspects (mostly up to and including the physical) function as subject. Further it became clear that technological artefacts function in all aspects as object, that is, are objects of the multifaceted activity of human beings. Designing therefore is devising such a technological artefact. The challenge to the designer is to design the artefact in such a way that it functions well in all aspects. It will therefore have to comply with all laws valid for the various aspects. In the physical aspect, for instance, the law of gravity is effective: things having mass fall. It would not be wise for a designer to disregard this law or to assume that it would not apply to his/her artefact. Likewise economic, juridical and moral norms also are valid for that artefact, and a designer has to keep this in mind.

Designing demands creativity. For the designer adds something new to reality. In this sense s/he is busy in a creative way. For that matter we have to put into perspective the creative character of technology, for a lot of designing is actually re-designing. Entirely new designs are relatively rare. Almost invariably there has been a predecessor for the designer or an already existing source of inspiration. In the literature a difference is made between '*normal design*' and '*revolutionary design*' (see Constant, 1980). The former expression denotes designs making a minor alteration to a predecessor and the latter denotes an entirely new design. In the practice of technological design one could speak of a broad spectrum that varies from minor alterations to entirely new designs. Designing therefore is not creating in the sense that the Bible says in the book of Genesis that God created: out of nothing. This is not what human designers do. They always start with materials at hand. The meaning that lies latent in the orderliness of reality is in a manner of speaking disclosed by designing. By doing this meaning is also added.

Designers are therefore first and foremost discoverers of meaning who disclose meaning in their process of designing and therefore give meaning (see Chapter 2). A designer prepares the material disclosure by investigating which elements of meaning one could disclose and in which way. And for this creativity is needed because the disclosing can happen in different ways. From one piece of wood one can make various objects with different destinations. One can also solve the same problem in different ways and with different materials. Further disclosure is closely linked with coherence and context. A part of a machine is only really disclosed when it functions within the greater whole of the machine. An apparatus only reaches disclosure when it functions within the greater whole of society. So when a designer thinks about his/her design the context will also be considered.

The consumer, too, plays a role in the disclosure. Only when s/he puts the technological artefact to use does it reach its destination. One could say: only then does the meaning become explicit. As long as the artefact gathers dust in a cupboard, full justice is not done to it. The fact that the meaning was a given beforehand, appears from the fact that the artefact cannot be used in just any way. Apart from the designer the consumer, too, has to consider the laws that apply in the different aspects in which the artefact functions. A consumer can instead of using a wooden spoon stir hot soup with a steel spoon but will soon become aware that steel falls in the physical aspect under the law that metal conducts heat. That is why that spoon was not meant for stirring hot soup. For that purpose the designer has invented another kind of spoon: the wooden ladle. Not only the designer and the manufacturer have to think about the laws of the aspects, but also the consumer. However, this does not mean that the consumer is bound by the purpose that the designer had in mind. When dealing with an artefact multiple forms of meaning disclosure are possible. Thus with a screwdriver one can screw or unscrew screws, but one can also use it to open a can of paint. The first we call the proper function the designer had in mind. S/he chose form and material in such a way that the artefact can fulfil its function properly. One can then say that the meaning of the artefact is being actualised: there is an actualising relation between consumer and artefact. We call other functions that a consumer can ascribe to the artefact 'non-proper' functions. One artefact can have many of these, depending on the creativity of the user. It can sometimes lead to a certain meaning disclosure, even one the designer had not thought of. In this way the use of simple pain killers can sometimes have unexpected positive effects on people suffering from heart and vascular diseases. This is a clear example of an unexpected meaning disclosure.

Not all non-proper functions lead to good meaning disclosure. When the use made of it is not compatible with the features of the artefact, it is no good and it can even be broken, whereby the meaning is then lost. If one tries to open a tin of paint with the accompanying stick meant for stirring the paint, it will break and then one would have neither an open tin nor a stick for stirring. In such a case one could speak of a non-actualising relation. A good designer will take into consideration the possibility that consumers may ascribe non-proper functions to the artefact. S/he can design the product in such a way that the artefact itself discourages the non-proper functions or makes them impossible. In the directions for use it will also be stated for what purpose the apparatus may or may not be used. Sometimes explicit directions are needed. There is the well-known story of a woman in the USA who sued the producers of a microwave oven because it did not say in the directions that the oven was not suitable for drying her dog that had become wet in the rain. This activity proved in a painful way to be a non-actualising relation for herself and her dog. The fact that her claim for damages was granted by the judge would have driven the designers to despair, for how should one know to prevent the use of all possible non-proper functions? Yet this case demonstrates that designers are challenged to do just this in practice. In the '*Dual Nature*' research at the Delft University of Technology mentioned in Chapter 4, the insight was acquired that designers actually always design two things: an artefact and an accompanying 'operating plan' (Houkes et al., 2002). By an operating plan is meant the actions that the consumer is supposed to carry out with the artefact, often including specific instructions on the circumstances under which it should be used – for instance temperature or humidity.

There are several ways in which the designer can make known that operating plan. Sometimes it can be done by means of the artefact itself. Norman (1988) gives the example of doorknobs that almost seem to cry out 'pull me' or 'push me'. Another example is electronic peripheral equipment which, under the motto 'plug and play', can only be connected and installed in one way: the different plugs 'communicate' how the apparatus should be connected. The total of all messages concerning the use being present in the apparatus itself, is sometimes called a '*script*', by amongst others, the technology sociologist, Bruno Latour. The designer can also make known the operating plan by means of explicit messages on the artefact. Thus the cable connecting a keyboard with the computer sometimes bears a label with a warning about health risks. On this label there also are references to the underside of the keyboard on which the user can read about the danger of repetitive strain injuries. Finally a designer can also make known the operating plan by means of a manual. The example of the dog in the microwave demonstrates that it may be important not only to make known the operating plan in a manual, but also to discourage openly wrong use (by a statement on the apparatus) or even to make it completely impossible (to prevent it technologically). An example of the latter option is a lorry with cruise control.

In designing and using there are therefore two elements of importance, best described with the terms 'structure' and 'direction' (see Wolters, 1985 and Mouw and Griffioen, 1993). Both designer and consumer allot a goal to the technological artefact (proper and accidental functions) and both have to take into account the laws and norms valid in the different aspects. All of this falls under the concept 'structure'. The term 'structure' also denotes properties that cohere with the laws (and norms) that are applicable to the technological artefact and the materials used, and that determine what one can and cannot do with the product. 'Structure' refers to the way in which the materials used are disclosed to realise a certain destination. The term 'direction' subsequently denotes the underlying motives with which the artefact is designed or used. For an understanding of designing and using technological artefacts there is a third important element, namely the context in which they are used. Thus there is a difference between a room thermometer and an outside thermometer. A room thermometer is designed in such a way that it gives information on the temperature in a domestic space. But this instrument is not suitable for an outside thermometer because its range is inadequate (low temperatures) and because it cannot withstand all kinds of weather (like rain and frost). For an artefact like a thermometer the influence of the context on the design process is relatively easy to determine. However, for new complex technologies the influence of, for instance, the societal context on the design process is much more difficult to assess.

7.3 Designing as a model

Due to the variety of aspects which a designer has to keep in mind, his/her work has become a complex matter. Every engineer will be able to confirm this from their experience in practice. Therefore there is a need to bring structure into the design process and methods of designing have originated. According to *Developments in Design Methodology* (1984) an overview of the history of the methodology of design by Nigel Cross it was especially in the 1960s that an attempt was made to summarise the design process in one general ideal scheme. Such a scheme described, irrespective of what was being designed, the steps that each design process should follow.

An example of such an ideal scheme can be found in the 'classic' book for mechanical engineers by Pahl and Beitz, entitled *Konstruktionslehre* (1977). The book was later published in English by the *Design Council* which shows how broad the interest in it was. It describes designing as a process that runs as follows.

1. *Analysing the problem:* departing from a given commission the problem is analysed, a programme of requirements is drawn up and separate problems are identified.
2. *Making a concept design:* for each separate problem a solution is in principle devised so that a concept design comes into being.
3. *Making the preliminary design:* the details of the concept design are worked out further so that a first preliminary design comes into being.
4. *Making the definite design*: the preliminary design is optimised so that a definite design originates.
5. *Preparing for production:* the documentation with directions for production is prepared.

(Pahl and Beitz, 1996)

The essence of this process can be expressed as follows: analysis–synthesis–evaluation. First the problem is analysed which leads to a programme of requirements (step 1), subsequently potential solutions are 'integrated', synthesis (steps 2 and 3), after which the designer evaluates the possible solutions with reference to the programme of requirements (step 4). The book by Pahl and Beitz explains that the solution is first devised in an abstract form and only later concretised. For instance, the designer first thinks that there will have to be a part of the system to convert electrical energy into mechanical, and only later s/he works this out by choosing a specific electric motor. This motor will probably also be used in other designs.

In the design process actually something remarkable happens, for a certain degree of abstraction takes place. However, abstraction is much more characteristic of science than of technology. According to Schuurman (2009) we here have an instance of the influence of science on technology. You can speak of a certain 'scientification' of technology. This term supposes that formerly abstraction was not really known in design. Although we can never know for sure what went on in the heads of designers from earlier times, we do not find any indications in their work that they made a step of abstraction the way it is directed by Pahl and Beitz. We find drawings of concrete objects only. Neither do we find evidence of a subsystem taken up in another system (as the electric motor). It is a matter of neutralising division of functions. Dividing the function of the total system into parts in abstracto in a manner of speaking neutralises such an electric motor. It is no longer characterised by its role in the one system for which it was devised but can be transferred from one system to the next. The electric motor hereby acquires the character of a scientific concept or a scientific theory that has, according to its nature, the pretensions of *common* validity and usefulness.

The general usefulness flowing from abstraction in designing can be seen amongst other things in standardisation and uniformity. In mechanical engineering this development has probably infiltrated furthest. Many components in that field have been fully standardised and it functions well. One also finds this in the architecture from the second half of the twentieth century. A row of terraced houses is in a manner of speaking also a design of general usefulness, one can copy it, which comes in handy for its production. In this way houses can almost be made in a factory as a product for the mass consumer (prefab). But some people raise the question whether this has only positive sides. If you were to ask someone which they found more beautiful, a row of houses in a new extension which are all alike, or a row of houses on a canal in Amsterdam where no two are alike, it is easy to guess what the answer will be. Uniformity also means not diversified. In science this is no problem, but in giving form to reality many people find that it leads to a boring and uninspiring environment. The scientific approach in design therefore has a downside, too.

After some years of research in design process it appeared that the scientific ideal did not function well in practice. This is not surprising. To understand this we have to keep in mind that the schemes of steps have the character of a model. They represent a complex reality in a simplified way. And that is a typical scientific approach, namely one of idealisation. While designing it became clear that reality and the model exhibited essential differences. In the first instance it appeared that design processes could vary considerably. Thus by researching the actual designing behaviour of, for instance, mechanical engineers and architects, it was found that they go about it in very different ways. Mechanical engineers often work with system representations of their designs while architects hardly ever do this. Besides, with architects fantasy and beauty play a much more serious part than with mechanical engineers. A second difference between the representation by means of models of design processes in the schemes of steps and the practice of design was that designers often did not run through a number of steps in a fixed order. Most of the schemes were, as we remarked when quoting the example of Pahl and Beitz, based on the order (a) analysis of the problem, (b) synthesis of possible solutions and (c) evaluation of the solution(s) by comparing them with the requirements emerging from the analysis. Many designers, however, proved to be switching back and forth all the time between working on solutions and getting more insight into the problem. This switching to and fro – currently designers are stimulated to do this regularly – is sometimes called 'back and forth' thinking. Reality often proves too complex to be represented adequately in a simple scheme of steps. Even worse, when designers were forced to adhere to the scheme things regularly went wrong. Examples are known of companies who worked with a handbook for designers in which each project was neatly divided up into phases, with a project leader for each phase. So the project was passed on from the one to the other and in the end resulted in a product. Only then did it become clear that the market had meanwhile changed so much that the product was not at all attuned to what clients desired. Since they were so focused on following the scheme of steps the designers had completely forgotten that the reality around them was in a continuous state of change. Failing to distinguish adequately between model and reality can therefore in such cases result in serious problems.

We can understand that one design process is not like the next when we bear in mind how different artefacts are. First and foremost, artefacts can have different qualifying functions. A passenger car has to convey people from A to B (social aspect), a computer has to process symbols in a controlled manner (technological aspect), an audio system should produce audible sound (perceptive aspect), and a hospital should promote the cure of the patient (moral aspect). This kind of difference cannot but have consequences for the design process. Besides, artefacts differ in their subject and object functions. A computer programme can only act as subject in the arithmetic aspect, a work bench is subject in the arithmetic, spatial, kinematic and physical aspect, and for an artificially cultivated plant the biotic aspect can be added. These differences, too, play an important part in the work of designing. For that matter all artefacts share the foundational aspect: they all emerged from the cultural activity of human beings. But the coherence between the different aspects entails that the way in which technological artefacts are designed is dependent on these aspects. Here we think in particular of the qualifying function and the different subject and object functions.

The philosopher of technology Andries Sarlemijn has highlighted still another difference between technological artefacts (Sarlemijn, 1993). That difference concerns the role of scientific knowledge in the design process. Designers of cork screws seldom or never use formulas from mechanics to calculate which forces are exerted when a cork is taken from a bottle. Designers of cascades do that all the time. Sarlemijn therefore differentiates between different kinds of technologies: experience technologies in which the knowledge concerned is especially the experience of the designer; macro-technologies in which models and formulas from classic natural sciences are used; and micro-technologies in which the even more abstract models and formulas of quantum physics are needed. The differences between these technologies have an effect on the way in which technological artefacts are designed. For designing a cork screw one does not need quantum physicists, but for designing the transistor one does.

7.4 Designing for the market: attention to the client

So far we have emphasised the design process from the perspective of the engineer: the steps s/he takes to make a new design. But we have hardly given any attention to the context in which this process takes place. Currently, however, the goal of designing new products is primarily seen as: the satisfied clients. This means that in the design process the 'voice of the client' has to be heard.

In the classic model for the design process a distinction is made between the technological strategy and the marketing strategy (Wheelwright and Clark, 1992). These strategies are developed in different departments of the company, are organised as different projects and communicate with each other via the formal channels. In many cases these strategies only meet at a high level in the organisation. We find this for instance in the early years of the beginning of Philips. It is told that the division between technological and commercial responsibilities was actually based on the specific qualities of the Philips brothers: Gerard was the technician and Anton the salesman. Since Gerard and Anton had a good understanding there was a productive relationship between technological and commercial responsibilities. Later on, when the company had crystallised into a much more extensive and formal structure this was often more difficult. In the various departments a technological manager was responsible for all technological activities (*development & manufacturing*) and a commercial manager for all commercial activities (*marketing & sales*). Both strategies were only connected at the level of management and that with fluctuating success.

The classic model for the development of new products works very well in a stable environment. The commercial departments make an analysis of the potential of a new product as possible applications, expected turnover and desired specifications. The technological departments draw up a plan for the development of the new product. They make an assessment of the possible technologies that can be used for its development. In cooperation with the factory they also make an appraisal of the possibilities of mass production. The commercial and technological assessments are evaluated at high level in the organisation. Further an integral plan is drawn up containing all aspects of the development, introduction to the markets and mass manufacture of the new product. Finally the necessary financial means are released and the various teams are put together. The classic model seems to be adequate. The knowledge of the commercial department guarantees an accurate assessment of the possibilities of the new product in the market. The knowledge and experience of the technological department are necessary for an evaluation of the technologies to be used to develop and produce this product successfully. It is also an advantage that the integration of the commercial and technological activities takes place at a higher level, because this level can take decisions on the priority of that new product, as well as on the budgets and manpower needed. Once the integral plan has been agreed upon, the commercial departments can work out their marketing plans and the technological departments can start on the development and production. In this phase each department can subsequently carry out its own activities reasonably independently without being disturbed.

In this model the new products also carry a promise in themselves: first a short period of investments and then a longer period in which the product is a moneymaker and the initial investments amply recover the costs. Alas! Reality is often completely different. Wheelwright and Clark give a description of the promise and the reality of the *A14 Stereo* project. In September 1989 Marta Sorensen of *Northern Electronics Company* submitted a plan for the development of a new stereo system. This system promised to offer sound of a superior quality by using the most sophisticated technologies. One of the reasons for developing this plan was the expected introduction of a comparable system by one of the rivals. According to the plan the development of this product would take about a year after which production would be ready to start. There would then be enough time to stock the shops before the important Christmas sales in December 1990. Directly after starting the project the first problems cropped up. As a consequence of a shortage of engineers and of differences of opinion on exactly what the product had to look like the concept phase took six weeks longer than planned. During this period it was decided, in response to a request by the engineers, to introduce a new amplifier technology. The consequence of this decision was that two extra months were needed for building and testing the prototype. More delays ensued as a result of chips and amplifiers having been delivered too late. In spite of these setbacks the planning of the project was not revised, since they were confident that the lost time could be made up. In May 1990 new problems with the design arose. These problems proved to have a serious effect on the whole planning. New components had to be ordered and machines had to be adapted. Once more the new plan proved to be too optimistic. All in all the problems resulted in a year's delay, taking up twice the time that had been planned to put the new product on the market.

What caused the failure of the *A14 Stereo* project? Was it due to the inability of the marketers to realise the wishes of the market successfully? Was it due to a lack of technological competence in the engineers to develop a new product? The answer to these questions is no. The marketers knew the market well and had excellent relations with the dealers and the clients. Nor was there any doubt about the technological competence of the engineers. The problems of the *A14 Stereo* originated in the inability of the organisation to manage the insight and expertise of the various disciplines in a coherent and effective way: i.e. they lacked the ability to *integrate*. Analysis shows that the commercial and technological departments had too much of their 'own agendas'. The marketers wanted to 'put a lovely product on the market' and the engineers wanted to 'develop a sophisticated apparatus with extraordinary quality of sound'. Moreover, the marketers took too little notice of the changes in the wishes of clients which occurred as time went by and the engineers focused too much on the technologically beautiful solutions without finding out what the consequences would be for the planning. There also was inadequate cooperation between the engineers that designed the new product and the engineers responsible for the production. The many alterations to the design also led to many alterations to the manufacture. Finally they lacked a strong leader to take responsibility for the whole project.

In the end the *A14 Stereo* received a good reception on the market. But the sales lagged behind because the rival had earlier launched even better products. In this case study we conclude that the promise of a short time for development and a long period of making profit was not fulfilled. Clearly the period of development was too long and the period in which profit could be made too short, causing the project to generate too little income.

The story of the *A14 Stereo* is not an isolated case. It is one of the many examples that prove that the classic model for the development of new products often does not do justice to the complexity of the design process. *Northern Electronics Company* is one of the many companies who have learned by bitter experience that a model-wise approach to the designing of new products in which each discipline makes its contribution in relative independence can lead to insufficient quality and huge delays in the introduction onto the market. The classic model for the development of new products gave too little attention to the changing markets and the changing wishes of the client. Philips also learned some hard lessons in this field. In the period 1970 to 1980 substantial investments were made in the development of the video disc, but it rendered hardly any income, because in the end the market did not accept the product. In the field of video Philips also failed to find a footing in the market with its own design – the *V2000* recorder. Another product that was less successful was the *P2000*, which proved not to be able to hold its own against the competition of IBM and Macintosh. Later similar 'catastrophes' occurred with the *Digital Compact Cassette* (*DCC*) and the interactive Compact Disc (the so-called CD-I). Philips was thwarted over and over by the same problem: the wishes of the client were not sufficiently known.

7.5 Designing for the market: an integral approach

The classic model for designing new products originated at a time in which business was relatively stable. The rate of technological innovations was low, the wishes of the client could well be surmised and competition was limited. The market was a typical '*sellers market*' in which the producers were in control of the market. One of the most important demands made by the market was a low price. But around 1970 this started to change. Since then more emphasis has been laid on the quality of the products, on the flexibility of the organisation in speedily delivering various types of products, and on the ability of the companies to put innovative products onto the market. In other words, the market gradually developed slowly but surely into a '*buyers market*' in which the wishes of the client took central position. It is clear that in such a changing environment the classic model for designing new products no longer sufficed, something that *Northern Electronics Company* learned from their own experience.

In this framework the changes in the development process made by *Southern Electronics Company*, the most important rival of *Northern Electronics Company*, are interesting to observe. Up until 1985 both companies biannually introduced a new series of products. They had the same price strategy and their production costs were comparable. But after 1986 differences became noticeable. *Southern* set itself the aim of shortening the time needed for setting a new product on the market from 24 to 18 months. In order to achieve this the entire development process, including the introduction to the market, was organised in a different way. The new strategy was characterised by strong leadership, effective integration of commercial and technological activities, a sharp focus on introduction to the market in good time, much attention to the '*manufacturability*' of the new design, better testing of the prototypes, and a renovation of the complete development process. By *manufacturability* we understand the ease with which the product can be made in a controlled manner. The consequence of these changes was that *Southern* could introduce a new series to the market six months earlier than was expected. It proved that for this new system that was more sophisticated and had a better performance the client was prepared to pay a higher price. Although *Northern* marketed a comparable audio system six months later, they had to concede with envy that they had lost a considerable share of the market to *Southern*. During the design process *Southern* had given more attention to the *manufacturability* of the new product and thus its production costs were lower than *Northern's*. Another 18 months later *Southern* introduced the next generation of products. This time they did not charge a higher price for it, but rather lowered the price. Because *Southern* brought down their production costs their profit margin remained stable. The consequences for *Northern* were dramatic: the company lost a further share in the market and the profit margin disappeared.

The above provides a good illustration of the 'modern' model for the development of new products. This model consists of four core elements:

a *Selection:* creating, defining and selecting a number of projects to be developed that lead to superior products and processes.
b *Integration:* integrating and coordinating functional duties, commercial duties and technological duties. Including the integration and coordination of the various departments involved in the development of the new product.
c *Aim of the enterprise:* managing the development process in such a way that the aims of the enterprise are achieved as effectively and efficiently as possible.
d *Skills:* developing the skills by which the development strategy contributes to a strengthening of the company's position as a rival in the long run.

This modern model for developing new products requires a completely different arrangement within the organisation. Instead of a strong functional organisation – i.e., an organisation in which the various functions like marketing and sales, development and manufacturing are housed in separate organisational units – there is a development towards an integrated organisation in which all the different functions are housed. In such organisations separate teams are usually set up for the development of new products, in which the various functions are represented so that the work is done in an integrated manner from the start.

We therefore see some characteristic differences from the classic model in the modern approach to the development of new products. First, it is evident that the wishes of the client receive greater emphasis. There is a greater awareness that a product is developed for a specific *context* and that the meaning the *context* has for the design is explicitly taken care of. Second we see that much attention is given to the integration of the various functions within a company. Here we think of the integration of marketing and development but at the same time also of integration of development and manufacturing. In fact we see here that the scientification of technology that led to functional barriers is more or less overturned in favour of an integral approach. Such an integral approach does more justice to the complexity of the development of technology and to the complexity of organisations in which the development takes place. Finally we see that even the process of development itself changes.

7.6 Designing for the market: the concept of quality

Since around 1970 there has been increasing focus on the quality of products. Under the influence of Japanese companies, in particular, a great number of methods and technologies were developed to improve the quality of products and processes.[1]

Initially the term 'quality' was used in particular for testing the products coming from the conveyor belt, to check whether they functioned well and would not break too soon when used. Methods were devised to take care that the number of products not meeting the quality requirements would be restricted to a minimum, for instance to less than one in 100. One method was taking a random test by which one would be able to make a statement about the quality of the entire production in a statistically sound way. The aim with these measurements was to keep the clients satisfied. In this phase the clients themselves were not involved in articulating their ideas on the concept 'quality'.

In a later phase there originated a broader definition of quality. It was based on a desire to please the client in all possible ways. To achieve this the product not only had to function properly but also had to display other features which the client would appreciate. Furthermore it had to be easy to maintain and quick to repair. In this way the complete life cycle of the product gradually became important. This cycle – from design to production, from sale to usage, and from repair to dispensing with it – would have to be considered in the design process. Therefore designers would have to design the product in such a way that the client would be satisfied in each phase of the life cycle. In this broad definition both the production phase and the recycling phase are therefore taken care of. Taking care of quality over the complete life cycle was indicated with a separate term: *Total Quality Management* (TQM). It is clear that for the designers the rise of TQM broadened the scope of their attention enormously. More so than before the integral nature and the complexity of reality were taken into consideration. To our way of thinking this is good, since designing is not an abstract activity but takes place in concrete reality.

By this change in reflection on designing, attention was once more given to the design meeting the individual needs of clients. Above we have seen that the influence of the scientific method on technological design led to a certain uniformity of products. Nowadays people again want less uniform and more unique products. But to retain the economic advantages of mass production something called '*mass customisation*' is used at times. In this approach the client can choose from a number of variations of a standard product or s/he can, within certain limits, even think up new variations him/herself. One no longer buys a standard computer. Via a website one determines which components one wishes to have in which configuration. One no longer orders a standard motor car, but independent from other buyers determines the colour on the outside, the colour of the upholstery and maybe also all kinds of other technological properties. The shoe manufacturer Nike allows one to compose one's running shoes on the website exactly according to wish. This development is eminently suitable to an era of individualisation.

The rise of client-oriented and life cycle conscious designing causes a new set of methods. However, often people still fall into the old pitfalls: the new methods are once again treated as if they were reality itself, instead of a simplified version thereof. And, just as in the initial phase of design methodology, TQM literature creates the impression that all these methods can be applied successfully, irrespective of what is being designed. We will be discussing a number of TQM methods of design to see what goes wrong when sight is lost of the fact that these methods have the character of models.

Various TQM methods involve feedback from the making process to the process of designing in itself. Examples of such methods are: *design for manufacturing, design for assembly* and *design for logistics*. Each one of these methods consists of a plan of steps. With the help of these methods designers attempt to make the design in such a way that it can easily be produced, easily be assembled and that the necessary logistics in production are as simple as possible. By means of all this the price of the product can come down, while retaining quality. Often it is possible to realise a higher quality with a lower price. The scope of these methods is limited to some aspects of the technological artefact. In particular we are concerned with the aspect of number (number of components), the spatial aspect (the form of the components, the design of the production line), the physical aspect (physical properties of the product), the juridical aspects (safety, environment) and the economic aspects (cost price, profit).

Life cycle conscious designing also concerns the end of the life cycle of the product. Many products are discarded and land on the rubbish dump. A great deal could still be done with the discarded product, however this was not taken into consideration in the design. A form of recycling conscious designing therefore means that the product is designed in such a way that it can easily be dismantled and that the various materials can be separated. This is also called *design for disassembly*. If one knows that it is hardly possible to realise this, one can still attempt to retain something useful when the product is ground into small pieces and all the materials are mixed. This is then called: *design for non-disassembly*. One can for instance avoid using different types of plastic in the product that work destructively on one another's properties when they are mixed. There are tables of synthetics showing which synthetics may and which may not be mixed with others. Naturally such tables can be used to support *design for non-disassembly*.

This attention to environmental aspects originates from an entirely different development than the quality approach. For the quality approach stems from Japan where environmental aspects do not play a great role. The environmental approach initially concentrated particularly on cleaning up the environment that had been polluted by technology (production, use or disposal). One also looked at possibilities of restricting pollution, but actually only in a late stage of the pollution itself. There was talk of '*end-of-pipe*' technologies because not much more was done than putting a filter on the chimney. Soon people became aware that in this way the problem is merely shifted: the air did remain clean, but now the dirt was in the filter. Only later did it dawn on people that the production and with it therefore also the artefact to be produced had to be designed in such a way that there would not be any dirty smoke at all. For prevention is always better than cure. Towards the end of the 1980s people in Europe in particular realised that this issue needed to be kept in mind while designing. In this way environmental aspects found their way into life cycle conscious design.

In other methods the client explicitly comes into focus. An example is Quality Function Deployment (QFD) by which the wishes of clients are translated into technological parameters. The essence of this method is a matrix in which each row represents the wish of a client and each column a technological parameter. In the cells of the matrix numbers are filled in that represent the extent of the relation between the specific wish of a client and the technological parameter. For instance, a client's wish regarding a pencil to be newly designed could be that it should feel good in the hand. Technological parameters accompanying this wish are the length of the pencil and the form of the diameter (round or hexagonal). A parameter that does not accompany this wish might be how much lead is given off while writing. This again goes with another wish of the client, namely that the point should last for some time. The form of the diameter is not only connected with its feeling good in the hand, but also with the client's wish that a pencil 'should not roll off the table'. The matrix helps to get an overview of the whole constellation of client wishes and technological parameters. By means of the scores for the relations one can calculate which technological parameter coheres best with the most important client wishes.

The question is whether the QFD method works in practice. No unequivocal answer can be given to this. The success of this method proves to be closely dependent on the situation. QFD is no cure-all for all situations. If that is what we expect we do not do justice to the complexity of reality and we are confusing model and reality. During the course of time it became clear that the QFD method can best be applied in situations where the product is well known to both consumer and designer. In general these are the technologies called 'experience technologies' by Sarlemijn and to a lesser extent the so-called 'macro-technologies'. When the product is completely new it often is not really possible for client and designers to state their wishes, as the story of Vincenti in the introduction to this chapter demonstrates. The use of QFD *presupposes* a certain learning process in which knowledge of clients and designers is made explicit. Unless these presumptions are met both consumers and designers will fill in just anything with the result that the outcome is undependable. It therefore amounts to a careful study of the reality in which the designing takes place and to its being tested against the presuppositions attached to the method. In particular the issue is what the clients should know and what the designers should know.

Another quality method used in designing for the market is that of drawing up so-called 'technology roadmaps'. These maps give the possible development of a material, product, process or technology. In such a road map one often works 'from back to front'. For instance, one could decide that in 10 years' time one would like to put on the market an entirely new kind of television or television screen. Then one investigates which part-technologies will be needed, when they will have to be ready and how they have to be integrated in the whole. Subsequently one investigates which sub-projects will be needed to realise the technologies and calculates which project has to be started when, so that in 10 years' time the total product can be put on the market. It will be clear that drawing up a roadmap implies numerous presuppositions on future market developments, social changes and technological innovations. All this, while our knowledge of the future naturally is limited.

The same applies to a method like '*technology assessment*' (TA) in which an attempt is made to document the possible consequences of the development and introduction of a new technology. The aim with TA is the development of a technology that will make a successful landing in society (Sollie and Düwall, 2009). Here, likewise, our knowledge of the possible effects is limited, even when – as the method prescribes – account is given of the actors involved, what their interests are and what means of power they have to realise their interests. The future is partly predictable because reality is ordered and subject to certain laws (the so-called descriptive laws, like the law of gravitation). But a part of the order is related to the intentions of people (as in the case of many economic laws). For people do not always act in the same way. If in a technology assessment study one were to forget that there are various prescriptive laws or norms to be reckoned with, one might seriously misjudge the reliability of the outcomes of the study. Here, too, it is important to carefully consider the application of the method and never follow it blindly. It is essential to involve critical reflection when designing and when applying methods.

7.7 Open innovation

In the view of Chesbrough (2003) the classic and modern models for designing products are examples of the so-called *Closed Innovation* paradigm. The essence of this paradigm is that successful innovation requires control of the innovation process. Companies have to generate, develop, produce and market their own ideas. The Closed Innovation paradigm states that (a) companies have to employ the best and most intelligent people, (b) devise and develop new products and services on their own, (c) a self-devised and self-developed innovation is the best guarantee to be first on the market, (d) the company who puts a product on the market first will mostly win, (e) high investment in research and development lead to the most and best ideas, and (f) the intellectual property has to be safeguarded so that rivals cannot make a profit from other people's original ideas.

Chesbrough states that the old paradigm of the Closed Innovation no longer works. One of the most important reasons is that highly trained knowledge bearers switch jobs easily thereby taking with them to the new company knowledge acquired by great effort. Another reason is that a lot of risk capital has to be available for financing new companies who try to commercialise completed research. It goes without saying that in such companies there are employees who moved from a huge laboratory to the small company and have brought with them their knowledge. A last reason is that more and more companies do no basic research themselves.

The new upcoming paradigm is called *Open Innovation*. In this the one-to-one relation between doing basic research and applied development on the one hand and the development of new business on the other hand has been rejected. Companies can make use of knowledge not generated by themselves and technologies they did not develop themselves. Open Innovation therefore makes use of ideas generated internally *and* externally, and of materials, processes and technologies developed both internally and externally. This paradigm also means that internal ideas not used by the company itself and technologies developed by the company but not applied by them, are actively 'marketed' and thus add value to the company. Open Innovation presupposes a new business model in connection with the role of innovation.

7.8 Evaluation

We have seen what changes have taken place in the practice of technological design. More than previously attention is given to the wishes of clients and, still more commonly, to the social effects of technology. In our opinion this broadening is a positive matter for it does greater justice to the variety and complexity of reality. However, one can pose some critical questions about it. Is it by definition a good thing to let the client's wishes take such a central position? Are these wishes always directed towards a good disclosure of the meaning of reality? Are clients, for instance, always so concerned about the consequences the realisation of their wishes have for the environment? Did not the great number of new products that emerged from the market-oriented design approach also bring various problems with them? A truly integrated approach to reality requires designers not to look and listen exclusively to clients and their wishes, but to take a wider look at the different aspects within reality and reflect carefully what would serve a true meaning disclosure. Such normative questions are dealt with in Part III of this book.

Note

1 See amongst others Schonberger (1996), Imai (1986), and Womack, Jones and Roos (1991).

References

Chesbrough, H. (2003) *Open Innovation. The New Imperative for Creating and Profiting from Technology*, Boston, MA: Harvard Business School Press.

Constant, E.W. (1980) *The Origin of the Turbojet Revolution*, Baltimore, MD: Johns Hopkins University Press.

Cross, N. (1984) *Developments in Design Methodology*, Chichester: Wiley.

Imai, M. (1986) *Kaizen. The Key to Japan's Competitive Success*, New York: McGraw-Hill.

Houkes, W., Vermaas, P.E., Dorst, K. and Vries, M.J. de (2002) 'Design and use as plans: an action-theoretical account', *Design Studies*, 23: 303–320.
Mouw, R. and Griffioen, S. (1993) *Pluralisms & Horizons*, Grand Rapids, MI: Eerdmans.
Norman, D.A. (1988) *The Psychology of Everyday Things*, New York: Basic Books.
Pahl, G. and Beitz, W. (1996) *Engineering Design: A Systematic Approach*, Berlin: Springer. Translation of *Konstruktionslehre* (1977).
Sarlemijn, A. (1993) 'Designs are cultural alloys', in Vries, M.J. de, Cross, N.G., and Grant, D. (eds), *Design Methodology and the Relationships with Science*, Dordrecht: Kluwer Academic Publishers.
Schonberger, R.J. (1996) *World Class Manufacturing: The Next Decade*, New York: Free Press.
Schuurman, E. (2009) *Technology and the Future. A Philosophical Challenge*, Grand Rapids, MI: Paideia Press. Translation of *Techniek en Toekomst. Confrontatie met wijsgerige beschouwingen* (1972).
Sollie, P. and Düwall, M. (eds) (2009) *Evaluating New Technologies: Methodological Problems for the Ethical Assessment of Technology Developments*, Dordrecht: Springer.
Vincenti, W. (1990) *What Engineers Know and How They Know It*, Baltimore, MD: Johns Hopkins Press.
Wheelwright, S.C. and Clark, K.B. (1992) *Revolutionizing Product Development. Quantum Leaps in Speed, Efficiency, and Quality*, New York: Free Press.
Wolters, A.M. (1985) *Creation Regained: Biblical Basics for a Reformational Worldview*, Grand Rapids, MI: Eerdmans.
Womack, J.P., Jones, D.T. and Roos, D. (1991) *The Machine that Changed the World. The Story of Lean Production*, New York: Rawson Associates.

Further reading

Dorst, C.H. (1997) *Describing Design – A Comparison of Paradigms*, Delft: Technische Universiteit Delft.
Petroski, H. (1985) *To Engineer is Human*, London: Knopf Publishing Group.
Schön, D.A. (1983) *The Reflective Practitioner*, New York: Basic Books.
Tenner, E. (1997) *Why Things Bite Back*, New York: Vintage Books.
Vries, M.J. de, Grant, D.P. and Cross, N.G. (eds) (1993) *Design Methodology and Relationships with Science*, Dordrecht: Kluwer Academic Publishers.

BRUNO LATOUR

(1947)

PORTRAIT 7 Bruno Latour

Bruno Latour was born in 1947. From 1982 to 2006 he was professor in sociology at the Centre de sociologie de l'innovation at the Ecole Nationale Supérieure des Mines in Paris. He also was visiting professor at the well-known London School of Economics and at Harvard University in the USA. He was awarded two honorary doctorates.

Is Bruno Latour really a philosopher of technology? Not everybody agrees on this. Although he also studied philosophy he is in the first place a sociologist. Yet he did more than empirical research. Many of his interesting reflections bring him very close to philosophy of technology. This is certainly the case if we bear in mind that after the empirical turn in philosophy of technology sociological studies are used with much greater emphasis than before.

Apart from philosophy Latour also studied anthropology. Undoubtedly this study stood him in good stead when he was doing research for one of his best known books, *Laboratory Life: The Social Construction of Scientific Facts* (1979), that he wrote together with Steve Woolgar. In that book he described daily life in a medical research laboratory on the basis of observations. It proves that science in practice sometimes runs completely contrary to the idealistic descriptions in handbooks. The human side appears to play a much more significant role than suggested by the books. It seems that scientists are human beings of flesh and blood in all respects, with all the good and bad characteristics associated with that. Jealousy among colleagues or clique forming contributes to the acceptance or rejection of research output. The mark of quality, 'scientific', is often given not only on the basis of objective norms. Different social and human factors also play a role. Scientific results are therefore 'socially construed' rather than 'discovered'. In a later book, *Science in Action* (1987), Latour treated this more systematically. In this he reaches a theory about what science looks like in practice, and also directs his attention to technology. In *Aramis, or the Love of Technology* (1996) the whole subject is technology. He describes how a promising project in the eyes of the engineers (in their 'love of technology') for a new transport system in Paris founders on discord between the parties involved.

In theoretical study Latour is known especially for the actor-network-theory that he developed in cooperation with Michel Callon and John Law. In this it is explained that technological developments take place in a network of actors, each having their own interests and own means of power (Latour, 2005). In Latour's philosophy of technology this element of power and strategy plays an important role. What comes out of a project depends only partially on the quality of the technology. The situation within the network, the power relations and the strategic skills of the parties take up an even greater part. To denote the fact that things can also be players in that game, Latour coined the term 'actant'. Both human actors and non-human 'actors' fall under this term.

Latour's ideas can be clearly discerned in the following quotations:

> The competition between scientists to turn one another's claims into subjective opinion leads to expensive laboratories equipped with more and more

black boxes introduced as early as possible into the discussion. This game, however, would soon stop if only existing black boxes were mobilised. After a time dissenters and authors – all things remaining equal – would have access to the same equipment, would tie their claims to the same harder, colder and older facts and none would be able to get an edge on the other; their claims would be thus left in limbo, in intermediary stages between fact and artefact, objectivity and subjectivity. The only way to break this stalemate is to find either new and unexpected resources or, more simply, to force the opponent's allies to change camp.

(Latour, 1987: 83)

I have sought to offer humanists a detailed analysis of a technology sufficiently magnificent and spiritual to convince them that the machines by which they are surrounded are cultural objects worthy of their attention and respect. They'll find that if they add interpretation of machines to interpretation of texts, their culture will not fall to pieces; instead, it will take on added density. I have sought to show technicians that they cannot even conceive of a technological object without taking into account the mass of human beings with all their passions and politics and pitiful calculations, and that by becoming good sociologists and good humanists they can become better engineers and better informed decision-makers.

(Latour, 1996: viii)

References

Latour, B. and Woolgar, S. (1979) *Laboratory Life: The Social Construction of Scientific Facts*, Princeton, NJ: Princeton University Press.

Latour, B. (1987) *Science in Action*, Cambridge, MA: Harvard University Press.

Latour, B. (1996) *Aramis, or the Love of Technology*, Cambridge, MA: Harvard University Press.

Latour, B. (2005) *Reassembling the Social: An Introduction to Actor-network Theory*, New York: Oxford University Press.

8

TECHNOLOGY AND PRODUCTION

From dehumanisation to human measure

Summary

In this chapter we consider the influence of the scientific approach on the design of production organisation. We will begin with an analysis of Taylorism and Fordism, approaches that laid the foundation for Western mass production. We demonstrate that their design is determined to a great extent by the method of science: breaking down the manufacturing process into small steps, studying each step in a scientific way and laying down the most efficient method. Research has proved that production organisations designed according to the principles of Frederick Taylor and Henry Ford are characterised by dehumanisation and alienation. During the course of time various alternatives have been presented. We will investigate in depth the so-called sociotechnology because this approach claims to break with the principles of Taylor and Ford. The core concept of sociotechnology is the so-called 'complete task'. We will also look at various quality techniques used in factories. An analysis of the phenomenon of power shows that these technologies can also be used to strengthen the power of management. Finally we will show that the engineer who designs production organisations faces the challenge to do justice to the broad complexity that we investigated in Chapter 3.

In Chapter 7 we saw that scientification of the design process led to specialisation. The result was that functions that actually belong together, like marketing and sales, design and production, were divided over time into separate departments. This development led to all kinds of problems. In the 1970s a reversal took place. By means of new methods and technologies an attempt was made to overcome the consequences of the separation of the various functions. Also, more attention was demanded for the requirements of the clients. In this chapter we analyse the development of the production process. We will show that since the beginning of the previous century the scientific approach has gained ever more influence on this process. We begin with a discussion of Taylorism and Fordism that are the foundation of Western industry. Subsequently we will discuss some new approaches.

8.1 Case study: Inspection of bicycle balls

'When the bicycle craze was at its high ...'. With these words the American engineer Frederick Taylor began his description of the rationalisation of the visual inspection of a factory for steel balls for bicycle bearings. Taylor was convinced that each activity of an operator in the factory can be analysed in a scientific way. On the basis of this analysis it had to be possible to ascertain the 'one best method' and to develop the 'one best implement' for performing the action. One of the examples that he describes in detail to praise the success of his method is the visual inspection of small steel balls.

Who was this American engineer? Robert Kanigel in his book *The One Best Way. Frederick Winslow Taylor and the Enigma of Efficiency* (1997) gives a good description of his life and work. Frederick Winslow Taylor (1856–1917) was the first to introduce scientific methods in the factory. He employed new management techniques to control the production process. Taylor's influence on the development of Western industry can hardly be overestimated. Many engineers applied his methods and developed them further. His concepts have formed several generations of managers. His view was exported to Europe, Russia and the Far East. And, what is more significant, his thoughts have stood the test of time. Taylorism is still dominant as Hans Pruyt has shown in his doctoral thesis *The Fight Against Taylorism in Europe* (1996).

Even as a child Taylor was taken up with the question of efficiency. One of his old friends related how Fred was constantly experimenting with walking to determine with which tread he could cover the greatest distance while using the least amount of energy. In the Midvale Steel Company, the company where Taylor started working after his training in mechanical engineering he landed in an environment where scientific methods were used to solve problems in manufacturing and processing steel. In 1885 Taylor first presented his ideas on the scientific study of industrial processes and management systems at a conference of engineers. From 1901 he devoted all his time to propagating his ideas. All over the world he held lectures and gave scientific advice. He was '*Mr Scientific Management*'. In 1903 he unfolded his ideas in his book *Shop Management.* Some years later he set out his ideas more systematically in his work *The Principles of Scientific Management* (1911).[1]

During his lifetime Taylor was the centre of bitter controversy. His ideas were welcomed by industry and universities but heartily rejected by trade unions and politicians. The opposition reached a peak when in 1912 Taylor was summoned to give evidence before a special committee of the *House of Representatives.* The reason for this summons lay in the fact that the introduction of the Taylor system in a government organisation had led to a huge strike. The course of this hearing was recorded in the annals with the title *Taylor's Testimony Before the Special House Committee.* In spite of all the opposition his ideas swiftly caught on. In 1919 the French poet Paul Valéry described the progress in post-war Europe as a march 'from Taylorisation to Taylorisation'.

Taylor discusses the case study 'Inspection of bicycle balls' both in *Shop Management* and *The Principles of Scientific Management.* This case study gives a good illustration of his approach. In the department 'Visual inspection' there worked about 120 experienced girls. The visual inspection was done by placing in the inside of the hand, in the 'crease' between two fingers, a row of small polished balls. They rolled the balls to and fro and inspected them carefully in the light of a bright lamp. By means of a magnet held in the right hand the defective balls were taken out. The visual inspection required a great deal of attention and concentration. In a number of steps Taylor arrived at a significant improvement in efficiency and quality. The first step was simply an observation of the checkers. This observation revealed that a considerable time of the working day was spent in 'idleness'. It was decided to shorten the working day step by step from 10½ hours to 10, from 10 to 9½, from 9½ to 9 and finally from 9 to 8½ hours, while the pay remained the same. To his astonishment Taylor discovered that with each shortening of the working day the output (the number of inspected balls) increased instead of decreased. The second step was the selection of the girls who carried out the visual inspection. The inspection required a combination of fast perception followed by fast action. On closer investigation it was found that the combination of these characteristics was only present in girls with a so-called low *personal coefficient.* This research provided the management with a method for selecting *first-class* controllers. A considerable number of the hard-working and motivated girls were dismissed because they did not meet the selection criteria. The third step was the improvement of the quality of the checking. This was done by measuring how many of the defective balls were passed over by the checkers. The results of this inspection were given in feedback to the checkers. Girls who did not meet the required quality were retrained. The last step was the introduction of rest breaks. By intensive observation it was proved that the checkers became tense after working for an hour and a half. It was clear that they needed a break to rest. So it was decided to introduce a break of 10 minutes after they had worked for an hour and a quarter, so that they could recover.

Taylor's approach led to a breakthrough in the inspection process. After the introduction of all the changes the checkers received a raise in salary of 80 to 100 % while their working day had also been shortened from 10½ to 8½ hours. During the day they had regular breaks of 10 minutes. They also got two days' leave per month (without loss of salary). The number of girls needed for the visual inspection fell from 120 to an amazing 35. In addition the quality of the inspection process improved. This example shows that a scientific analysis of a step in the production process can lead to increased efficiency, a rise in quality of the products and an improvement in the labour conditions.

8.2 Scientific management

In his book *The Principles of Scientific Management* Frederick Taylor gives a systematic analysis of the responsibilities of the 'modern manager'. He distinguishes four principles: (1) the development of a science, (2) the scientific selection and training of workers, (3) hearty cooperation between management and workers, and (4) an equal distribution of work and responsibilities between management and workers.

According to Taylor management is responsible for replacing the old rules of thumb of the craftsman with a genuine science. In other words, one can speak of converting *tacit knowledge* into *codified knowledge* (see Chapter 6). To do this, management has to collect the traditional knowledge of the employees and translate it into rules and instructions. In the scientific study of the production process, division of labour and time studies play an important part. Division of labour means that a complex production process is divided into a number of steps. Each step is studied in a scientific manner: the time needed for performing this step is determined, unnecessary movements are eliminated and the correct tools are devised. In this way for every step the '*one best method*' and the '*one best implement*' are determined which leads to a complete standardisation of all work processes. As we will see in the next paragraph, each employee is trained in only one production step. In the case study 'Inspection of bicycle balls', we clearly see the different elements of the development of the '*one best method*'.

Taylor was not the inventor of the division of labour. As early as 1776 Adam Smith in his book *The Wealth of Nations* spoke about the great advantages of dividing labour. He saw it as one of the main causes of increasing wealth. Following in Smith's steps, Charles Babbage stated in his book *On the Economy of Machinery and Manufactures* (1832) that division of labour leads to a cheaper production process. Further he emphasised that dividing the work process into a great number of steps makes it easier to discipline individual workers and maintain order.

The second responsibility of management is the scientific selection and training of workers. In the case study 'Inspection of bicycle balls' this element was treated explicitly. Taylor was of the opinion that the scientific approach would lead to *first-class* workers. According to him many workers would be able to move up to a 'higher level' of work and would be able to earn more money.

The third responsibility of management is to ensure that all work is done according to the principles and laws of science. To realise this would demand a 'hearty cooperation' between management and workers. Taylor realised that this demand required a mental revolution in the work process. He observed that in this area there was much opposition from the side of management. Therefore he denounced factory managers who did not want to invest in the development of a relationship based on mutual trust with their workers.

The fourth and last responsibility concerns an equal division of work and responsibility between management and employees. Taylor remarks that under the old management systems the workers were responsible for everything: planning, materials, tools and execution. According to him, however, managers should take upon themselves the duties in which they are good: drawing up plans, determining the best method, developing the best implements and training and guiding the worker. The equal division of work and responsibility basically means a separation of manual work and brainwork, of doing and thinking: 'All possible brain work should be removed from the shop floor' (Taylor, 1947: 98). Taylor gives an extensive analysis of all duties management should undertake. He remarks that these duties are so diverse that it is impossible to find *gang bosses* who have all the necessary qualities. Therefore he proposes introducing functional management, that is, managers who are responsible for *one* functional aspect of managing. In an ideal situation there are four managerial staff members for the direct execution of duties (*gang boss, speed boss, inspector* and *repair boss*) and four managerial staff members form the planning department to guide the workers (*order of work and route clerk, instruction card clerk, time and cost clerk* and *shop disciplinarian*).

So what is the worker's responsibility? It is the worker's duty to leave behind the old manner of working, to learn the new fast methods, always keep learning, think ahead and to take care that every minute is used productively. In other words s/he should work according to the '*one best way*' and use the '*one best implement*'. In Taylor's experience most workers take on their new responsibility without resistance. However, some miss the bus because they cannot meet the requirements for the new system and others because they are 'lazy' or 'balky'.

8.3 Between science and ideology

How scientific was Frederick Taylor's work? Do his opinions represent a certain ideology? Did he defend the interests of one certain class? It is difficult to answer these questions unambiguously. First we would like to point out that the contribution Taylor made to the scientific study of industrial organisations was recognised by universities more or less right from the start. Thus in October 1906 he received an honorary doctorate from the University of Pennsylvania. In the motivation his approach and merit were stated as follows: 'Thorough and industrious in preparation for his life work, patient in investigation and experiment, logical in analysis and deduction, versatile in invention, his labours have brought system out of disorder in the organization of industrial establishments' (Kanigel, 1997: 386).

But this does not detract from the fact that several times Taylor was criticized on the scientific basis of his standards. So he was given a hard time by the *House of Representatives* about a factor of 20 to 27 per cent for 'unavoidable delays and interruptions' that he introduced in his time studies for the automobile industry. He did not succeed in convincing the committee that this factor had been determined in a scientific manner (Taylor, 1947: 160 et seq.). There also are questions about a similar factor that he had introduced earlier during his studies on loading crude iron. In this case he set the factor at 40 per cent. The many pages of numbers and calculations indeed suggest a scientific approach, but still do not prove that this factor had been determined in a scientific manner (Kanigel, 1997: 320).

From about 1904 rhetoric played an ever-increasing role in his work. In the so-called 'Boxly lectures' (Taylor's estate lay in Boxly) he defended his approach with the necessary fervour. He became the zealous propagator of 'the new gospel of the industrial enlightenment' (Kanigel, 1997: 412). His defence before the special committee of the *House of Representatives* was also one great rhetorical piece. He tried to silence all critics by explaining what his system was not. He stressed that it was *not* an efficiency system, *not* a paying system and *not* a time study either. According to him the essence of his scientific system on the contrary was a complete mental revolution. It was a revolution that led to a new attitude of workers towards their work and of management toward their responsibilities. A revolution that would even lead to brotherly cooperation between workers and management and to the replacement of war by peace. Taylor emphasised that scientific management ceased the moment injustice made its entrance (Taylor, 1947: 26–33, 188–194). It need not be said that in this defence the speaker was an evangelist and not a scientist.

Finally it has to be stressed that Taylor's ideas had a definite ideological character. Only according to his system (and not according to other management systems) could the interests of employers and employees be managed in a balanced way. But then the mental revolution must have taken place first. This demand for a mental revolution does not yet occur in Taylor's earlier work. At the end of his later work *The Principles* he first formulated this demand. And in his *Testimony* before the special committee the idea of the mental revolution is emphatically foregrounded. Opinions differ on the question of whose interests Taylor was actually defending. Erik Bloemen (1988: 42) concluded in his doctoral thesis *Scientific Management in Nederland 1900–1930* that Taylorism can be seen as 'the ideology of the middle class, as an ideology that is functional *for* the ruling class and as an ideology that contains instigation towards opposition *against* the ruling class'. He points out that the contradictory elements in Taylor's work make it possible for different groups to be drawn to it. For our part we would like to mention that Taylor's conduct carried the message that management, power and scientific principles are all closely connected. For he used his power to implement his principles: he forced workers to work according to his instructions and dismissed them if they did not obey swiftly enough.

We conclude that there are several features of the empirical sciences to be found in Frederick Taylor's work. But when the crucial factors are determined, it transpires that the scientific basis is weak. Besides, rhetoric and ideology also play a great part.

8.4 The conveyor belt

James Womack, Daniel Jones and Daniel Roos in their book *The Machine that Changed the World* (1991) describe the development of the automobile industry. Up to the end of the nineteenth and the beginning of the twentieth century the production of automobiles took place according to traditional methods. A typical feature of this way of production is firstly the presence of experienced craftsmen who have the skills in technological design, operating machines and assembling components. They acquired these skills during a long period as apprentices. Second, the total organisation was highly decentralised. Most of the components and important elements of the design came from small workshops. The whole system was coordinated by an owner who also was the entrepreneur who stood in direct contact with all interested parties. Third, *general purpose* machines were used, that is, standard machines that could produce all kinds of components. Finally one could speak of a low production per type per year, often less than 1,000 models and sometimes it was 50 or fewer. The consequence of working according to traditional methods was that no two automobiles looked exactly the same.

An important result of the organisation of automobile production according to traditional methods was that fundamental innovation did not occur easily. The smaller workshops simply did not have the means for it. During this period Henry Ford (1863–1947) developed some methods that enabled him to surmount the problems inherent in working according to traditional methods. These new methods would result in an enormous drop in the cost and a substantial improvement in the quality of the automobile. Ford called his innovation system '*mass production*'.

In 1903 Ford began with his first efforts to assemble an automobile. In 1908 he began production of the famous model T. The most significant breakthrough for facilitating mass production was not, as is generally believed, the introduction of the conveyor belt but the interchangeability of components. The standardisation of components (and means of measuring) enabled them to assemble an automobile without having to make the components to fit every time.

Initially one assembling operator of Ford assembled *one* complete automobile on *one* assembler platform. But later when all components were fully interchangeable this way of working was discarded. In the new set-up each assembling operator did only one assembling action. The consequence of this was that the operator had to move from automobile to automobile. Ford realised that this solution also entailed disadvantages. First, walking from one platform to the next takes up time. Furthermore congestion occurred in the factory, because the slower workers were overtaken by the faster ones. In 1913 Ford realised his second breakthrough: the introduction of the conveyor belt. He got the inspiration for this from the pork abattoirs of Chicago where such a belt was already in use. The essence of the conveyor belt is that the worker stands still while the product moves along. There is no doubt that Ford's engineers were influenced by Taylor's ideas. Most probably they derived their method of time studies from him. There are no known statements made by Ford which indicate that he was directly influenced by Taylor.

Apart from technological innovations Ford also introduced a number of social and organisational innovations. First of all he gave much attention to improving the work ethics. He drastically raised the wages: from less than two and a half dollars a day to five dollars a day. This measure resulted in a considerable decline in turnover of employees and absenteeism. Another advantage was that so many workers applied that Ford could vigorously select whom he wanted to admit. Yet Ford did add a condition to the high wages: a good moral lifestyle. In order to stimulate this, he founded a Social Department consisting of psychologists and sociologists who checked out the lifestyles of the workers. Therefore the influence of the company in the long run extended even to the private lives of the workers.

Mass production had an enormous influence on the technological and business economics parameters of an automobile factory. In the initial stage in which one operator assembled an automobile mostly on their own, the time for a cycle came to at least 500 minutes. The cycle time is the time one operator works before s/he starts doing the same activity again. From the moment when every operator performed only one activity on each platform this time fell to 2.3 minutes and afterwards at the conveyor belt to 1.2 minutes. The introduction of the conveyor belt caused the time for assembling one automobile to decline by a factor of 2 or 3. All these changes led to a substantial fall in the price of the model T. Around 1920 a new record was reached in production at the Ford factory: more than 2 million identical (!) specimens of the model T per annum.

8.5 Limits to control

The conduct of Frederick Taylor and Henry Ford heralded a new phase in industry. Work according to traditional methods was replaced on a great scale by mass production. The craftsman was replaced by the worker. Management became a separate discipline with its own duties and responsibilities. The principles of Taylor and Ford formed the basis of American industry. Companies like Ford, Singer, Du Pont and AT&T became the leaders in mass production. This new phase in industry was facilitated to a great extent by a scientification of the production process. The methods devised by Taylor and Ford have certain essential features in common, which we summarise by the term 'technological control'. The systems developed by Taylor and Ford consist of two basic components. The first is the technological control of the material processes and the second the technological control of the organisational and social processes.

The technological control of the material processes comes about by dividing the manufacturing process into a great number of individual steps and by researching each one of these steps. In this way the problems of the industry could be managed by science. This led to the standardisation of components, implements and operations. In the works of both Taylor and Ford we see that the method of analysis of the manufacturing process (dividing into small steps) lays the foundation for the method of organisation of this process (every worker performs *one* small task). What is striking is the naturalness with which the scientific approach of industrial problems leads to solutions which are characterised by the same scientific approach.

The technological control of material processes had an enormous influence on the organisation of the work. The moment the production process was set out in a great number of sub-steps and the duties of management were divided into various functional areas, the fragmentation of labour became a fact. No worker any longer had knowledge of the whole product and no manager knew the full connectedness between the different functional areas. We can here speak of an irreversible causal relation. The moment the 'architecture' of the material processes was fixed, the 'architecture' of human labour was also fixed. And with the architecture of labour we also lay down the profile of workers and managers. In the terminology of critical management studies, the Scientific Management 'produces' its own workers and managers (see Jacques, 1996).

The second basic component of Taylorism and Fordism is the technological control of the organisational and social processes. This we see in the scientific selection and training of employees, the separation of work requiring doing and thinking, the idea of functional management and the activities of the social departments. In these organisational and social processes the worker and the functional manager therefore have to be controlled by means of these technologies, or rather modelled according to the requirements of the technological system. We deliberately use the word 'modelled' because these organisational and social processes are not meant to do justice to the human being as a unique individual but to 'deliver' a worker or manager who meets the requirements of the system. And we call it 'technological control' because the notion of a worker or a manager who can be moulded is the dominant idea. In the approach of Taylor and Ford the scientification of the production process therefore leads to a technological approach to the human being as a social creature and a technological control of the organisation as a social community. In the introduction to *The Principles of Scientific Management* Taylor writes that in the period of traditional methods the human being took central position but that in the scientific approach the system forms the centre: 'In the past man has been first; in the future the system must be first' (Taylor, 1947: 7). The idea of the control of employees by social processes is to be found particularly in Ford's work in the duties of the social departments. By means of psychologists and sociologists the private life of the worker was controlled to promote a moral lifestyle.

The Taylorist and Fordist labour organisation is a typical product of technological-scientific society. It is a product that was developed to offer freedom to a human being. But exactly because of the technological control it degenerated into something that eventually jeopardised this freedom. Here, we see an area of great tension: the development of the modern organisation was meant to give human beings more power over the world, but instead it now exerts power over human beings. Indeed, Taylorist and Fordist organisations form a power that threatens the freedom of human beings. When Taylor defended his system before the special committee of the *House of Representatives* he stated that the '*very essence*' of Scientific Management consists of two principles. First a mental revolution on the side of the workers and management, second a scientific control of the whole production process. These two principles are not compatible. A full mental revolution excludes technological control of the production (organisation), and complete control of the production (organisation) excludes a mental revolution. It seems as if Taylor's talk of the mental revolution was little more than rhetoric to defend his system against attacks from society. In his system the technological control takes central position and the mental revolution he mentioned is subordinate to it.

The threat posed by the system of Taylor (and Ford) was observed fairly quickly. As early as 1911, Samuel Gompers, head of the American Federation of Labour wrote that the Taylor system turned every person into 'a cog or a nut or a pin in a big machine' who need do nothing more than 'a few mechanical motions' and need not use more 'brain power' than what is necessary for the single mechanical movement (Kanigel, 1997: 446). And in 1936 Charlie Chaplin made the film *Modern Times* in which he exposed all the negative effects of Taylorism and Fordism (see Box 8.1).

BOX 8.1 *MODERN TIMES*

In the film *Modern Times* (1936) Charlie Chaplin plays the role of a factory worker at the conveyor belt. In each hand he has a spanner with which he is supposed to tighten a nut on a component. The belt moves at a considerable speed: if Charlie loses concentration even for a moment, he lags behind. As long as the belt moves at its normal speed, he just manages. But when the order is given to move at top speed everything goes wrong. Charlie cannot keep up any longer, he lands on the conveyor belt and is 'swallowed up' by the machine. Charlie is 'rescued' but mentally breaks down completely. He sees nuts everywhere that he has to tighten. Thus he runs after a secretary to tighten the dark buttons on her dress. The end of the story is that Charlie lands in hospital with a nervous breakdown.

Modern Times has a very powerful first scene. In the first images on the screen we see a huge flock of white sheep – in a tight pack – which are driven in the same direction. A striking detail is that there is *one* black sheep among them. In the following scenes we see a great 'flock' of workers – in a tight pack – hurrying from the underground stations to the factory. Subsequently the film exaggerates the misery of the conveyor belt in a satirical way. The first is that the monotony of the work leads to psychological degeneration and even to a complete breakdown. The second is the negative effect of such a system on the relations between people. The workers are so dependent on one another that each disturbance or problem leads to friction, irritation and tension. Subsequently there are the dehumanising aspects of technology as a hierarchy: people are reduced to extensions of the machine. And finally the threat posed by mass production: human beings are literally swallowed up by the cogs of the machine.

Modern Times is an accusation against the socio-economic situation of the 1920s and 1930s, in particular in North America. In *My Autobiography* (1964) Chaplin says that he was inspired to make the film by a young journalist who told him about the system of conveyor belts in the factories in Detroit where healthy young men over four or five years were changed into physical and psychological wrecks. David Robinson stresses in his book *Chaplin: The Mirror of*

Opinion (1983) the element of social criticism. He writes that for the first time an American film dared 'to challenge the superiority of an industrial civilization'. For him the film pictured 'the story of a pathetic little man trying bravely to uphold his end in this mad world' (Robinson, 1983: 107–108).

The tension between control and freedom is a consequence of a technological approach by the production organisation. As soon as 'technological control' starts dominating the design of production organisations there no longer seems to be any reason to limit this control to the material processes.

8.6 The birth of a new paradigm

The insight that the approach of Taylor and Ford took place at the expense of the humanity of human beings grew soon after the introduction of their principles. During and after the First World War a number of research projects were started in England and America on these approaches which would lead to the so-called *Human Relations Movement*. The rise of this school meant a paradigmatic change in the thinking about organisations. For the focus shifted from technological aspects to socio-psychological aspects. A first breakthrough in thinking about human beings and organisation was introduced by Elton Mayo and his colleagues. They did research on the influence of labour conditions on the productivity in the Hawthorne factory of General Electric in Chicago (1927–1932). They were of the opinion that an improvement in physical conditions would lead to an improvement in productivity. Mayo researched the productivity of a group of six female workers. During a certain period he introduced a great number of changes in the labour conditions, like better wages, shorter work days, longer breaks and free refreshments. Each change led – as could be predicted according to the theory – to an increase in productivity. In the following stage of the research all introduced changes were reversed simultaneously and the original conditions were reinstated. Everyone expected that the productivity would subsequently drop again and return to the original level. But to everyone's astonishment the productivity rose even further.

Mayo concluded that the rise in productivity could not be explained by the changes in the labour conditions. To understand this phenomenon another explanation had to be sought. Mayo demonstrated that the researchers – unconsciously – had not only altered the direct labour conditions but also the social environment in which the women worked. The explanation that the researchers gave was that the occupational satisfaction of these women had increased hugely during the research because they enjoyed greater freedom than their colleagues and had more influence on the execution of their work. The group of women had become a social unit. They had established good relations with the researchers. Because each alteration in the labour conditions had been thoroughly discussed with them, they had conducted themselves according to the wishes of the researchers. Therefore Mayo concluded that the occupational satisfaction was highly dependent on the informal social patterns of and within the group.

The most important contribution of the research in the Hawthorne factory was actually the (re-)discovery that the working human being is also a social creature. This research laid the foundations for industrial sociology and led to the Human Relations Movement. Under the influence of this movement employee departments were founded, labour conditions improved, training and instruction were stimulated and the participation of employees was encouraged. Later on, under the influence of the work done by Maslow (1970), McGregor (1960) and Herzberg et al. (1959), attention shifted to the individual development and personal growth of employees.

A second breakthrough in organisational thinking occurred after the Second World War. Eric Trist and Ken Bamford from the Tavistock Institute of Human Relations in London did research on the psychological, social and economic consequences of various technologies in coal mining. Ken Bamford, a former mine worker, visited the Elsecar mine in South Yorkshire (England) where he had worked for years before, and discovered that a new way of working was now being applied there. Together with Eric Trist he received permission to do closer research. Before the mechanisation the miners worked in groups of three to seven men who were responsible for the whole mining process. By the introduction of machines it had become possible to mine coal on a broad front. The consequence was that they had to work together in a completely different manner. They worked in groups of 40–50 men, divided into three teams. Each team was specialised in certain activities. The first team prepared the coal front, the second made new main and side passages, and the third blasted loose the coal and transported it. The full cycle lasted 24 hours (each team worked eight hours). The group of 40–50 men was divided into seven specialised fields. The new mechanisation demanded a careful tailoring of the work within and between the teams. Exactly because the work in the mines was unpredictable it proved to be difficult to realise this tailoring well. The method of working proved to be particularly vulnerable. If the work of one specialised field fell short the others could not proceed. Or if *one* team had not finished the next team could not continue successfully. At worst a whole cycle of 24 hours was lost. This dependence caused much tension between the mine workers themselves and between mine workers and management. It was a tension intensified by the fact that the employees were rewarded individually according to the amount of work they had finished. And when a whole cycle was lost it certainly led to anger and frustration.

In a new vein of the Elsecar mine this method of working – the so-called *long wall* method – could not be used because the conditions of the ceiling did not permit long fronts. In this vein the mine workers themselves developed another method of working. By an adaptation of the technology they were able to mine coal along a short front. Further they had adapted the working method in such a way that all activities could take place within a cycle of eight hours: each team prepared a small length of coal front, made some new passages, blasted loose the coal and transported everything. The miners also learned to perform more than one task. It became clear that as a result of this new way of working the teams could cope well with the unpredictability of the mines. Also they were able to help one another within the team if some particular duty appeared to be difficult. Finally there no longer was any dependence between the teams because each team could proceed where the previous had left off. The result of this was that the tension within and between the teams dropped sharply. The pleasing point was that the new way of working showed great similarity with the method used before the mechanisation. Under the leadership of Trist and Bamforth the new way of working – the so-called *composite long wall* method – was introduced with some refinements into all mines in South Yorkshire. At the same time the principle of group rewards was introduced so that each miner and each team was equally concerned about the whole process running smoothly.

The Yorkshire case is a nice illustration of the necessity of studying production systems in their entirety. Only then can the different aspects be understood in their interdependence. The case shows that the problems in the social system were caused to a great extent by the technological system. The transition from the *long wall* method to the *composite long wall* method not only was a technological adaptation, but facilitated all kinds of changes in the organisation of the mining process. The case further demonstrates that a technological system cannot function well without giving attention to the social system. Trist and Bamforth therefore concluded that the production process has to be characterised as a *socio-technological* system. They also concluded that employees should have the authority to coordinate their own work. With these conclusions a new paradigm came into being: *Socio-Technical Systems Design*.

Socio-Technical Systems Design, also called sociotechnology, claims to offer a dignified alternative to Taylorism and Fordism.

8.7 Socio-Technical Systems Design

One of the main objectives of sociotechnology is to end the extreme division of work that has dominated industrial labour since the introduction of Taylor's and Ford's principles. The socio-technological approach implies a fundamental redesigning of organisations. This redesigning is focused on the entire production process, the layout of individual workplaces, the management of employees and communication. In this approach the objective is an improvement in the quality of work, quality of the organisation and the quality of labour relations.

The characteristic feature of the sociotechnological approach can be expressed in the concept 'complete task'. In Taylorism and Fordism the employee performs only one executive task. In sociotechnology, one employee, besides performing a number of executive tasks, also does various regulating tasks. S/he preferably makes a complete product or a whole module. If that is not possible s/he performs a number of coherent processes. The employee also performs a number of regulating duties, such as carrying out quality measurements, performing minor reparations and regulating logistics. Thus we see that sociotechnology is different in that the extremely rigid division of work in the executive labour is diminished as far as possible and employees are allowed to perform a number of coherent activities. Sociotechnology also attempts to diminish the rigid division between manual and brain work by letting the employees do as many regulating tasks ('brain work') as possible.

In modern sociotechnology various different schools can be distinguished.[2] In this section we will focus on the Dutch approach as developed by Ulbo de Sitter and others. This method offers a detailed design theory, a carefully described methodology of design and a participatory process of change. Later on we will deal briefly with the Australian approach which focuses on the use of participatory technologies and with the Scandinavian approach which stresses the importance of democratic dialogue (Section 8.9). The various socio-technological schools make a significant contribution to the participatory or democratic tradition in organisation theory. The word 'democratic' in this context means the right to take decisions on matters pertaining to one's own workplace and to take part in decisions regarding the areas of knowledge and experience of the employee.

The Dutch approach to sociotechnology has four points of departure. The first is that organisations should be designed in an integrated way. The design has to consider the relations between the different departments (subsystems) like manufacturing, development, sales, engineering and finances. But there should also be attention to the different aspects (aspect systems) like the technological, the economic, the social and the psychological. The second point of departure is that the design should specifically explore the manageability of the system. Employees should be given enough space to take care by themselves of unexpected situations such as illness of a colleague, a breakdown of a machine or a problem of quality. Third, the design process should explicitly pay attention to the so-called production structure and the control structure. The production structure comprises the executive tasks in the production process, and the control structure the various regulating tasks. We observe that later on another so-called information structure is also introduced which contains all the necessary information for performing the regulating tasks. The last point of departure concerns the concept of the structural parameters. The idea is that for a good design one should quantify the fundamental parameters of the different structures and should state the relations between the different parameters.

The four points of departure of the Dutch approach in sociotechnology resulted in a number of principles, strategies and sequences in the design process. The golden thread in this process is formed by the idea of reduction of complexity (simple design) and of an increase in managing capacity (more authority for the worker). Figure 8.1 gives an overview of the design sequence. We start with the design of the production structure and subsequently discuss the design of the managing structure. The production structure is designed in the order: Macro >> Meso >> Micro. The managing structure is designed in the reverse order: Micro >> Meso >> Macro.

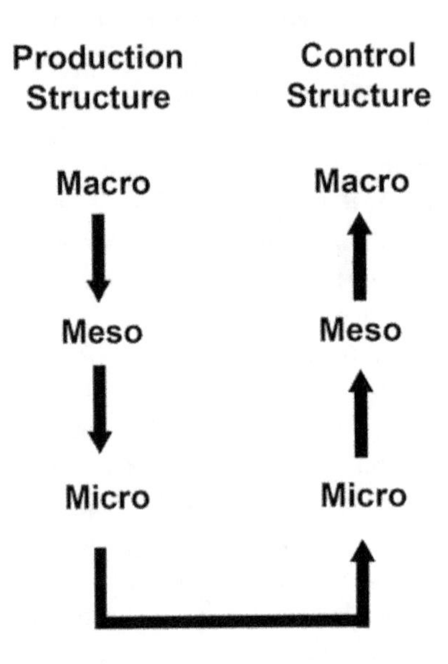

FIGURE 8.1 Design sequence

Figure 8.2 shows how the redesign of the production structure takes place. We here presuppose a factory founded on Taylorist principles. In this factory six different products are manufactured: a, a', b, c, d and e. To manufacture these products seven different processes are needed: 1 up to and including 7. For making product a that would be the processes 1 and 4, for product a' the processes 1, 3 and 4, for product b the processes 2 and 3 and thereafter once more 2 and 3, etc. For maximal efficiency (division of labour) all machines doing process 1 are placed together. Subsequently this process was standardised ('*one best method*'), tools were developed ('*one best implement*') and the operators were trained in carrying out this one process. The same applies to the processes 2 up to and including 7. In this set-up each operator therefore only performs one process. The consequence of this is that no operator has an overview of the whole process and that no employee knows the product thoroughly.

The redesigning of this factory to a socio-technological design takes place in three steps. The first step is the design at macro-level. Here an attempt is made to simplify the factory by dividing it up into a number of smaller factories. This process is called parallelisation. In this example we depart from three parallel factories as can be seen in Figure 8.2. In each of these parallel factories we place the machines needed for making the products concerned. The first factory only makes the products a and a'. For this are needed the machines performing the processes 1, 3 and 4. The second factory only makes the products b and c. For this are needed

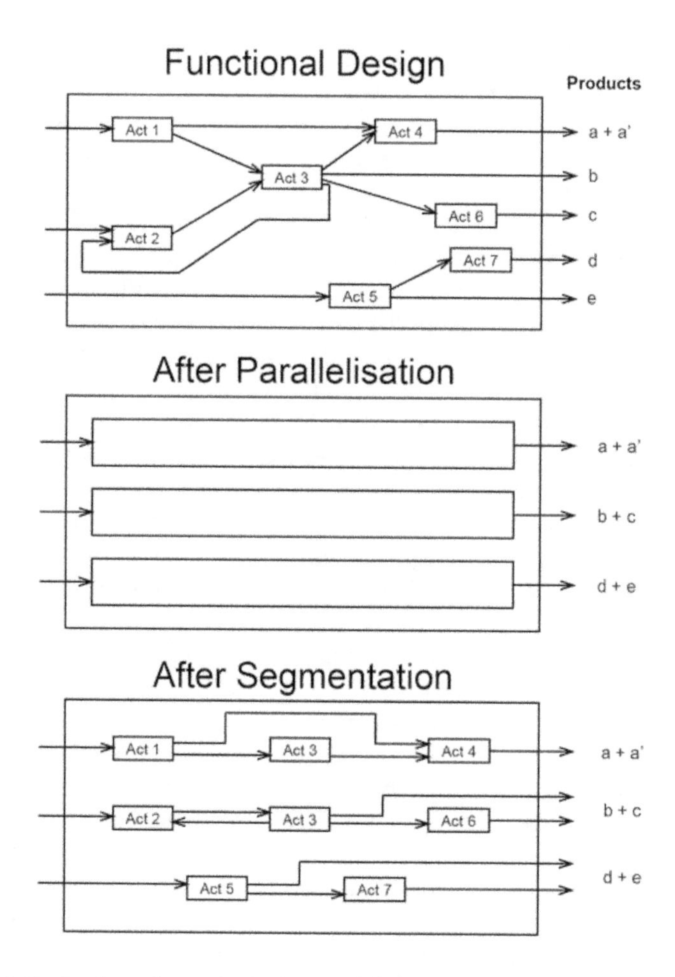

FIGURE 8.2 Reduction of complexity by parallelisation and segmentation

the machines performing the processes 2, 3 and 6. The third factory only makes the products d and e. For this are needed the machines performing the processes 5 and 7. By dividing the factory into three smaller factories an enormous reduction of complexity took place, because (1) the *routing* of each product is clear, (2) attention is focused on a small number of products and (3) less discussion is needed because the employees perform various managerial tasks.

The second step in the redesigning according to socio-technological principles is the design at meso-level. For each one of the smaller factories we consider whether a further division into segments is necessary. Say, for instance, that relatively few of products a and a' are sold. Then total production can be realised by a group of 10 operators. A group of 10 has a size that can be comfortably steered. On the other hand the products b and c are *runners* (meaning products that sell particularly fast). For the performance of the total production about 30 operators are needed: 20 for processes 2 and 3 and 10 for process 6. It then is obvious that we can divide this small factory into two segments, namely processes 2 and 3 in one segment and process 6 also in one segment. If the products d and e are also runners, a division into two segments seems obvious. It will be clear that segmentation likewise leads to reduction in complexity.

The third step in the redesigning is the design at micro-level, i.e. the workplace of the operator. In the socio-technological approach the design of the workplace always takes place in close consultation with the operator. Which executive tasks do the operators perform in the smaller factories? In the first factory they would be trained as far as possible to be able to carry out all processes so that in principle all operators can make a complete product. In the second factory the operators in the first segment will preferably perform both activities 2 and 3, while the operators working in the second segment will in principle only master activity 6. To reach sufficient flexibility several operators from the first segment will learn the activities of the second segment, and vice versa. In the third factory the operators will learn the processes from their own segment. Here, too, for reasons of flexibility a number of operators from the first segment will learn the process from the second and vice versa.

We said earlier that the design of the managing structure takes place in a reversed order. We start at micro-level (workplace), proceed to the meso-level (segment, group) and end at macro-level (parallel factory). At workplace level an effort is made to give the operator the competence s/he needs to let the production run smoothly. Expressed in socio-technological terms, this means that there should be a balance between the given regulatory capacity and the regulatory capacity needed. Here important elements are the performance of quality measurements so that the operator can determine for themselves whether they have done a proper job and also the performance of minor repairs and maintenance. At the level of the segment, the group is granted the power to manage the logistics and planning in the segment. At the level of the parallel factory, the power relates in particular to the logistics and the planning of the whole line. The managing structure takes place intentionally from micro-level to macro-level because the idea is that at the higher level only those duties should be performed that the lower level cannot cope with.

In summary, socio-technological redesigning leads to a completely different factory. By dividing up the factory into a number of smaller factories a considerable reduction in complexity has taken place. Each smaller factory makes its own products. The idea of the whole task brings with it more complex jobs. The operators not only perform executive tasks but also a number of regulative tasks.

8.8 The Mini-Company Concept

A variation on socio-technological thinking was worked out by the Japanese Kiyoshi Suzaki in his book *The New Shop Floor Management. Empowering People for Continuous Improvement* (1993). Suzaki sharply criticises the Taylorist organisations because they focus too little on the quality of the products and on the client's wishes. As an alternative he proposes the so-called 'Mini-Company Concept'. Suzaki's approach reminds one very much of the Dutch variety of sociotechnology. He also divides a huge factory into small parallel factories and regards it as important for operators to perform a complete task. But Suzaki goes one step further by defining a production department as a 'mini-company'. That is to say, a small company within a company. Essential to such a mini-company is that it has its own mission, keeps up contact with clients and suppliers on its own and has its own improvement programme. In Suzaki's view the operators of a mini-company draw up their own mission and define their own improvement programme. Within the framework of the policy of the factory we could here speak of a high degree of self-rule by labourers.

In the Netherlands experiments were also done with Kiyoshi Suzaki's method. For this we refer the reader to the doctoral thesis *Trust and Power on the Shop Floor* (2004) by Maarten J. Verkerk. From this thesis it emerges that there are huge parallels between Suzaki's Mini-Company Concept and the train of thought of the Socio-Technical Systems Design.

8.9 The human standard

How should we evaluate all the developments discussed up to now? Does the Human Relations Movement actually constitute a genuine breakthrough? And what about the participatory and democratic approaches like sociotechnology and the Mini-Company Concept? Quite soon after Ford and Taylor's principles had been widely applied, it became clear that this approach also had its darker side. The mass production in many cases led to high absenteeism and a great turnover of employees. Further it became evident that the employees were often dissatisfied, unmotivated and unhappy. Finally there occasionally occurred serious conflict between management and employees. We once more refer to the Chaplin film *Modern Times* that portrays all these effects. Most of the scientific research on these phenomena concludes that dehumanisation and alienation are inherent to the principles of mass production. In popular terms: organise the industry according to Taylor and Ford's insights, and one gets high absenteeism and low motivation among the employees. (See further Walker and Guest, 1952 and Blauner, 1964.)

In Section 8.6 we called the Human Relations Movement a first breakthrough because it played a great role in the improvement of the social aspects of industrial organisation. Attention was given to the employee as a social creature and to the organisation as a social community. Training and education, the participation of employees and attention to the personal development of each individual were important. But was this first breakthrough enough to prevent the phenomena of dehumanisation and alienation? Supporters of sociotechnology answer in the negative. In their view the problems of mass production are caused by the principles of division of labour and functional management. They are of the opinion that the introduction of a social policy will not lead to a fundamental solution to the problems of dehumanisation and alienation. At most it will lead to mitigation of the pain. They say that a fundamental solution demands a different design of the organisation, something that the Human Relations Movement has not given.

Sociotechnology – the Dutch variation – breaks with the principles of Taylorism which is the reason why we spoke of a second breakthrough. The difference between Taylorism and Fordism on the one hand and sociotechnology and the Mini-Company Concept on the other emerges in the rules for the design of the organisation, in particular by means of the phenomena of isolation and abstraction (see Chapter 3). In Taylorism and Fordism the phenomenon of isolation is manifested in the division of the production process into sub-steps. By means of this division each sub-step is abstracted from the relational coherence with the complete production process. Isolation moreover returns in the analysis of each sub-step that is limited to technological and economic aspects. As a result of this limitation the technological and economic aspects are abstracted from the modal coherence with all aspects. The scientific analysis subsequently leads to rules for the design characterised by the same phenomena of isolation and abstraction. This we first find in the division of labour and the functional management (abstraction of the relational coherence) and further in technological control of material and social processes (abstraction of the modal coherence).

Sociotechnology and the Mini-Company Concept take a different route. Both the scientific analysis and the redesigning take place in an entirely different manner. This we see in the design sequence and the rules for the design, directed at as many complete tasks as possible, consisting of both coherent executive and regulating activities. We could also speak of a more integral approach, for example because consideration is given to psychological and social aspects of the production process itself. In this connection we would also like to refer to the Australian approach of sociotechnology in which there is much focus on the participation by employees and to the Scandinavian approach that stresses dialogue between management and employees.

Sociotechnology helps us to get a better view of the structure of a production organisation. The Dutch approach differentiates between three structures: the production structure, the management structure and the information structure. The production structure comprises all aspects of the production line, like apparatus, the processes, the tools and the lay-out. The management structure comprises all actions and controls carried out to manage the production line. The information structure comprises all information needed to take the correct decisions. These three structures are characterised by different qualifying functions. In the production structure, the technological control is the qualifying function and hence this function is in the formative dimension. In the management structure the concern is for the cooperation and coordination of all managing activities, and the qualifying function is the social dimension. The information structure concerns the information that employees need to take decisions; this information has to be meaningful. The qualifying function of this structure is the lingual dimension. Taylorism and Fordism do not recognise that the management and information structures have their own peculiar character. The technological approach which proved so successful in the control of the manufacturing process was also applied to the managing of the social processes and the information processes. However, these last two processes are characterised by the social and lingual dimension. To put it differently, in Taylorism and Fordism a category is incorrectly labelled. In the participatory and democratic approaches justice is indeed done to the particular character of the management and information structure.

The sequence of design of the different structures also plays an important part. In the socio-technological approach one begins with the production structure, then the management structure and lastly the information structure. Here the importance of the design of the production structure takes the most prominent place. For when laying down the production structure one 'automatically' lays down the management structure and with the management structure the information structure. If one chooses a production structure according to Taylor and Ford's principles, one also chooses 'division of labour' in management and information. This means a low rate of involvement, disrupted social relations and economic loss (just think of the *long wall* method in the Yorkshire case). These negative effects can indeed be lessened by a good social policy but they cannot fundamentally be prevented, as the Human Relations Movement proved. But if one chooses a socio-technological design and a participatory process, one chooses a 'complete task' in management and information and with it conditions for eliminating dehumanisation and alienation.

We conclude that the human standard begins with the design of the production structure and has to be consolidated in the management and information structure. If the human standard is an integral part of the design process it is understandable that this type of approach in general leads to better motivation and greater involvement and thereby to higher quality and efficiency.

8.10 Other alternatives?

The question is whether besides sociotechnology there are other alternatives to Taylorism and Fordism. As a first alternative we mention 'Lean Production' that originated in the production systems developed by the engineer Taiichi Ohno of Toyota. In his book *Toyota Production System. Beyond Large-Scale Production* (1988) he gives an overview of this system. Lean Production departs from a high degree of division of labour, uses a number of new manufacturing technologies and logistical concepts and is supported by the philosophy of continuous improvement. Some researchers are of the opinion that Lean Production could do away with the shortcomings of Taylorism and Fordism. Besides, they point out that this approach can also be implemented successfully in Western factories (Besser, 1996). Other researchers say that Lean Production offers no fundamental solution to the problems of dehumanisation and alienation since there still is a high degree of division of labour.

A second alternative is 'Business Process Reengineering' of which the most important champions are Michael Hammer and James Champy, known for their book *Reengineering the Corporation. A Manifesto for Business Revolution* (1993). They advocate a radical redesigning of 'business processes' with the help of modern information technology. Business Process Reengineering, according to them, leads to a 'new world of work' in which employees focus on the wishes of the client instead of the wishes of their boss. The features of this new world of work are very similar to the participatory and democratic tradition but the great difference lies in 'the way in which'. In sociotechnology the design and the implementation of the new organisation take place in a participatory manner and in Business Process Reengineering in a hierarchical manner.

How should we evaluate these alternatives? In the literature there are indications that the different traditions are growing towards one another. Supporters of the Socio-Technical Systems Design come to the conclusion that application of the different Japanese manufacturing and improvement technologies are needed to improve the quality of the products and the satisfaction of the client. Within the Taylorist and neo-Taylorist approaches the design of the conveyor belt has to be adapted in order to overcome its negative effects. Evaluation of a great number of Business Process Reengineering projects have shown that hierarchical methods have serious limitations. It seems as if design and implementation processes have a greater chance of success when participatory and democratic methods are used.

8.11 Power, language and organisation

In Section 5 of this chapter we pointed out the relation between power, control and organisation. We would like to give greater depth to the analysis by posing the question whether the alternatives offered like sociotechnology and the Mini-Company Concept actually do imply a breakthrough or whether we are here still dealing with more subtle varieties of Taylorism and Fordism. Could it be possible that participatory and democratic concepts are in essence also directed at controlling the employee, that these concepts in a subtle way exercise power since in the end it is management that designs the processes of change and formulates the frameworks? Could it be that the concept 'responsibility' is an illusion that is based on the rhetoric of management? These questions cannot be answered by pointing out the good intentions of scientists, consultants and managers when developing the new concepts. The point is the capacity that managers have in essence to use participatory concepts and democratic technologies for the technological control of an organisation.

In this context we refer to the masterful study *Surveiller et punir. Naissance de la prison* (*Surveillance and punishment. The birth of the prison*) (1975) by Michel Foucault. Foucault did historical research on the birth of the prison and studied in depth the way in which power is exercised over (the body of) the prisoner. In the discussion of one of these techniques, namely discipline, he goes into elaborate detail on other organisations like schools, armies, monasteries and factories. Foucault focuses his attention on the so-called 'microphysics' of power. In doing this he is not concerned with the power someone has, but with the strategy someone uses in order to exercise their dominance. He covers the whole spectrum of 'dispositions, manoeuvres, tactics, techniques, functionings'. The essence of this kind of strategy is that power is not just enforced on people that do not have it but that power 'invests' them, 'is through' them and even 'exerts pressure' upon them. Foucault states that perhaps we should abandon 'a whole tradition that allows us to imagine that knowledge can exist only where the power relations are suspended' and that 'knowledge can develop only outside its injunctions, its demands and its interests'. According to him we should admit that 'power produces knowledge (...) that power and knowledge directly imply one another; that there is no power relation without the correlative constitution of a field of knowledge, nor any knowledge that does not presuppose and constitute at the same time power relations' (Foucault, 1995: 26–27). One of the most important techniques of power is discipline. Foucault demonstrates that in this 'specific technology of power' the body of the person involved is not appropriated in a brutal way as happened, for instance, in organised slavery but that this appropriation happens with 'elegance' by means of a 'codification that partitions as closely as possible time, space, movement'. By this he understands the intentional division of time and space and the painstaking prescription of movement (activities). By portioning the human body is taken up into a 'machinery of power' (Foucault, 1995: 137–138).

Foucault does not want to describe the effects of power exclusively in negative terms like exclude, repress or conceal. 'In fact', he writes, 'power produces; it produces reality; it produces domains of objects and rituals of truth. The individual and the knowledge that may be gained of him belong to this production' (Foucault, 1995: 194). Foucault convincingly proves that power is organised in spoken and written language ('*discours*') and in the way in which activities are organised in actual situations ('discursive practices'). He explains that every element of technology and organisation has an aspect of power. And this aspect of power can be both positive and negative. It will be clear that in the participatory or democratic approaches the aspect of power in both technology and organisational processes is disclosed in a more accountable way than in Taylorism and Fordism. But Foucault's analysis puts into perspective the participatory or democratic approaches as a positive idealised picture. Technology, the interaction between human beings and technology, the interactions between people and the different mechanisms are too complex and too subtle to prevent abuse of power. In the practice of management and organisation we see that the same approach, method or technology can indeed be used to grant power to employees but (at the same time) also to control and oppress employees. For example, a technology of quality, such as Total Productive Maintenance gives clear insight into the achievement of a machine. In a participatory culture the application of this technology will be intended to support the operator to raise the availability of a machine and to improve the quality of the products. But the same method can also be used by management to exact greater achievements from the employees by intensifying the control and raising the goals for production (in a one-sided manner). In modern organisations the difference between 'supporting employees' and 'exacting greater achievements' can be very flimsy.

8.12 Quality technologies

In Chapter 7 we pointed out that from about 1970 various methods were developed to improve the quality of the design process. In manufacturing science the same thing happened. Under the banner of *Manufacturing Excellence* and *World Class Manufacturing* a great number of methods were developed to improve the quality of the production process. Examples are the *Statistical Process Control* for the control of processes, *Failure Mode and Effect Analysis* for calculating risks, *5S* for obtaining a cleaner workplace, *Seven Tools* for continuous improvement, *kanban* for managing logistics, *Design of Experiments* for solving problems, *Total Productive Maintenance* for the realisation of a high rate of availability of machines, and *Design for Assembly* for minimising the number of components. Methods have also been devised for improving safety, minimising the impact on the environment and decreasing energy consumption.

The methods above are focused on *one* aspect of the manufacturing process. However, quality systems have been developed that claim to encompass the most important elements and processes of the whole organisation. The best known quality system is ISO 9000. In this system 20 aspects of an organisation are investigated, including the accountability of management, the quality procedures, the control of the design process, the control of the manufacturing processes, the training of employees and the performance of audits.[3]

Seen from the perspective of the human standard quality systems can be problematic. In certain cases it seems as if these systems are used as an 'invisible hand' to exert power over the employees. Thus Joyce Rothschild and Marjukka Ollilainen (1999) in their article 'Obscuring but not reducing managerial control: Does TQM measure up to democracy standards?' show that the modern *total quality management* (TQM) technologies do not lead to a lessening of the controlling power of managers; on the contrary, they contribute to making the power invisible. They state that concepts like *participation, empowerment* and *self-management* have great rhetorical value and eventually acquired a different meaning from what was originally meant in the democratic tradition. According to them no real redistribution of power takes place; on the contrary, the quality systems contribute to the refinement of the power apparatus.

The use of quality systems as a strategy of controlling organisations 'automatically' leads to their being applied as instruments. The idea of the organisation as a community of workers is thereby reduced to the idea of the organisation as an object to be controlled. In practice it has been proved that instrumental use of quality systems leads to instrumental conduct in employees. For the consequence is that in many cases the employees 'follow the rules' instead of actively thinking and working together for the improvement of the quality of the production process so that in the end the effect of this kind of system proves to be limited. It seems as if we are dealing with an organisational variety of the '*revenge effects*' of a technological approach to reality.[4] These effects are caused in particular by reduction of the quality systems to *one* aspect (the formative aspect) and the denial of the production organisation's own normative structure.

8.13 Conclusion

In the introduction of this chapter we raised the question about the influence of scientification on the manufacturing process. We have seen that scientification led to a technological control of this process by means of which mass production became possible. The design principles of Frederick Taylor and Henry Ford which formed the basis of the mass production bear the main features of scientification in themselves, namely isolation and abstraction. An alternative to Taylorism and Fordism is offered by sociotechnology which attempts to understand the various aspects of manufacturing in their interdependence. In particular in the idea of 'the complete task' there is a break away from Taylor and Ford's design principles. A philosophical analysis has demonstrated that the participatory and democratic tradition attempts to do justice to the relational and modal coherence of the manufacturing process. In this context we use the term 'the human standard'. Comparing sociotechnology with alternatives like Lean Production and Business Process Reengineering leads to some interesting conclusions. First, this comparison confirms the statement made by sociotechnology that division of labour has a negative influence on organisations. Second, there are definite indications that the use of participatory technologies in design and implementation has a positive effect on all aspects of organisations. Finally the different professional technologies should be applied in their context. Therefore the challenge in the design of production organisations is to do justice to the different kinds of complexity: the different modal aspects, the different parties and the different technologies. Yet we should temper our optimism about the participatory and democratic tradition because there often is a fine line between 'granting power' and 'controlling and oppressing'. That is to say, concepts of organisation which in principle were developed to allow employees to act in freedom and responsibility can also simply be used to dominate and oppress them. The same applies to modern quality techniques.

Notes

1 *Shop Management, The Principles of Scientific Management* and *Taylor's Testimony Before the Special House Committee* were reprinted and incorporated into *Scientific Management* (1947).

2 Van Eijnatten (1993) gives an overview of the birth of sociotechnology. For a description of the Dutch approach we refer to De Sitter, Hertog and Dankbaar (1997). The Australian approach was described by Emery (1993) and the Scandinavian by Gustavsen (1992). Verkerk (2004) in his doctoral thesis draws a comprehensive comparison between sociotechnology and other approaches.

3 An audit is an official check to see if the organisation in practice works according to the quality procedures agreed.
4 The expression 'revenge effects' of technology is derived from Tenner (1997). He shows that technological artefacts 'bite back' if too little consideration is given in the design process to the context in which a technology is used. A well-known example is the air-conditioning of underground trains which is meant to lower the temperature in the coaches. An unwanted side-effect is that the warm air pushed out from the air-conditioning causes the temperature on the platforms to rise.

References

Babbage, C. (1832) *On the Economy of Machinery and Manufactures*, London: Charles Knight.

Besser, T. (1996) *Team Toyota*, New York: State University of New York Press.

Blauner, R. (1964) *Alienation and Freedom*, Chicago, IL: University of Chicago Press.

Bloemen, E.S.A. (1988) *Scientific Management in Nederland 1900–1930*, Amsterdam: NEHA.

Chaplin, C. (1964) *My Autobiography*, London: Bodley Head.

Eijnatten, F.M. van (1993) *The Paradigm that Changed the Work Place*, Assen: Van Gorcum.

Emery, M. (ed.) (1993) *Participative Design for Participative Democracy*, Canberra: Australian National University.

Foucault, M. (1995) *Discipline and Punish. The Birth of the Prison*, 2nd edn, New York: Vintage. Translation of: *Surveiller et punir. Naissance de la prison*, (1975).

Gustavsen, B. (1992) *Dialogue and Development*, Assen: Van Gorcum.

Hammer, M. and Champy, J. (1993) *Reengineering the Corporation. A Manifesto for Business Revolution*, New York: Harper Business.

Herzberg, F., Mausner, B. and Snyderman, B. (1959) *The Motivation to Work*, New York: Wiley.

Jacques, R. (1996) *Manufacturing the Employee. Management Knowledge from the 19th to 21st Centuries*, Londen: Sage Publications.

Kanigel, R. (1997) *The One Best Way. Frederick Winslow Taylor and the Enigma of Efficiency*, London: Little, Brown and Company.

Maslow, A.H. (1970) *Motivation and Personality*, New York: Harper & Row.

McGregor, D. (1960) *The Human Side of Enterprise*, New York: McGraw-Hill.

Ohno, T. (1988) *Toyota Production System. Beyond Large-Scale Production*, Portland, OR: Productivity Press.

Pruyt, H.D. (1996) *The Fight Against Taylorism in Europe*, thesis Rotterdam.

Robinson, D. (1983) *Chaplin: The Mirror of Opinion*, London: Indiana University Press.

Rothschild, J., and Ollilainen, M. (1999) 'Obscuring but not reducing managerial control: Does TQM measure up to democracy standards?', *Economic and Industrial Democracy*, 20: 583–623.

Sitter, L.U. de, Hertog, J.F. den and Dankbaar, B. (1997) 'From complex organisations with simple jobs to simple organisations with complex jobs', *Human Relations*, 50(5): 497–534.

Smith, A. (1776) *An Inquiry into the Nature and Causes of the Wealth of Nations*, London: W. Strahan.

Suzaki, K. (1993) *The New Shop Floor Management. Empowering People for Continuous Improvement*, New York: The Free Press.

Taylor, F.W. (1947) *Scientific Management*, London: Harper & Row.

Tenner, E. (1997) *Why Things Bite Back*, New York: Vintage Books.

Verkerk, M.J. (2004) *Trust and Power on the Shop Floor. An Ethnographical, Ethical, and Philosophical Study on Responsible Behaviour in Industrial Organisations*, Delft: Eburon.

Walker, C.R., and Guest, R.H. (1952) *The Man on the Assembly Line*, Cambridge, MA: Harvard University Press.

Womack, J.P., Jones, D.T. and Roos, D. (1991) *The Machine that Changed the World*, New York: Harper Perennial.

Further reading

Imai, M. (1986) *Kaizen*, New York: Random House.

Japan Human Relations Association (1995) *Improvement Engine. Creativity & Innovation through Employee Involvement*, Portland, OR: Productivity Press.

Hayes, R.H., Wheelwright, S.C. and Clark, K.B. (1988) *Dynamic Manufacturing. Creating the Learning Organization*, New York: The Free Press.

Monden, Y. (1998) *Toyota Production System*, 3rd edn, Norcross, GA: Engineering & Management Press.

Pascale, R.T. and Anthos, A.G. (1981) *The Art of Japanese Management*, New York: Simon & Schuster.

Schonberger, R.J. (1982) *Japanese Manufacturing Techniques*, New York: The Free Press.

Schonberger, R.J. (1996) *World Class Manufacturing: The Next Decade*, New York: The Free Press.

LARRY HICKMAN

(1942)

PORTRAIT 8 Larry Hickman

Larry Hickman is a representative of pragmatism in the philosophy of technology. He is a professor of Philosophy and Director of the Center for Dewey Studies in the USA. John Dewey is the philosopher who serves as his primary source of inspiration. Hickman's major publication on the philosophy of technology is his book *Philosophical Tools for Technological Culture* (2001). Following Dewey, Hickman takes a very broad perspective on technology. Any use of tools and artefacts to solve practical problems can be seen as technology. It is a cognitive activity that emerged in the process of human evolution. Technological developments should be realised in a process of experimenting and thus finding out what works and what does not. Therefore it cannot be a matter of experts defining a priori what would be good, but a matter of systematic experimentation. Dewey even suggested that this systematic experimenting in engineering should be taken as a model for all human activities, including politics. Even philosophy is a sort of technology, according to Dewey, when its aim is to have a practical impact rather than only the theoretical investigation of the nature of reality. Technoscience should not be the search for ultimate truth according to fixated procedures, but this open-ended search for what works and what does not. Dewey saw this approach also as the most appropriate for education. Rather than transferring fixated truths from teacher to pupils, education should be based on inquiry by the learners themselves. Hickman applies this to what he calls 'Technoscience education for a lifelong curriculum'. In *Philosophical Tools* Hickman uses Dewey's pragmatist concept of technology to show how the information superhighway developed in a process of evolution from previous technologies but with all kinds of 'hiccups', like temporary exclusion of certain population groups, stress coming along with increased non-physical labour and the formation of various subcultures. The pragmatic stance causes Hickman to be critical about other philosophical 'streams' in the philosophy of technology. According to him, analytical philosophers make the mistake of ascribing truth value to the outcome of a conceptual analysis, the Critical Theorists' approach is too limited to small-scale actor-network analyses and therefore not fundamental enough, the phenomenologists do not offer a way forward and religious approaches (like theologian Reinhold Niebuhr's plea for 'agape' as a leading principle, that he mentions in *Philosophical Tools*) work with a priori truths that hamper open-ended search. Dewey's 'productive pragmatism' according to Hickman is the most appropriate alternative for all these philosophical 'streams'. In this 'productive pragmatism' responsibility is not a move away from technological control (like Ellul and Heidegger suggested), but a means to deal with it in a proper way. There is no reason to be afraid of technology, as long as we take an open view and accept that it is not possible to define beforehand what shape technology should take.

Some quotes that present Hickman's view on technology and the way he builds on Dewey:

> In the hands of the pragmatists, then, theory and practice become equal partners as phases of inquiry. Working together, they orient themselves not just to the analysis of the past or present, but to plans for the future. Like good business partners, they are always negotiating with one another about the feasibility, design, cost, and marketability of potential products. Theory keeps an eye on practice, making sure that options are kept open, that imagination enters into the design stage, and that potential products are coherent with the larger goals of the firm. Practice keeps an eye on theory, making sure that design and production goals are not too ambitious or too fanciful, that products correspond to the needs of the market, that inventories of products and spare parts are maintained, and that the cash flow is sufficient to start the next project. Together, theory and practice engage in a conversation that constantly adjusts means to ends-in-view, and ends-in-view to the means at hand. The

> goal of the partnership is not merely action, but production. The goal of the partnership is continual adjustment to changing situations by means of the development of enhanced tools and new products.
>
> *(Hickman, 1996)*

Simply put, for Dewey it was never 'technology' that was the problem. It was always:

> faulty tools and techniques, or intransigence in the face of new ideas and methods, or overriding class and economic interests, or failure either through ignorance or through force of will to avail oneself of the best of tools at one's disposal, or combinations of these and myriad other factors that are so easily and frequently arrayed against efforts to promote human growth and flourishing. For Dewey, technology – as the experimental involvement with our tools, techniques, traditions, and so on – is intelligent. It is therefore the antithesis of ignorance, greed, intransigence, and ideology.
>
> *(Hickman, 2008)*

References

Hickman, L.A. (1996) 'Techne and politeia revisited: pragmatic paths to technological revolution', *Techne*, 1: 3–4.

Hickman, L.A. (2001) *Philosophical Tools for Technological Culture*, Bloomington and Indianapolis: Indiana University Press.

Hickman, L.A. (2008) 'Truth deflationism, technology, and classical pragmatism', paper presented at the First Nordic Pragmatism Conference, Helsinki, Finland, June.

CASE STUDY II

A new factory

In the summer of 2000 Philips sold some factories in which *multilayer ceramic capacitors* (MLCCs) were made for the Taiwanese firm Yageo. In the light of the strong growth in the market for electronic components the new owner decided on a substantial expansion of the capacity of the establishment in the Netherlands. For this a new factory had to be built. The production would be increased in one year from twelve billion to forty-eight billion units per annum.

In this case study we will be taking a closer look at considerations that played a role in the process of designing the new factory.[1] Various stakeholders are involved in the design of a factory: the employees, the clients, the parent company, the municipality and the province but also the immediate environment (neighbourhood). It is a complex process aiming to synchronise social, economic and technological considerations. To give insight into this process we start with a short description of the manufacturing process. Subsequently the various requirements to be met by the design will be discussed and the choices made during the process of designing. Finally we will say something about the completion.

1 Manufacturing process

An MLCC is a modern type of capacitor composed of a great number of very thin layers consisting of alternatively conductive and insulating material. The conductive materials are metals with a thickness of about one micrometre and the insulating materials consist of a mixture of oxides with a thickness of ten to eighty micrometres. The smallest MLCC is hardly one millimetre in size and the largest is about five millimetres.

The production process is made up of five steps, every step in turn made up of a number of sub-steps. The first step is *casting the foil*. Diverse materials are added to ceramic powder (an oxide). This is ground until the particles have a specific size.

The whole is then mixed with an organic binder (a kind of liquid plastic) and some chemicals until a white liquid slib emerges. This slib is then 'spread' very thinly over a metal band that goes through a tunnel oven. The moisture evaporates and the result is a very thin layer: the ceramic foil. This thin white foil is wound around a spool.

The second step is the *screen-printing, pressing and cutting*. During screen-printing the ceramic foil is cut up into sheets, screen-printed (application of the conductive materials) and stacked on top of one another. Under high pressure these loose stacks are pressed into small thin plates of ten by ten centimetres. These are then cut into loose products in the cutting machine. This is done first in the length and then in the width. By carefully going over the cut plate with a roller, the individual products come loose.

The third step is to get rid of the binding material by heating and baking. First of all the organic binding material added during the mixing in the first step is removed by a thermic process. This is done by heating the product slowly to a high temperature so that the binding material is 'combusted' in a controlled manner. Then follows the baking. During this process the ceramic powder particles are baked onto one another at a high temperature. The products now obtain the desired mechanical and electrical properties.

The fourth step is applying the electrical contacts, baking and galvanising. Each product is provided with two contacts: one on each side of the capacitor. These contacts are needed to connect the electrodes with one another and to enable the product to make contact with the electric circuit in which it will later be placed. The installation of the contacts is done by dipping each product in a special 'paste' of precious metals and subsequently baking them. After this each head is galvanically equipped with a nickel and tin layer. To affect this the products are collected in baskets and moved through diverse galvanic pools.

The last step is *measuring and packing*. Finally the electric properties of all products are measured and some quality measurements done. Then the products are packed in a tape.

2 Technological and organisational requirements

What would the new factory have to look like? A copy of the old one, only four times the size? Or would a new concept be needed? In the initial phase of the project these questions were discussed in detail. For it is known from the literature that the design of a factory has a great influence on its eventual performance. We are not here concerned only with technological performance such as quality and yield but also with social 'performance' such as satisfaction and absenteeism. Besides the immediate environment had to be taken into account (the factory stood in a residential neighbourhood) and also the natural environment (removal of heavy metals and the emission of gases).[2]

The design of the existing factory was inspired to a great extent by the principles of Taylor. The machines belonging to a specific step were grouped together in a separate space. In that factory an order would go through all the processes of a certain step and was next transported to another space where the processes of the following step were carried out. Most of the operators knew only the process of their own sub-step and could only operate one kind of machine. There were indeed a few operators who knew all processes of a certain step, but no operators with knowledge of all the production steps.

The existing design, however, had a number of disadvantages. First, throughput time was quite long. This is the time a product needs to pass through the whole manufacturing process: from casting the foil to packing. A long throughput time has as its most serious disadvantage that it takes a long time before problems are discovered. These often arise in the first steps of the process but only become evident when measuring is done. The consequence could be that all products still in the pipeline suffer from the same defect. The second disadvantage was that the yield was not particularly high. This is the relation between the number of products that are set up (at the screen-printing) and the number that are found to be good (after having been measured). A lower yield is detrimental because even the quality of the good products is often then lower. Third, in the existing factory products were made from different kinds of materials. But some materials 'bite into' one another. This means that when small particles of one material come into contact with the other material it can lead to scrap and a lower quality of the products. In the fourth place the existing design was not environmentally friendly enough. If production using the existing process was to be increased by a factor of four it would lead to enormous pressure on the environment. In particular problems would ensue with the removal of heavy metals via the drains and with the emission of gases. To this would be added that the removal of waste material would come at an increasing cost. Finally the atmosphere in the factory was average, the absenteeism relatively high and the cooperation between the different teams and between management and employees not at its best.

On the basis of these considerations the following technological requirements were formulated: a short throughput time for the products, high yield for all processes, statistically controlled processes (which also lead to a high quality), drastic reduction of pressure on the environment and a considerable improvement in efficiency. In addition, the following organisational requirements were formulated: great satisfaction among employees, enough variation in activity, participation by employees in important decisions about their work place, decision-making capacity about the process for employees in specific situations, good cooperation among employees and between management and employees, good communication among employees and between management and employees, a culture of constantly improving and great carefulness in setting priorities and solving problems.

3 Designing the new factory

What would the design of the new factory have to look like? What concept would meet the various requirements for the design? Would this concept fit into the situation in the Netherlands? The team responsible for designing the new factory had two resources. The first source was the *flow*-concept of a sister factory in Taiwan and the second source was organisation theory and manufacturing science.

In the sister factory in Taiwan, acting on the advice of a consultant, they had investigated so-called 'mini-factories'. This is a relatively small unit that manufactures a limited number of products. In general about eighty to a hundred and twenty employees are employed here, divided into four or five shifts. The number of products manufactured is restricted to one or two main types. From this investigation it was concluded that the throughput time in a mini-factory is much shorter than in an ordinary factory. It is as if the products just 'flow' from one process to the next (hence *flow-production*). In this way problems are detected earlier, the yield increases and the quality improves. There was also a positive influence on efficiency. Since the employees of the mini-factory regularly discussed the results involvement grew considerably. On the basis of these experiences the management in Taiwan had decided to break down the 'huge' factory step by step and replace it with mini-factories.

The experiences of the sister factory in Taiwan were theoretically confirmed by sociotechnology, modern quality methods and Japanese manufacturing technologies (see Chapter 8). In Dutch sociotechnology and in the Japanese concept of a Mini-Company pleas are heard for replacing the Tayloristic concept that has a functionally oriented lay-out (where machines with the same function are grouped together) with a so-called product-oriented lay-out where machines with different functions are grouped together for making a product or a family of products. In this set-up it is possible to synchronise the capacities of the various machines very well so that a short throughput time can be realised (*flow production*). This almost automatically leads to mini-factories. Another example is the modern quality methods and the Japanese manufacturing technologies whereby the whole production process can be systematically analysed and the whole manufacturing design can be optimised. The issues at stake here are yield, quality and cost. In the last instance the participatory or democratic tradition claims that involving employees in the routine and in the improvement process leads to a growth in motivation and involvement. This, too, once more becomes possible in mini-factories.

On the basis of the above-mentioned considerations the following design was proposed:

1. The new factory would consist of a number of mini-factories (seven in total).
2. A rigid separation would be made between mini-factories using the so-called BME-like materials (five in total) and the mini-factories departing from so-called NP0-like materials (two in total). These materials 'bite into' one another. To preclude 'cross contamination' these mini-factories would be housed in separate buildings.
3. Separate mini-factories would be erected for making foil: one mini-factory for making BME foil and one for NP0 foil.

4. Four mini-factories would be erected for multilayer ceramic capacitors based on BME materials. One for the types 0402 and 0603, one for the types 0603 and 0805, one for the type 0805 and one for the type 1206.[3]
5. One mini-factory would be erected for all multilayer ceramic capacitors based on NP0 materials.
6. The factory premises would be redesigned to facilitate the delivery and transport of materials and products and to ensure sufficient parking space on the premises.

For each mini-factory the following technological choices were made:

1. The mini-factories for making foil would be redesigned. The most important reasons for this were: (a) raising efficiency, (b) improving the quality, (c) substantially reducing the number of lifting movements ('back breakers'), (d) reducing the pressure on the environment and (e) reducing the amount of waste.
2. Several production steps would have to be improved to reduce odour and noise for the neighbourhood.
3. The lay-out of the mini-factories for multilayer ceramic capacitors had to be logical, facilitate a short throughput time and minimise the number of walking movements.
4. The supply and transport of all apparatus would be analysed anew to minimise the number of activities.
5. Modern quality technologies and Japanese improvement technologies would have to be incorporated into the design.

For every mini-factory the following organisational choices were made:

1. The mini-factory is an organisational unit. That means that each mini-factory would have its own operators, group leaders, technological specialists and production manager. Only a few specialist functions would be housed in one central section. In each mini-factory about 75 to 125 employees would be employed.
2. Each operator would have to be able to carry out more than one process and thus to fall into the line in different places.
3. Each team would be responsible for the number of products they make and for their quality.
4. Each mini-factory would begin the working day with 'morning prayers' during which the production manager, technological specialists and one or two employees of the team would run through the results and problems of the past 24 hours.
5. The teams of each mini-factory would meet once in four to six weeks to discuss the results and raise problems.

6. The improvement process – in which operators take an active part – would be organised per mini-factory.
7. In a participation process with employees from all layers of the organisation a new code of conduct would be developed. This development rested on the realisation that a new manufacturing concept also demands a new way of working and working together. The central values and norms would be summarised in an easy to use document.

As we have said, what had been experienced in Taiwan was integrated into the whole process of designing. Different operators, technological specialists and managers visited and analysed mini-factories there. In addition they used the different approaches from the participatory or democratic tradition, different improvement and quality technologies and Japanese methods of manufacturing. During the whole process of designing close consultation was kept up with the higher management about the budgets that would be needed and the leadership of the project, with the municipality and the province about environmental permits and with the residents of the neighbourhood about problems with parking, odour and noise.

4 Completion and further developments

Fifteen months after starting the first mini-factories were completed: one mini-factory for making BME foil and one for making BME products. The start did present some problems. The machines had to be synchronised meticulously and the production processes had to be set very carefully. Operators played an important role in the training of new employees. Two to three months later the third mini-factory was completed. And another two or three months later the fourth mini-factory, and so forth.

However, shortly after the opening of the fourth mini-factory a huge setback was experienced: the market for electronic components collapsed. The consequence was that the prices of products like capacitors and resistances plummeted. For this reason the fourth mini-factory was eventually not put into use and the contracts for employees were not renewed. The recession in the component industry lasted a long time and losses mounted enormously. At the end of 2004 Yageo decided to close the factory in Roermond and transport the apparatus to China.[4]

Notes

1 This case study is a 'view from within'. One of the authors (Maarten J. Verkerk) was the project leader for the design and building of the new factory.
2 This is comparable to what in industrial ethics is called the '3P approach': *People, Planet* and *Profit*. The idea behind this approach is that responsible action demands a balance between these three values. '*People*' refers to the various human aspects of the company, '*Planet*' is concerned with the use of feedstock and pressure on the environment, and '*Profit*' refers to the financial results of the company.

3 The type 0402 comprises a family of products that all have the same length (0.04 inches) and width (0.02 inches) but that can vary in the number of layers and the thickness of the layers. The same applies to the other type numbers. It is difficult to synchronise properly the capacities (in the number of products per hour) of the various machines. For some of the machines the capacity is dependent on the surface of the products, for others on the number of layers and the thickness of the layers, for still other machines on the circumference of the products and finally for still other machines on the number of products. From an extensive computer simulation it proved that the best way of getting a balanced production line is to 'run' one or two types per mini-factory.
4 See also the references and further reading of Chapter 8.

References

Schumacher, E.F. (1973) *Small is Beautiful. Economics as if People Mattered*, London: Blond & Briggs Ltd.

Schumacher, E.F. (1979) *Good Work*, New York: Harper & Row Publishers.

PART III

Designing and thinking

9

THE RULES OF THE GAME

Technology as a social practice

Summary

This chapter offers an analysis of the activity of engineers taking as its point of departure the concept of practice as developed by the American philosopher Alisdair MacIntyre. By referring to a case study we will work out this concept in more detail. We differentiate between the structure, the context and the direction of a practice. We will show that this differentiation does justice to the complexity of technological practices, giving insight into the influence of power, values and norms, and life-view on the development of technology. From the analysis it emerges that different kinds of rules or norms apply in technological practices: foundational, qualifying and facilitating. An investigation is done into the nature of these different kinds of rules and norms. Finally we will show that a technological practice is in a class of its own and is characterised by the fact that the qualifying rules have a technological or formative character.

9.1 Introduction: technology as a practice

In Chapter 2 we discussed the coherence between meaning, activity, technology, culture and history and we defined activity as conduct involving meaning. We start from the assumption that activity supposes the conscious intention of those who act. The person who acts, attempts to reach certain goals or make special things. The philosopher Ludwig Wittgenstein (1889–1951) occupied himself extensively with the relationship between meaning and activity. According to him meaning is not only peculiar to language, but also to the human way of life as such. Language, activity and life are intensively interconnected according to him. In his view 'meaning' always has to do with rules. Something has meaning because it conforms to certain rules. For that matter we are not always concerned with written rules. Moreover, most rules for our conduct are unwritten.

When people talk to one another, they exchange meaningful sentences. The sentences have to be formulated according to certain rules, in order to be understood by others. There are rules for the meaning of words (semantic rules), for the structure of the sentences (syntactic rules) and for the correct declension and conjugation of words (grammatical rules). These rules make it possible for the speech act of one person to be interpreted by another. Should somebody just think up and apply their own rules, the speech acts of this person would become unintelligible to the other person who would not know how to interpret them.

Because all human activity is directed by rules, one could define activity as *rule-directed conduct*. In order to be able to act people have to internalise the rules that direct their activity. Wittgenstein points out that these rules are always *inter-subjective*. A rule can only be tested if it is shared by a number of people. For rules enable us to discuss what people do. The basic example, to which Wittgenstein returns in the exposition of these matters, is the game. With reference to the metaphor of the game he attempts to explain how meaning, language and activity are interconnected.

Chess, for instance, is played on a board with 64 squares. Each player has 16 pieces with six different 'characters' (king, queen, castle, bishop, knight and pawn). The players take turns to make a move with the intention to checkmate one of the pieces of the opposition – the king. In addition each character has its own rules that determine which moves are admissible or not. When teaching someone to play chess, we seldom do it by reading out the rules or giving a booklet in which these are written. Mostly we would demonstrate the moves that can be made with the different characters and then we would play a game to practise. When the pupil makes an incorrect move, s/he is corrected. In this way s/he learns the game. Of course one can also write down the rules for chess but for chess to be played this is not essential. What is important is that there are people who know the rules and who can teach others to play the game.

What applies to language and to a game, actually applies to every form of activity, according to Wittgenstein. Even speaking or communicating can be taken as a form of activity that can only be properly understood when one sees the use of language as a part of a much broader context of activity. A person who says something is doing something, often with the intention that it will lead to some kind of activity. Take for instance a request: 'Pass me the hammer please.' Such a sentence is seldom spoken with the intention of reflecting on the deeper meaning of the spoken words. A person saying something like this usually has a plank and a nail in his hand and needs somebody to pass him the hammer. In this example it is assumed that there is someone who understands this request and connects it with the correct import: the request has been understood when the person addressed passes the hammer as requested.

Understanding (speech) acts, according to Wittgenstein, consists of having mastered the rules laying down what to do in situations as they occur. Therefore the use of language should always be seen within a context of speaking as well as acting. He speaks of *language games* which are embedded in *ways of life*. It strikes one that Wittgenstein gives no hard and fast definitions but demonstrates what he means by referring to examples and situations. For instance, by the concept 'language game' Wittgenstein refers to the metaphor of the game. The question what all games have in common is difficult to answer unambiguously. Some games you play by yourself (Sudoku), others in pairs (chess) and still others in a team (football). Often there is an element of contest in a game (only one is the victor) but not always. Thus there is no common feature applying to *all* games but at most we can speak of *family resemblance*. Wittgenstein uses the comparison of a rope. A rope does not derive its strength from *one* unbreakable fibre that runs through the whole length of the rope. No, the strength of the rope is derived from the interweaving of numerous shorter fibres. Likewise with games: there is no *one* feature that is common to all games, but there is a sheaf of characteristics of which all games have at least a few.

For all games it is imperative that the participants have mastered the rules. By this much of the concept 'language game' of Wittgenstein's and of his ideas on language and the coherence between language and acting can be made clear. Yet the image has its shortcomings. Contrary to other games, in 'language games' one cannot just quit, neither is the playing time over after a while. Language games are less distinctly outlined. Of course a game within the lines of the playing field and within the appointed playing time is a 'serious' matter. Everyone 'goes for it', for as long as it lasts. When the game is over, one returns to the normal routine of the day. But language games are an integral component of our daily lives. We cannot simply 'quit' and do something else. A language game is a component of being busy. The context, within which the use of language and activity are connected in daily life, is denoted by the term 'ways of life' by Wittgenstein. Among these he mentions examples like the following: issuing commands and acting on commands, making a certain object by referring to a description or diagram, reporting on a specific event, inventing a hypothesis and testing it, playing theatre, telling a joke and more. So here we are concerned with situations occurring in daily life in which people use language and perform acts in a way that is mutually coherent. We can also mention sundry examples for these ways of life that are to an important extent connected with technology. In these 'technological' ways of life we could speak of speech utterances and acts that have a substantially technological character. For an analysis of that kind of life form we link up with the work of the British philosopher Alisdair MacIntyre, in particular with his use of the term 'practice'.

9.2 Deepening of the concept practice

In his book *After Virtue* (1984) MacIntyre introduces the concept 'practice' to do something about what he calls the contemporary crisis of ethics. According to him the viewpoint of 'emotivism' dominates the ethical debate. By this he understands the conviction that all our evaluative and moral judgments – meaning statements that say nothing about how something *is* but about how it *should* be – can in the final instance be reduced to preferences, personal predilections or expressions of attitude or emotion. Because it is difficult to have a rational discussion of these preferences or predilections, ethics limits itself mostly to procedural matters within this context. In the case of differences of insight into moral issues how can one still come to a correct decision in a rational way? In this case ethics no longer focuses on making statements about the contents (unless it is an undisputed matter) but actually only on the formal rules and procedures by means of which one could arrive at settling on ethical statements concerning content.

To break out of this crisis in ethics, MacIntyre falls back on the tradition of virtue ethics. In the ethics of virtues it is not so much the *rules* and *guidelines* for acting properly that take central place, but *virtues* taken as characteristics of persons. A virtue actually is nothing other than a good characteristic of a person – reduced to a pattern – in other words, an acquired quality. For instance, a virtue like trustworthiness denotes an (enduring) characteristic of somebody. When one asks such a person a question s/he gives an honest answer; should you entrust something to them it will be in safe hands. When s/he promises something, s/he keeps her/his promise. Neither is this something that happens once only, it is a recognisable pattern that characterises the particular person: s/he is trustworthy. Not only her/his statements, her/his promises or her/his deeds are trustworthy, it is the recognisable pattern behind this, *s/he her/himself* is a trustworthy character. To be trustworthy in this sense, a person should indeed be disciplined. One could say s/he has to practise it in her/himself. Of course this does not mean that a virtuous person can never have a slip. Naturally a trustworthy human being can once in a while make a mistake or know a moment of weakness. But then their trustworthiness should remain recognisable, for instance by their willingness to be reprimanded and by admitting that they have failed in that situation. In other words, a virtue is a moral disposition. By the word *disposition* we here denote a basic attitude of a person that has become a habit. With this concept of virtue MacIntyre joins a longstanding philosophical tradition that began with Aristotle. However, the application of this concept as visualised by MacIntyre is new. According to him virtues are important to people within 'social practices' because they contribute to reaching the goods inherent to practices. When people do not have these dispositions the relevant 'internal goods' will not be able to be realised well or perhaps not at all.

This application of the concept of virtue evokes the question of what MacIntyre understands by a *social practice*. By this he indicates a meaningful coherence of human activity in which certain goods are realised.[1] In MacIntyre's definition there are four aspects of importance. First of all he characterises a practice as a 'cooperative human activity'. In other words, he sketches the acting person not as an individual but as a member of a community. Second, a practice is 'socially established', that is, it concerns activities that are rooted in the community, as for instance in the context of societies and in organisations. Therefore these activities do not merely exist in the minds of people, but are *institutionalised*. Third, this practice realises 'goods internal to that form of activity'. In other words, the realising of these goods is specific to the kind of practice involved and also characteristic of that specific practice. Take for example the practice of jurisprudence. In this practice many aspects play a part, but the essential part of the practice of jurisprudence is that justice is done. The concept 'external goods' is reserved by MacIntyre for the goods that are external to the goal of a practice. Examples of external goods are the remuneration of judges and advocates. Finally, in a practice the activity is evaluated with reference to 'standards of excellence'. By these MacIntyre means the rules and procedures which determine whether the realised goods are meeting the requirements that hold for that particular practice: the *standards of quality*. People who attempt to realise the internal goods of the particular practice unequivocally pursue these standards of quality. They will do their best to excel in their field. Nowadays one would say: they have a passion for their work. To MacIntyre virtues are acquired qualities or attitudes that enable people to pursue the internal goods of practices and excel in these.[2] In other words: virtues are attitudes that people have acquired and that equip them to realise the *internal goods* of *practices*.

What MacIntyre says is not always clear. For instance, it is not easy to determine exactly which contexts of activity are social practices and which are not by referring to his definition of the concept *social practice*. Questions also arise when he differentiates between internal and external goods (see Section 9.5). But in spite of these limitations, his concept of *social practice* offers a starting point for a complete reflection on human activity. According to MacIntyre a practice is a *coherent and complex form of socially accepted, cooperative human activity*. Practices are therefore socially stabilised structures of meaningful human activity and the sense of that activity is very much relevant to the destination of the practice.

9.3 A case history

What is the connection between the above arguments about language, activity and practices and the world of technology? Most of the engineers who finish their studies at university go into *practice*. It is therefore a matter of importance to take a closer look at this *practice*, and this we will do with reference to the following case study (Verkerk, 2004).

BOX 9.1: CERAMIC MULTILAYER ACTUATOR

Around 1985 the Business Group Passive Components of Royal Philips Ltd started development of the ceramic multilayer actuator in a laboratory in Eindhoven. The engineers developed the basic materials, designed products for specific applications, determined the physical characteristics and conducted an extensive market study. At the beginning of 1990 it was decided to concentrate the pilot production in Kaohsiung (Taiwan). This decision was prompted by the presence of the multilayer technology and the availability of cheap labour. Further the company considered that the industrial process was already stable enough to conduct trial production in a foreign country. Shortly after starting the production in Taiwan a serious problem presented itself. Some production runs resulted in products with a high internal resistance and other runs in products with a low internal resistance. No research into the cause of this problem rendered any distinct results. Simultaneously, however, a contract was signed to supply actuators to a Japanese client for the inkjet printer market. In view of the fact that this contract could well mean a breakthrough for the new product it was decided in the spring of 1992 to transfer all actuator activities to Roermond (The Netherlands). In this factory knowledge of the multilayer technology was also present. At more or less the same time it was decided to transfer the laboratory of Passive Components where the actuators were developed to Roermond. Due to the new strategy of the business group, Roermond got the status of the 'birthplace' of new businesses based on the multilayer technology. Ultimately the 'Taiwan adventure' had made it clear that the production process had been insufficiently developed to justify settling it in a foreign country. In addition the new contract set such high demands for cooperation that realising them was only possible when all disciplines – marketing, development and production – were concentrated in Roermond.

Figure 9.1 shows a cross section of an actuator. This product is made up of thin layers of piezoelectric material, separated by an even thinner metallic layer. One of the characteristics of this material is that under the influence of an electric charge it becomes slightly longer or shorter. This feature can be used for designing the head of an inkjet printer. Thin needles of piezoelectric material are placed against an ink cartridge. As a result of an electric pulse the length of such a needle increases, a shock wave ensues in the ink cartridge and shoots out a small drop through the aperture in the cartridge. During the years 1992–1997 the Epson company applied this technology in their printers.

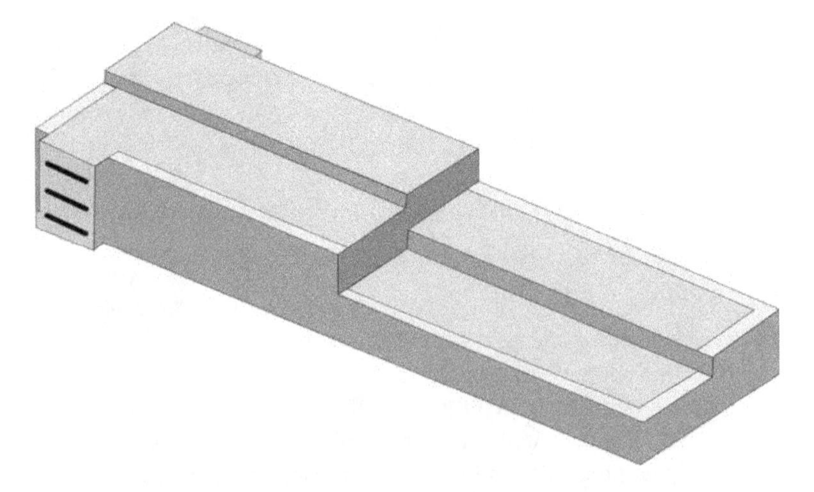

FIGURE 9.1 Cross section through an actuator (20 mm x 8 mm x 1.5 mm)

The decision to concentrate all activity on actuators in Roermond (The Netherlands) meant a far-reaching change in the manner of working and collaborating. Formerly the marketing and developing activities were concentrated in various buildings in Eindhoven (The Netherlands) and the production activities in Taiwan. These lines met at a high level of the organisation. But now marketing, development and manufacture were brought together in one building and under the leadership of one business manager. The communication lines were short. The decisions were taken integrally: with the input of and supported by every discipline. The result was that engineers no longer focused on practising their own discipline but rather on contributing to the business from their discipline.

In the above case study engineers played an important role: they developed new products (*developers*), negotiated with the client (*marketers*), supervised the production process (*factory engineers*) and gave guidance to the business-unit (*managers*). This case study is representative of the different questions and problems which confront engineers in practice. We now analyse this case study with reference to the four criteria of MacIntyre's concept of practice:

1. An industrial practice is a *cooperative human activity* in which different professional disciplines work together to develop, produce and sell a product. In the case study this cooperative activity took form specifically in the organisational structure: the divisions marketing and sales, development and manufacture were brought together in one organisational unit, in one location and under the leadership of one responsible manager.
2. An industrial practice is *rooted in society* and formed in a long historical process. It is the end result of a great number of decisions and is characterised by specific knowledge, certain methods and its own networks. The decision to harbour the disciplines marketing, development and manufacture in one organisation promoted the coherent development of the diverse disciplines.
3. The concept *internal goods* expresses what should be the main concern in the industrial practice: realising products that contribute to the development of society. The different disciplines have no autonomous position but have to be focused on the encompassing objective of the practice. The internal goods of this industrial practice are producing actuators of good quality for the printer industry.
4. The *standards of excellence* of the industrial practice are to a great extent determined by the quality systems which are applied to it. One of the systems most often used is ISO 9001. The objective of this system is to 'achieve client satisfaction by preventing deviations in all stages from design up to and including aftercare' (ISO 9001, Section 1). In order to realise this objective 20 requirements or norms are formulated which the organisation has to meet. These requirements relate amongst other things to the responsibility of the

management, the quality system, the appraisal of contracts with clients, designing of new products and control of the manufacturing process. Each of these requirements is 'translated' into a number of procedures giving detailed instructions regarding the activity of the particular staff members in the organisation. In 1995 the Business Unit *Ceramic Multilayer Actuators* was officially certified for ISO 9001. Table 9.1 gives a few examples of ISO norms and their corresponding procedures. Some of the norms and procedures apply to all disciplines and others only to one specific discipline.

In the business unit they had started using a separate manual with *standards of excellence* for the fields of safety, health, well-being and environment. The development of these standards was supported by some *improvement programmes* presented in-house by Philips, like *Energy Consumption Analysis, Ergonomics/Human Factors*

TABLE 9.1 Examples of ISO norms (4.1 t/m 4.4 and 4.9) and ISO procedures for the various disciplines

ISO norm/Procedure	*Marketing and sales*	*Development*	*Manufacture*
Responsibility of Board of Directors (4.1)			
Organisation manual (no. 4002)	X	X	X
Quality system (4.2)			
Manual product quality (no. 4001)	X	X	X
Appraisal of contracts (4.3)			
Market analysis (no 4049)	X		
Acceptance of production orders (no. 4050)	X		
Treatment of clients (no. 4026)	X		
Designing new products (4.4)			
Drawing up a roadmap		X	
Project management (no. 4014)		X	
Client-friendly designs (no. 4015)		X	
Control of the manufacturing process (4.9)			
Production planning (no. 4002)			X
Process control (no. 4022)			X
Operational instructions (no. 4028)			X

and *Managing Health and Safety Risks*. Finally the business unit used the quality system 'Philips Quality Award for process management in the nineties' (PQA-90). This system has six categories which are evaluated: the role of management, the process of improving quality, the quality system, relationship with clients, relationship with suppliers, and finally the results. Each category is divided into a number of subcategories, each subcategory consists of a number of elements to be tested. PQA-90 describes, just like ISO 9001, a great number of norms for delivering good quality products to a client, but goes even further. It describes the way managers and staff members should conduct themselves in order to realise these norms. The introduction of PQA-90 in Philips was supported by parallel improvement programmes for marketing, development and manufacture.

9.4 The rules of practice

Activity, we have said, is meaningful or rule-directed behaviour. Acts derive their meaning from the rules with which they comply. These rules apply within the context of social practices as, for instance, a football association or a business unit.

Football begins with a relatively simple set of rules of the game that determines when we can call a certain game 'football'. With this formulation we lay down what football is: acting according to the rules of the game. But at the same time this formulation affords a criterion for judging and evaluating different kinds of activities connected with football. It could happen that certain components of the football practice deviate so much from the rules of the game or begin to dominate it, that they pose a threat to the practice. We see this in particular in professional football where all the financial activities (buying and selling of players, gate takings, broadcasting rights, etc.) become so important that they enjoy preference above and easily clash with the set of rules defined for the game of football. The result of the game as the outcome of the exertion to play a football match according to the *standards of excellence* then risks becoming a means towards something else, namely a higher profit or a higher share price. The relation between the pursuit of internal and external values then becomes unbalanced.

We find a comparable meaning of rules in the case study we discussed above. There, too, we can discern a number of elements: management, marketing division, development division, apparatus, the building, cafeteria, cleaning services, etc. Here it also holds that: 'the business' begins with a simple set of 'rules of the game'. In this case it is to deliver products meeting certain specifications according to the terms and conditions laid down in the contract. The whole of specifications, terms and conditions determines whether we can denote a 'game' as 'doing business'. If during the production process – for whatever reason – the rules of the game are infringed, it can lead to defective products. So in the final instance all activities in the organisation should be focused on delivering actuators that meet the technological specifications and are delivered as arranged. Certain elements can end up being so far removed from the rules of the game that they constitute a threat to the business. If, for instance, the financial section should decide on their own authority to forbid staff members to travel in the framework of cutting costs, it could have a negative effect on the business, since developers and marketers would no longer be able to visit clients.

9.5 Structure, context and direction

In our discussion of the rules of a practice we come up against the distinction we mentioned earlier made by MacIntyre between the *internal* and *external goods*. The *internal goods* are values specifically realised in practice. In football the focal point would be an attractive match and in the production practice discussed it would be delivering properly functioning actuators. The *external goods* on the other hand, are values which are 'external' to the practices, for instance maintaining the field or cleaning the production hall. The tickets for the football match are sold and the bills are sent to the client. And every month the salaries of the players and the employees have to be paid out. All these activities are important for the functioning of a football club or a business unit while not being part of the primary process.

The distinction between *internal and external goods*, however, does not do adequate justice to the complexity of practices. When the board of directors of the business unit decides that the factory may be expanded in the following year because the profit exceeded 10% of the turnover, can we still clearly separate the manufacture of actuators (*internal good*) from making a profit (*external good*)? Is this not too simplistic a representation? This distinction suggests that one can easily ascertain what belongs to the core and what to the periphery of a certain practice. Let us illustrate this by means of an example from a different sector. Promoting the cure of a patient belongs to the internal goods of medical practice. The fact that the doctor is paid for this pursuit – according to MacIntyre an external good – can hardly be interpreted as a matter that is merely coincidentally connected with the practice. That is how professional practices are. But the medical practice is not absorbed in the economic meaning of this for the professionals. On the contrary: the *primary process* of the medical practice remains the care for human beings that need it.

As the difference between internal and external values is still not fine enough, we will try another angle of approach to gain a good understanding of technological practices, by an analysis done from three different angles of approach: *structure, context* and *direction*; subsequently we will demonstrate that different kinds of rules give direction *to the* activity *of* engineers. Such a combination of lines of approach and rules enables an integral approach to technological practices.

The concepts 'structure', 'context' and 'direction' are particularly productive for an analysis of societal relationships in a plural society (Mouw and Griffioen 1993). 'Structure' pertains to the *formal* aspects of an organisation. Thus we find great differences between, for instance, a technological practice, a medical practice and an agricultural practice. These differences are to a great extent caused by the nature of the primary process. The concept 'context' refers to the influence of the environment within which a practice is developed. Thus there are differences between a laboratory in the Netherlands and a laboratory in Japan. These differences cannot be connected exclusively and directly to the structure of the research process but have to do with the country and the culture in which the laboratories are situated. Finally, the concept 'direction' refers to the differences in the basic convictions underlying the design, manufacture and sales of technological artefacts. A clear example is the difference between the Rheinland (German) and the Anglo-Saxon (English) systems of commerce: do you see the company mainly as a community of people who, from different positions and interests, make a concerted effort to bring about one joint product or service package, or do you see a company mainly as an investment that has to yield adequate returns for the share holders?

The rules applying to a particular practice determine the structure. We will see that these rules (can) obtain a specific colour and content as a result of the context and the direction of a practice and also that the direction of a practice has great influence on the way in which these rules function.

9.6 The structure of a practice

Each practice is characterised by its own rules of conduct. In the case of the structure of a practice we think of the rules that are constitutive, that define and *constitute* the practice, in other words, that delineate the structure and its boundaries. We differentiate three kinds of *constitutive* rules: *foundational, qualifying and facilitating.*

Social practices are *cultural phenomena* formed by human activity that exist by the grace of that activity. They are *based* in the form-giving or formative aspect, the patterns or regularities that apply to them we call the *foundational* rules. This 'having-been-formed' is expressed in the fact that social practices are realised in *organisations*, institutionalised forms of cooperative activities set up by human beings. Foundational rules have to do with the formative aspect of practices, their *organisational* character, in business terms: you find the foundational rules in the organisational structure, the quality systems and the improvement programmes of the organisation. We are here concerned with what is often called the *formal* structure of an organisation: the rules and instructions that see to it that an organisation functions. In the description of the case study we came up against different foundational rules. An organogram shows who reports to whom and how the different sections are connected. The details of responsibilities, mutual relations and diverse consultative structures are laid down in the quality manual and in procedures (see Table 9.1). In addition foundational rules are found in the different quality systems. In this way the section Marketing and Sales is bound by different rules relating to market analysis, acceptance of orders and the treatment of clients; the section Development by the various instructions for drawing up an annual strategy, the managing of projects and client-friendly design; and the section Manufacture by detailed procedures for planning production, controlling processes and drawing up instructions (see Table 9.1). Also the improvement programmes contain a great number of instructions on the way processes should run. In short, technological practices are made up of rules.

The *qualifying* rules in a practice characterise the *primary process* of the practice. In the case study we find three different primary processes: the development, production and sales of actuators. Each one of these processes has its own qualifying rules. In the process of development and manufacture we are concerned with technological rules as described in the data-sheets of products, standards of excellence for measuring instruments, specifications for machines, norm sheets of institutes, safety and environmental instructions from the government, etc. In the sales process we are concerned with economic rules as written in procedures, laws and contracts. The nature of the rules depends on the primary process. The rules of a developmental and manufacturing process are primarily technologically qualified and the rules of a sales department primarily economically.

Finally we distinguish the *facilitating* rules. These contribute to enabling a practice without being characteristic of the primary process of the practice. We are here concerned with conditions essential for the functioning of the practice that do not, however, characterise or qualify the practice. In the case study discussed above we mention the rules for the financial administration, the payment of tax, hiring of staff, training of staff members, securing the complex, etc. These are rules we also find in other practices, in companies particularly in what is sometimes called the *staff sections* of an organisation: Purchases, Finances and Administration, Staff and Organisation and Services.

Technological practices have a normative character: the various kinds of rules lay down how professionals should act. An engineer comes up against different kinds of norms, like formative norms (specifications for products), social norms (cooperation in the primary process), economic norms (purchase of materials, product sales), juridical norms (contracts, safety and environmental instructions) and moral norms (quality of work, care for the environment) in their everyday work. We call this a *simultaneous realisation of norms.*

9.7 The context of a practice

The concept 'structure' refers to the principles of ordering or constitutive rules that similar practices have in common: What is it that makes a company a company and a laboratory a laboratory? The concept 'context', however, refers to the individualising circumstances which affect the unique individual form of a particular practice. Why does one actuator factory differ from the next? Surely, they are determined by the same structural principles and the same primary process. And do they not both have to follow the same qualifying rules? However, it makes a difference whether this factory is in Taiwan or in the Netherlands, for there always are social, economic, geographical and cultural differences. By 'context' is understood all these different temporal and spatial circumstances, in different situations, that have an effect on the concrete forms a particular practice takes on.

The case study demonstrates that the context of a practice is influenced by diverse specific factors. Here are a few examples. In the initial phase the developing activities were concentrated in The Netherlands and the production activities in Taiwan. The communication between these two localities was not easy. Not only because of the distance but in particular because of differences in culture. Later it was decided to bring together all actuator activities in *one* organisational unit, in *one* location and under the responsibility of *one* manager. This decision led to a far-reaching change in the way of working and cooperating. A second variable having great influence was the Japanese clients. They set high requirements for communication, the organisation's ability to react and the quality of the process. These requirements also significantly affected the way of working and working together. Finally, under the influence of the general policy of Philips the management of the business unit decided to introduce ISO 9001, manual for Safety, Health, Well-being and Environment, and PQA-90. These systems emphasise the relations with important stakeholders like clients, suppliers, government, local community and environment. This decision had a great influence on the context of the practice because the different relations were actively managed.

9.8 The direction of a practice

The concept 'direction' refers to the *regulative side of practices* with respect to the attitude, the intention, the values and the basic convictions which prompt human beings to act. Under this fall, for instance, the attitude towards the client ('the client is always right'), the view of subordinates ('they are only working for the salary'), the prejudice towards dark-skinned people ('they are lazy'), the opinions on the style of leadership ('you have to give your staff as much freedom as possible'), the view of technology ('technology should serve society') and the life-view ('you have to enjoy as much as you can in this life') or religion ('you should live to the honour of God') that a person subscribes to. All these attitudes, intentions, values and basic convictions influence the way in which human beings conduct themselves in practices and interpret, apply and obey the rules within those practices.

Let us apply the above reflection on the regulative side of practices and the concept of direction to the case study. In this case study the main concern is the *values* and *convictions* that motivate the board of directors and staff members of the business unit in their work and that influence the execution of the primary and the supporting processes. First we give consideration to the values and convictions shared by the organisation as a whole. In the case study under discussion the company values of the mother company Philips play a great part (see Box 9.2). Within the business unit these guided the development and the implementation of the diverse organisational structures, quality systems and improvement programmes. The effect of this can be seen not only in rules and procedures but also in the way in which people do their work.[3] The quality system PQA-90 (see Section 9.3) focuses on the conduct of staff members and the example set by the conduct of managers. This system presupposes that they share certain values and act out of conviction. In this context we will pay some attention to the research by James Collins and Jerry Porras on the influence of values on the functioning of a company. In their book *Built to Last* (2000) they point out that the success of companies in the market is determined mainly by the extent to which the values of the company are lived by all staff members and the way in which the values are made explicit for all activities of the company.

BOX 9.2 THE PHILIPS WAY – THIS IS HOW WE WORK

1. Make **clients** enthusiastic
2. **People:** our most important means
3. Do **quality** work and excel in everything you do
4. Get the highest possible **output** from your own capability
5. Stimulate **entrepreneurship** at all levels

Tony Watson in his book *In Search of Management* (1994) shows that every manager has her/his own vision of human beings, labour and organisation and that this determines the way they do their work. For this he uses the term '*work orientation*'. Watson discovered that there can be huge discrepancies between the values of the company and the work orientation of individual managers. This can result in a certain practice developing in a different direction from what is envisaged in the various quality systems. Tension then develops between the 'language' of the quality systems and the 'language' of the manager's conduct. Maarten Verkerk in his doctoral thesis *Trust and Power on the Shop Floor* (2004) shows that in a culture characterised by fear and distrust employees on the shop floor see the values of the company mainly as a means used by management to increase its power over them. The employees on their side try to protect their own interests as far as possible in their daily activities. It goes without saying that in such a situation the values of the company have little influence on the daily routine. Finally we would like to point out that personal opinions like 'clients nag', 'staff members only pursue their own interests' or 'I have to secure my bonus' can easily lead to conduct that is at odds with the values of the company.

We conclude that the concept of direction referring to the *regulative side* of practices points to intentions, values and convictions that are given form in a practice, to the question by what the engineer is oriented in his/her professional activity and how s/he envisages responsible development of our society and his/her view of 'the good life'. And ultimately it points to the convictions one has about one's own possibilities to contribute successfully to further development of the practice in which one is active and if so on what (spiritual) grounds.

9.9 Technology and practice

The relationship between science and technology is often typified with reference to the word 'practice'. Technology is then seen as the application of scientific theory in everyday practice. This view can be called rationalist since it means that human beings can control reality ever more effectively on the basis of scientific insights, for instance by the application of insights in the design of technological artefacts. This view, however, does not sufficiently take into consideration the specific limitations of theoretical knowledge (its abstract, modelling character) that impede the direct application in the process of designing. Formulated differently: this is too simplistic and one-sided a way of thinking about the process called 'application of a theory'. In this discussion the word 'practice' can definitely elucidate something. But then it will have to be given form in a different way, namely by seeing both science and technology as cooperative human activities, each with its own character and rules. Therefore not only technology is a practice. Science itself can be regarded as a practice in a class of its own.

We speak of a technological practice when we are dealing with a distinct, socially established form of cooperative activity qualified by formative technological rules. In other words, we are dealing with practices guided by rules for technological form-giving. One could also distinguish between *technological* and *technical* practices. *Technological* practices are distinguished by a scientific approach to technology: the research on *technological rules.* This forms the field of technological sciences. *Technological* practices are characterised by the *application* of the rules in concrete processes of design and execution (see Chapters 7 and 8).

The view presented above enables an integral approach to the practice of human activity and the place of technology in it. By having understanding of science, technology and technique as social practices, one can do justice to the integral character of the reality in which we live and the place technology has in it. This approach also offers a viable point of departure for the development of an ethics of technology (see Chapter 13).

9.10 Conclusion

In this chapter we have seen that science and technology can be seen as social practices that can be analysed from the perspectives 'structure', 'context' and 'direction'. This analysis demonstrates that practices are never neutral: norms, values, life-view and questions of meaning all play an important part. The rules that constitute a practice are normative rules: one can comply with them but one can also ignore them. The normativity also applies to the direction in which a practice is developed. For it is supported by convictions and life-views of human beings: opinions on the good life.

Notes

1 'By a "practice" I am going to mean any coherent and complex form of socially established cooperative human activity through which goods internal to that form of activity are realised in the course of trying to achieve those standards of excellence which are appropriate to, and partially definitive of, that form of activity, with the result that human powers to achieve excellence and human conceptions of the ends and goods involved, are systematically extended' (MacIntyre, 1984: 187).

2 'A virtue is an acquired human quality the possession and exercise of which tends to enable us to achieve those goods which are internal to practices and the lack of which effectively prevents us from achieving any such goods' (MacIntyre, 1984: 191).
3 The case study was investigated extensively by the University of Twente, the Netherlands (Leede, Looise and Verkerk, 2002) and by Verkerk (2004). From this research it was proved that company values were internalised to a great extent and gave direction to conduct in the organisation.

References

Collins, J.C. and Porras, J.I. (2000) *Built to Last. Successful Habits of Visionary Companies*, London: Century.

Leede, J. De, , Looise, , J.K. and Verkerk, M.J. (2002) 'The mini-company. A specification of sociotechnical business systems', *Personnel Review*, 31(2): 338–355.

MacIntyre, A. (1984) *After Virtue. A Study in Moral Theory*, 2nd edn, London: Duckworth.

Mouw, R.J. and Griffioen, S. (1993) *Pluralism and Horizons. An Essay in Christian Public Philosophy*, Grand Rapids, MI: Eerdmans.

Verkerk, M.J. (2004) *Trust and Power on the Shop Floor. An Ethnographical, Ethical, and Philosophical Study on Responsible Behaviour in Industrial Organisations*, Delft: Eburon.

Watson, T.J. (1994) *In Search of Management. Culture, Chaos and Control in Managerial Work*, London: Routledge.

LANGDON WINNER

(1944)

PORTRAIT 9 Langdon Winner

Langdon Winner became known especially for his studies of the social and political aspects of modern technological development. He was born and grew up in San Luis Obispo in California and obtained his doctorate in Berkeley. He is professor in political sciences in the Department of Science and Technology Studies of the Rensselaer Polytechnic Institute in Troy, New York. He has lectured at diverse other universities and institutes, among them the University of Leiden. From 1991–1993 Winner was president of the Society for Philosophy and Technology.

The most influential piece written by Winner is the article 'Do artifacts have politics?' The article was published in 1980 in the journal *Daedalus* and was incorporated in Winner's book *The Whale and the Reactor. A Search for Limits in an Age of High Technology* (1986). In 'Do artifacts have politics?' Winner opposes both a 'technological deterministic approach', which supposes that technological artefacts have inevitable social consequences, and a 'social deterministic approach' by which artefacts are regarded as neutral utilities that can be used for good and for evil. Winner states that we have to take a much closer look at the technological artefacts themselves, should we wish to know their social impact. For in the structure of technological artefacts and systems there is an intentional or unintentional expression of political convictions. One of his examples is the low viaducts built by the architect Robert Moses which had the intended effect that black people who had to resort to transport by bus could not reach the beaches of Long Island in New York. Another example, but one having an unintended effect, is the mechanical tomato harvester developed in California. This resulted in job loss for many agricultural labourers. The structure of technological artefacts and systems is not something to be changed in less than no time. Yet both in the case of the viaducts and of the tomato harvester an alteration to the design succeeded in bringing about a different political-social reality. But it is a different matter in the case of technological designs like a nuclear power station. For instance, in order to prevent abuse of radioactive material or a terrorist attack, a strictly hierarchical organisation is essential so that – and this Winner finds important – a democratic business management by definition becomes impossible.

In spite of the influence Winner had in philosophy of technology he wrote few books and articles. His first book is *Autonomous Technology. Technics-out-of-control as a Theme in Political Thought* (1977). In this book he takes a critical stand on Ellul's conviction that technology has to a great extent become autonomous (see Section 10.3.4). The second and also last book by Winner is the above-mentioned *The Whale and the Reactor*. He also edited the book *Democracy in a Technological Society* (1992) and wrote diverse articles, amongst others on the public responsibility of engineers and on social constructivism. In these articles he opposes an approach based on case studies which is the accepted thing in both ethics of technology and in social constructivism. It finally leads to a narrowing down of one's view since there is insufficient consideration for the broader political-social issues accompanying technological development.

> 'Autonomy' is at the heart a political or moral conception that brings together the ideas of freedom and control. To be autonomous is to be self governing, independent, not ruled by an external law or force. In the metaphysics of Immanuel Kant autonomy refers to the fundamental condition of free will – the capacity of the will to follow moral laws which it gives to itself. Kant opposes this idea to 'heteronomy', the rule of the will by external laws, namely the deterministic laws of nature. In this light the very mention of autonomous technology raises an unsettling irony, for the expected relationship between subject and object is exactly reversed. To say that technology is autonomous is to say that it is nonheteronomous, not governed by an external law. And what is the external law appropriate to technology? Human will, it would seem.
>
> *(Winner, 1977: 16)*

No idea is more provocative in controversies about technology and society than the notion that technical things have political qualities. At issue is the claim that the machines, structures, and systems of modern material culture can be accurately judged not only for their contributions to efficiency and productivity and their positive and negative environmental side effects, but also for the ways in which they can embody specific forms of power and authority. Since ideas of this kind are a persistent and troubling presence in discussions about the meaning of technology, they deserve explicit attention.

(Winner, 1986: 19)

Despite these admirable aims, however, the approach often used to teach ethics to scientists and engineers leaves much to be desired. In the way the topic is usually presented, personal responsibilities are situated in extremely limited contexts and ethical choices made to seem something like extraordinary, unwelcome intrusions within a person's normal working life. Rather than lead students to evaluate the most basic, most practical features of their career choices – the kinds of work they select and the social conditions in which that work is done – courses on professional ethics tend to focus upon relatively rare, narrowly bounded crises portrayed against an otherwise happy background of business as usual.

(Winner in Durbin, 1990: 53)

References

Durbin, P.T. (1990) *Broad and Narrow Interpretations of the Philosophy of Technology*, Dordrecht: Kluwer.

Winner, L. (1977) *Autonomous Technology. Technics-out-of-Control as a Theme in Political Thought*, Cambridge, MA: MIT Press.

Winner, L. (1986) *The Whale and the Reactor. A Search for Limits in an Age of High Technology*, Chicago, IL: University of Chicago Press.

Winner, L. (ed.) (1992) *Democracy in a Technological Society*, Deventer: Kluwer.

10

SYMMETRIES

Between pessimists and optimists

Summary

The technological sciences, just like the natural sciences, are classified under the exact disciplines. These sciences are supposed to be 'harder' or 'more objective' than the social sciences and the humanities. However, in this chapter we demonstrate that science and technology are likewise social realities. Technology is a product of human cooperation, therefore there are always social factors involved as well. The awareness of this social aspect of science and technology has increased markedly as a result of recent studies. In this chapter this development is sketched by referring to the *empirical turn* in the philosophy of technology. First of all we deal with two classical philosophers of technology: Heidegger and Ellul. Then we cross the bridge to the actual, more empirically oriented philosophy of technology. Via Winner we get to Latour and the SCOT approach.

We begin with a situation in everyday life. A scientist goes to her study, switches on the light, adjusts the heating, presses the 'on' button of her computer, makes herself a mug of coffee with an apparatus, puts a CD of *Symphonie Fantastique* by Berlioz into the CD player, opens a file and starts rattling on her keyboard. This scene which seems not worth reporting, demonstrates how great a role technology has in our lives. Even in a relatively simply furnished study the number of technological apparatuses can hardly be enumerated on the fingers of one hand, let alone in the whole house.

Let us look at the scene above in somewhat more old-fashioned terms: a scientist goes to his study, draws open the curtains so that the daylight falls on his desk, takes up a pen and paper, and starts writing. On the fire in the hearth a small kettle of water is simmering and now and again he pours some water into the coffee filter. In this description of the situation there also are technological elements but they are much less obvious. When we compare these two examples it seems that technological developments have at most made life somewhat easier and more pleasant. One can make coffee in the old-fashioned way, but one can also use an apparatus. Or one can write a book with a pen and paper, but one can also do this with a computer. Or one can work by daylight, but one can also switch on the electric light.

But is technology really as innocent as depicted above? Does technology only make life a little more comfortable and pleasant? Can one choose to make use of it or not? Or does technology play a much greater part in our lives? What would happen in our country if there was a power failure for a few days? Or if electronic transactions in the financial world could not be done for weeks on end? If modern telecommunications no longer functioned? This type of question worries people. Yet since the beginning of the nineteenth century these questions have been asked with increasing urgency. This disquiet concerns the technological development which in the modern world has a complete dynamics of its own. People acquire ever more scientific insight into the natural reality. These insights are translated into technological applications at an increasing rate. Human beings are constantly looking for new solutions, new technologies or new medicines to master problems as yet unsolved or illnesses which are still untreatable. Technology more and more develops into an autonomous system which is not easy to manage. Although technology has not been a constant theme in philosophy for a long while, one can nowadays speak of a tradition of philosophy of technology. It is striking that this tradition from the start has been critical towards modern technology and has been guided by a degree of mistrust towards its development.

10.1 Classical philosophy of technology

In the classical philosophy of technology negative and critical utterances about technology abound. In contrast to the instrumental view that sees technology as an 'innocent aid in a specific practice' (Smits, 2001), the critical philosophers of technology point out that technology radically changes human beings and society. Technology is not merely a neutral aid, but determines the way in which people look at reality. We find ourselves more and more in a world in which technology has become the dominant system affecting our entire lifestyle. We are living in a technological biotope. Why are the voices of traditional philosophy of technology so sceptical towards the technological culture and why do they resist surrender to the seemingly inevitable dominance of technology?

10.1.1 Technology pessimists and technology optimists

In his doctoral thesis *Techniek en Toekomst* (1972) (*Technology and the Future* (2009)) Egbert Schuurman differentiates between two important tendencies in philosophy of technology. The *positivists* and the *transcendentalists*. As the name indicates the *positivists* feel positively towards technological development.[1] Schuurman typifies their view as follows:

> The positivists derive their view from technological development, which they regard as the motor of cultural progress. The significance of this conviction for futurology is that the positivists place technology at the centre of human thought. They give the advance of technology such a central, all-controlling place that technology becomes the model for controlling even human relations, society, and the future. The utilization of technological possibilities is made decisive for the development of society. The future as a technological future is regarded by the positivists as a meaningful future.
>
> *(Schuurman, 2009: 205, 206)*

The positivists see little evil in technology. Progress and technological development are more or less synonymous and therefore they do not feel threatened by technology as a dominant factor in modern society.

Against these positivists Schuurman sets the *transcendentalists*, who 'all share a conscious concern that proper consideration be given to transcendental directedness of all experience, including the experience of technological factuality'. They 'strive to understand the origin, development, and future of technology. In their critical review of technology, they appear to set themselves against any worldview that is both natural-scientific and technological.' Schuurman summarises their view as follows: 'the transcendentalists oppose philosophical technicism, that is, the view that absolutises technological-scientific thought and results in practice in a technocracy bent on subjecting everything to the power of technology' (Schuurman, 2009: 61). But what is the meaning of this 'proper consideration to be given to transcendental directedness of all experience'? Contrary to the positivists who lock themselves up in a 'worldview that is both natural-scientific and technological' the transcendentalists state that experience relates to a reality that 'goes beyond' (transcends). According to them the world of the positivists is an embedded world. They have a 'worldview' without realising it themselves. They confuse their 'worldview' with the real world. That also applies in general to modern human beings: they see the world from a natural-scientific-technological perspective.

It is questionable whether one sees the world as it is. Does one not always observe the world from a specific perspective? And does the positivist worldview offer a productive approach? The transcendentalists think it does not. Although they have few objections against this worldview in itself, they are of the opinion that people forget that they look at the world from a certain – positivist – perspective while believing that they see the world as it really is. According to the transcendentalists the mistake made by the positivists is that they forget about their own perspective and act as if it is the only way of looking at the world. And it is exactly when one is no longer conscious of one's own one-sidedness, that one violates reality.

10.1.2 Heidegger

Martin Heidegger is a philosopher who looks at technology from a transcendentalist perspective. Without doubt he is one of the most important philosophers of the twentieth century; at the same time he is one of the most controversial. At the beginning of National Socialism in Germany Heidegger misjudged the aspirations of this movement and during his period of vice-chancellorship at the University of Freiburg was a member of the Nazi party.

The philosophy of Heidegger can be typified as a philosophical view of the way modern human beings are alienated from themselves. One of the central insights of Heidegger's is that the history of philosophy up to the present has been a history of 'forgetting', forgetting about the question of the meaning of 'being'. Philosophy has been searching for as long as people can remember for 'the last cause of everything that is'. The search for the last point of departure or Archimedean point of thought is sometimes called *metaphysics*. This means that philosophy is in search of what is hidden 'beyond' ('meta') the 'natural phenomena' ('physics'). Heidegger states that for the greater part of the history of philosophy these *metaphysical* questions were answered in a particular, and in his view incorrect, manner. Over and over an attempt was made to answer the search for the 'last cause of everything that exists' by looking for a being that is the foundation of the existence of all other beings. This being, according to Heidegger, is often denoted as god or as the divine. Hence Heidegger also calls the history of metaphysics *onto-theology*.

For Heidegger there is something still more fundamental than 'a being that is the foundation of the existence of all other beings'. For all beings, including the last-mentioned, have in common that they 'are' or 'exist'. This 'being' is forgotten in the quest for the last being. Hence, according to Heidegger, history of philosophy suffers from *Seinsvergessenheit* (forgetting about being). This is the pitfall that Heidegger wishes to avoid. He wants to give central position to the quest for the meaning of being. And this he does in the most literal meaning: he asks himself what we say about things when we state that they are. What is the meaning of the word 'being' in this context? What does it mean to say about something that it is?

Posing the question, however, is something different from answering it. Heidegger first works out his answer in a bulky volume: *Sein und Zeit* (*Being and Time*) (1927). When the book was first printed he put on the title page that this was only Part 1. Part 2 would never see the light. But this note showed clearly that Heidegger believed that with this book he had not adequately answered the question he himself had put. So there would have to be a sequel. And in a sense one could defend him by saying that a great part of his later work was an effort to master the issues of this never-written second part. In *Sein und Zeit* Heidegger is searching for the 'beginning' in particular: How exactly should one answer the general and abstract question about the meaning of being and where should one start? According to him one first has to know the origin of this question: Who asks it? The search for the origin of the question soon leads to a specific being, the one who poses this question, namely the human being. To denote this specific being Heidegger introduces the word 'Dasein' (there-being). It is the 'Dasein' who distinguishes itself as being from all other beings by asking the question about the meaning of being. In contrast to a stone or a block of wood – beings who simply are – the 'Dasein' is in a relationship with its own being. So the peculiarity of a human being is that in its being it has a relationship with its own being and tries to form an opinion on it.

In order to give an answer to the question about the meaning of being, Heidegger is of the opinion that one should first create more clarity in the particular way in which the 'Dasein' – the human being – exists in the world. Therefore Heidegger devotes a long dissertation to the way of existence of the human being. His work is too extensive to go into in detail. We merely highlight some main lines. The most striking aspect to the human way of existence, according to Heidegger is that the human being is a mixture of passivity and activity. He begins with the point that the human being 'has always been' in the world. When we think about the human way of existence, we are already there and living our life. We always think from a place in the world. A world that we have labelled or interpreted in many ways: a table is a table and bread is bread. Our world has 'always been' organised and is familiar to us to a certain extent. We live there and feel at home. We have not chosen this 'being in the world': there is no alternative. We have been 'thrown' into it. But this is only part of the story. For we can very well organise our existence and make many choices. People do not live passively, but *lead* an existence. Even if nobody chose to be born, one can still to some extent live according to one's own plan or 'design' ('Geworfenheit': 'being thrown in' over against 'Entwurf': 'design'). The life of a human being is not typified by a flat 'being' but to a great extent by a 'can-be': a being that has to be given form. This given is intrinsically connected with the fact that for Heidegger a human being is at bottom an interpreting being: a being that ascribes sense and meaning to beings. Man is responsible and thereby is compelled to give account.

Now the way in which a human being goes out to meet the being, depends on time. Being has a history. As we have said, philosophy at the moment finds itself in a time of forgetting about being. Instead of thinking the beings from the being, the being is constantly approached from the beings: from a highest being. In modern times (after Nietzsche's statement that God is dead) it has increasingly been the (human) subject who claimed this place for him/herself: the autonomous subject that deals with reality the way s/he chooses. People consider reality more and more as a collection of beings that are at the disposal of the subject who determines itself. This is exactly the view that to a great extent characterises the scientific-technological, positivist thinking of modern times.

In later writings Heidegger is very critical of technology. He then seems to reconsider his view in *Sein und Zeit.* In two texts which were later published in one volume, *Die Technik und die Kehre* (*Technology and the Turn*) (1962), Heidegger attempts to think even more radically, departing from being.[2] According to him in *Sein und Zeit* he was still too much bogged down in a *Daseinsanalyse* (an analysis of the human being). Heidegger speaks about a *Kehre,* a turning. Still more so than in his earlier work he now departs from being. He no longer thinks of it as static but as history. Being is an event. 'Being' consequently is a verb, which Heidegger expresses in more specific terms by speaking of *Wesen* (which means essence, but is used by Heidegger in its original German sense as a verb). He conjugates this word like a verb. The being as event is a continuous covering and uncovering: the being shows itself in a specific way, but by the way in which it shows itself (uncovering) it can in other respects simultaneously be hidden from sight (covering). With the term 'uncovering' Heidegger declares that truth is a process. He speaks about *alètheia*, uncovering. Truth is a process of 'coming to light'.

Heidegger considers the issue of technology as a quest for the *essence* (Wesen) of technology. This essence should be sought in the way in which we go out to meet nature. Technology is a form of 'uncovering'. In modern technology this 'uncovering' is not the classical bringing-to-the-fore (*Hervorbringen*) in the sense of *poièsis* (the ancient Greek word for 'making' or 'creating') but a compelling-to-the-fore (*Herausfordern*). The uncovering of modern technology therefore is more violent: the world is dealt with in the way man chooses. Modern technology consequently is not simply a rational input of means to reach specific goals, but is a way of compelling-to-the-fore beings in a technologically determined context. Human beings cannot freely choose their place in this, but are themelves controlled by this way of approaching reality. This uncovering, which is typical of modern technology, Heidegger calls *bestellen*. That which is brought about by *bestellen* he calls *Bestand*. With a third word – *Gestell* – Heidegger wants to make it clear that the *bestellen* is blind to itself. This also applies to the way in which modern human beings relate to reality. People cannot choose freely for or against technology. Modern human beings have been sentenced to a technological way of dealing with beings. Indeed, the *Gestell* was brought about by human beings, but they are no longer free from it. Thereby *Gestell* has become *Geschick* (Fate).

Heidegger's perspective is a sombre one. Modern technology as *Gestell* means nothing less than that modern people have been sentenced to *bestellen*. The only other way out is to become conscious of the *Gestell*. However, Heidegger remained vague on where this could lead. At the end of his reflection on technology Heidegger quotes the vaguely promising words of the poet Hölderlin:

> But where danger lurks, grows
> the chance of rescue.

Sometimes one gets the impression that he sees no other solution than a supernatural intervention. For instance in the famous interview in *Der Spiegel* he says that only God can still save us and that philosophy and poetry can only help to prepare us for his advent or teach us to cope with his absence.[3]

From the above interpretation of Heidegger's thinking it is clear that to him technology is no neutral instrumentation, but represents a fundamental way of being-in-the-world. Modern human beings interpret the reality in which they live in a technological way: as an object that can be dealt with by means of technological instruments. This way of being-in-the-world is one-sided and alienating. But most significant is that the human being plays only a minor role in the event of being in which technology is important.

10.1.3 Ellul

A second philosopher who is representative of the traditional philosophy of technology is the French philosopher Jacques Ellul. He is not strictly a philosopher, but a jurist, sociologist and theologian. Consequently his approach to technology can be called cultural-sociological rather than philosophical. Yet his theory is so comprehensive that many people with a cultural-philosophical approach easily recognise their own views in it.

The foundation of his contribution to the debate on philosophy of technology is laid out by Ellul in *The Technological Society* (1964). In this book he shows how dominant technology has become in our society. By technology Ellul means more than only apparatus and machines. It is true that in our modern society technology has gained an enormous weight of influence by the use of machines and the mechanisation that follows from this, however, its influence extends much further than the sphere of the factory. Technology has also penetrated to the sphere of academia, the economy, organisation theory, health services and communication. Actually technology is nothing but the means or the collection of means that people employ when practising activities. So, for instance, there is a technology of organising, of entering into relationships, or of swimming (Ellul, 1964: 22). Winning a swimming contest not only depends on the strength of a swimmer, but also on the question whether s/he has a good technique. In the sport of swimming the development of such a technique is subject to constant analysis and *monitoring* in order continually to raise the swimmer's achievement. But according to Ellul we do meet technology in its purest form in the machine.

Ellul rebels against the view that technology is applied science (Ellul, 1964: 7–11). Technology was there before there was science and science needs the products of technology to develop itself. Think for instance of lenses in microscopic research, Erlenmeyer flasks in chemical experiments or barometers when determining atmospheric pressure. Without technology no science would be possible. Therefore the dominant position of technology in our society is not merely the consequence of the scientific revolution at the beginning of the modern era. However, there is a huge difference between traditional and modern technology. It is true that hunters and gatherers do have their methods and that a blacksmith handles all kinds of technologies but in these cases one could hardly speak about a conscious and rational analysis of their particular method of working. Modern technology, however, abstracts from the complexity of the given situation and searches for the best method of realising its goal. In this way there develops an abstract system of technological knowledge and expertise which to a great extent is unrelated to the concrete reality as experienced by people. Traditionally technology conforms to the life of the human being, now the human being has to adapt to the technological system.

What distinguishes modern technology according to Ellul is that there is a conscious and rational search for the most efficient method. He distinguishes six features of modern technology: (1) the automatism of the technological choice,

(2) the fact that it reinforces itself, (3) its monism, (4) the interconnectedness of different technologies, (5) its universalism and (6) its autonomy. An important feature of modern technology is the *automatism of the technological choice*. Technology manifests itself as the most efficient way of reaching a goal and thereby excludes other alternatives as being less rational. In this way an automatism comes about: technology enforces its own use. A closely connected characteristic is the *reinforcement of itself* by modern technology. In modern technology there is an intrinsic driving force towards completion and perfection. Technological solutions can always be made better and more efficient so that there never is a definitive solution. Besides, the various technologies flow over into one another more and more and in this way contribute to the solution of problems in other fields. This also is an automatism: technological solutions are sought for problems caused by technology.

A third feature is that modern technology is *monistic:* it has the character of a system. Technology gives form to acting, and thus the possibility is lost for a human being to decide between good and bad applications. For when accepting a certain technology one also has to accept everything this technology inherently entails. Driving a motor car is inevitably connected with the fact that it generates pollution and that every year people die in accidents. And no one choosing nuclear power can exclude the possibility that the knowledge of this could also be used for warfare. However, not only does one great monolithic system come into being, there also ensues an *essential connection* between various fields of application. The introduction of the steam engine leads to the industrial factory which increasingly complicates the production process, making the development of technologies for managing such an organisation necessary. This subsequently leads to various changes in the field of town and country planning and the economy, as well as to numerous new commercial activities, changes in the structure of education, etc. The result is an intricate system of relations in which one form of technology calls for the next. By the *universalism* of modern technology, the fifth feature, Ellul then means two things. In the first instance he contrasts it to the local character of traditional technology: modern technology is geographically universal. But in a qualitative sense, too, modern technology is universal: technology looks for a universal language, independent of the subjective experience of human beings. This language is independent of place and time, and takes no notice of the borders between different cultures. Technology develops into a totalitarian power that penetrates, destroys and replaces traditional cultures.

Ellul gained a permanent place in the debate on technology with his sixth characteristic: *autonomy*. By this he means that modern technology is comprehensive and develops according to its own mechanisms, laws and norms. Modern human beings can no longer think outside technology. Technology has become independent with regard to social, political and economic aspects of life. Neither do world-viewish or normative values exert any influence on its development. Technology is not merely a toolbox full of neutral instruments which human beings can choose to use or not to use. Technology determines the whole culture in which we live and forms the frame of interpretation from which we think. It has become the natural and no longer criticised point of departure of modern human life. We can only free ourselves from the grip of technology if we have the nerve to stay free and independent from technology. We have to free ourselves from the drive to manage and control. It is exactly by attempting to get control of the power of technology by developing contra powers that we are once more encapsulated by it.

10.1.4 Winner: the transition to current philosophy of technology

The postulate made by Ellul that technology has become autonomous has raised a lot of dust. This is probably more due to the recognition than to the denial it elicits. For is it not true that much can be said in favour of the idea that the development of modern technology is autonomous to a great extent? Cannot one simply state that technological possibilities that become available are applied per definition, however grave the moral objections may be? Still, many objections have been raised against this statement. For no matter how difficult it may seem to influence technological developments, there are always human decisions involved. And however taken up people are with a technological way of looking at things, at the same time there always are critical voices. One cannot speak of a complete naturalness of technological thinking and the technological culture.

Although he has much appreciation of Ellul's philosophy, the American political philosopher Langdon Winner distances himself from it. He does consider it important to see, as Ellul does, that human beings lose control of technology, but this does not happen outside human beings. In his book *Autonomous Technology – Technics-out-of-control as a Theme in Political Thought* (1977) he visualises the political character of technology. He rejects the liberal idea that sees technology as a neutral instrument having a few negative side effects at most. But Winner is not a defeatist philosopher either. In his view Ellul who says a lot about modern technology that is worthy of consideration neglects the political dimension of technology. But what exactly does he understand as the political dimension? In the now famous article 'Do artifacts have politics?' from the volume *The Whale and the Reactor* (1986) Winner shows that the one-sided, monolithic account about the autonomy of technology is inadequate: the power of technology lodges in concrete technologies and the place they take in our lives. The issue is not 'technology' in the abstract sense but the concrete power and influence exerted by technologies on the actions of people and their mutual relationships. This exertion of power is promoted by two mechanisms: the *technological imperative* and *reversed adaptation.* The technological imperative entails that technology creates human needs which can only be met by technology. Reversed adaptation means that technology changes people – their convictions, habits, institutions – so that they are better adapted to the demands of technology (Winner, 1977: 208 et seq.).

Winner emphasises the fact that the 'autonomous' power of 'the' technology is concrete power exerted by technologies because of human choices and decisions.

This power is mostly the consequence of the fact that people forget 'to educate' technology. Winner explains this by referring to the narrative of Mary Shelley about Frankenstein. Victor Frankenstein manufactures his own monster but runs away in fear the moment his creation comes to life (Winner, 1977: 306 et seq.). Winner does not refer to this story to show what a monster modern technology is but to point out that technology becomes a monster as soon as people no longer give attention to its functioning and its conditions. The people then forget to incorporate their inventions in culture and society.

According to Winner we should not consider technology as an autonomous 'opposition' to our own freedom of acting and deciding but as a product of human creativity that has to remain under political control permanently. The power of technology should be managed democratically. Langdon Winner can be seen as a transitional figure between traditional and current philosophy of technology. What differentiates him from traditional philosophy of technology is that he does not speak about 'the' technology as an abstract and autonomous 'opposition' of human subjectivity (like Heidegger and Ellul). On the contrary, he pays attention to the concrete political functioning of techniques and technologies. Something that differentiates him from the current philosophy of technology and in particular from *social constructivism* is that he makes more normative statements about the development of technology. He recognises the merits of this philosophy like its proposal to open up the black box of various technologies but at the same time finds that they are hardly interested in normative or evaluative statements about the contribution of technology to the quality of human life and society (Winner, 1993).

10.2 The empirical turn

The social constructivism mentioned above has contributed significantly to the *empirical turn* in philosophy of technology.[4] This reversal entails that less is spoken about 'technology' and that more attention is paid to the specific development of concrete techniques and technologies: for example the rise of the aeroplane, television or a certain medical treatment technology. And instead of showing how technology as a kind of hostile power threatens culture and society from outside, the new batch of philosophers of technology demonstrate how technology and society 'create' and influence each other. In order to be able to place the empirical turn and the role of the social constructivists in it properly, however, it is important first to take a brief look at the changes in the view of science and rationality that took place during the twentieth century.

10.2.1 Developments in the philosophy of science

During the first half of the twentieth century a decisive debate was raging in philosophy of science between the logical positivists (also called the *Wiener Kreis* or Vienna Circle) and the philosopher Karl Popper. The issue was how one could justify scientific hypotheses (statements on the state of affairs in reality). The logical positivists were of the opinion that justification of hypotheses was a matter of *induction*. By repeated observation of the unvarying state of affairs one can *generalise* this state of affairs. When one observes that a certain pattern keeps repeating itself, one can make a statement like the following: y applies to all x. Strictly speaking one can never be absolutely certain of this since it cannot be excluded that someday an observation may be done in which the statement is refuted. So in this *inductive* method there always is a logical leap to some degree. The logical positivists wanted to determine more definitely under which circumstances this leap was justified.

Then Popper published his book *The Logic of Scientific Discovery* (1959). According to him the logical positivist approach had to lead to general statements having as little risk as possible of being refuted. However, s/he who attempts to formulate such statements will soon find that s/he is constantly trying to avoid contradiction and thus tends to remain as general and indefinite as possible. Thus induction leads to loss of information. Therefore Popper proposes that scientists should not so much pursue the maximum factual confirmation of their statements but a maximisation of the refutability of their statements. A scientific hypothesis would have to be formulated in such a way that it is immediately clear which observation would be able to refute it. Such an approach leads to the formulation of clear-cut, informative hypotheses. By this approach Popper thought he could entice scientists to have a more intrepid attitude. They would be able to formulate highly speculative theories on condition that they *deduce* refutable statements from them which would give direct insight into the conditions and observations under which they would be true or false. If the observations refuted such a statement it would be a serious argument against the plausibility of the underlying theory. Popper's view is sometimes called *deductivism*.

Actually Popper tries to overcome the logical leap of induction without having to let go of the intention of logical positivism. For he is also of the opinion that refuting or confirming hypotheses on the grounds of empirical material guarantees the rationality of the development of science. So Popper is also concerned with a rational development of science.

One important critic of Karl Popper is Thomas Kuhn. His best known book is *The Structure of Scientific Revolutions* (1962) which has exerted enormous influence on philosophy of science. With him a new phase in reflection on science took off. Kuhn investigated to what extent 'progress' in the history of science really was directed by rational processes as Popper represented them. In practice the research revealed that hypotheses which are emphatically refuted by observation mostly do not lead to a rejection of the underlying theory. For there is no alternative available. Because refuting a hypothesis does not mean that the observed phenomena can directly be interpreted in a different framework. Therefore such observations generally lead only to an adaptation of the theory so that the observed phenomena can still be explained within the existing framework. The consensus on the underlying theoretical perspective often is so hard to break out of that the theory is adapted rather than rejected. Such a phase in scientific development Kuhn calls 'normal science'. Scientists then occupy themselves in particular with fleshing out and refining a certain theoretical perspective. Kuhn calls this a paradigm: a complex entity of already acquired theoretical knowledge, philosophical suppositions, values that define and direct the research, methods of measuring and typical examples that serve as models of how one should find solutions for problems within a certain field of science.

A scientist adopts a paradigm during the time s/he is being moulded as a scientist. This to a great extent determines the direction in which s/he searches for solutions to problems. This is so determinative for the way in which scientists approach their field of research that generally they are not easily persuaded to abandon it. Even when there are numerous observations that seem to refute the theory they will keep clinging to it. Besides in practice it often is very difficult to decide precisely which part of the paradigm is being refuted by the contradictory observations. Often such observations (which Kuhn calls *anomalies*) can be brought into line with the theory by intentional adjustments.

It is only when a paradigm is tottering that we enter into another phase in scientific development: the *scientific revolution*. In the phase of *normal science* every anomaly is solved by an adjustment to the theoretical structure, but in the phase of a *scientific revolution* this is no longer successful. Then a situation ensues in which the theoretical perspective can no longer bear the weight of the anomalies and threatens to fall apart. Obviously even then many scientists attempt to fit anomalies into the old paradigm. However a small group of scientists will start looking for alternatives and will try to design new paradigms. In this phase of the development of a field of science there then occurs a conflict of *paradigms* between the old and the new paradigms. If the old paradigm promptly loses its authority it could lead to a crisis in the science. It is hard to discern which voices still have authority. A new situation comes into being in which it is not clear what will become the new, authoritative paradigm.

Kuhn's study means a significant turn in the research on science. The development of science is no longer seen as a purely rational process. There also are numerous other factors playing a part. In other words it becomes possible to regard science as a social practice. Popper wanted to demonstrate with his theory that the process of discovery by which people reach their scientific hypotheses is often whimsical and far from rational, but that the testing and justification of the theories are a rational process. This opinion is contradicted by Kuhn who pleads for a more historic study of the way in which paradigms originate in science and are subsequently reviewed. There are processes on two levels which are not unequivocally rational. The way in which consensus comes into being on the authoritative paradigm of '*normal science*' is irrational. This also applies to the way in which this authoritative paradigm is subsequently disputed in the phase of '*scientific revolution*'. Indeed the consensus on the authoritative paradigms is not so rational at all. Social, cultural and political factors play important parts. The same applies to the way in which this consensus is subsequently disputed.

With his book Kuhn made a breakthrough. The development of science could now also be studied in a historical and social-scientific way. There followed a boom of social-scientific and historical studies confirming that the development of science is not an unequivocal, straightforward rational process. The consequences go without saying. If the progress in science is determined to a great extent by social, cultural and political factors, what about the truth and the rational validity of scientific theories? Is a particular theory considered true because it is socially acceptable, culturally desirable or politically enforced? Or is a theory true because it tallies with the phenomena it explains?

One of the most extreme viewpoints discussed here is the so-called Strong Programme. According to the supporters of this paradigm variables like *truth* and *rational validity* are determined by social, cultural and political factors. They therefore are the resultants of non-rational processes. However improbable this viewpoint may look at first sight, research has shown that it can actually be made plausible. But does the Strong Programme tell the whole story? This question in itself reveals a problem: if this approach is correct it enfeebles itself. If research shows that the claims of the Strong Programme are justified (and that its theory therefore is *true*) then this theory is no more than a social or political phenomenon.

10.2.2 Looking back

In classical philosophy of technology one sees, apart from the positivist view in which technology is a neutral means of realising goals in a rational way, a cultural-critical approach in which pessimism dominates. The examples of Heidegger and Ellul present a global and abstract view of technology. However, a science scholar could say: Heidegger and Ellul were not hampered by any knowledge of technology. They are scholars of the humanities making statements about a scientific reality and that cannot fail to result in a catastrophe. They are talking about something of which they have absolutely no understanding. This opinion again is regarded by the scholars of the humanities as a compliment, for in their opinion it shows that they have not been contaminated by a natural-scientific way of thinking.

The philosophers of technology working in the wake of Thomas Kuhn have, however, broken out of this futile discussion between scholars of natural science and those of the humanities. They argue that one has to have an understanding of technology to be able to evaluate it, but that it need not be the *same understanding* as that of the scientists who developed the technology. It seems as if technology is based on nothing other than *hard* facts, but on closer inspection they prove to be not all that *hard*. Technology, too, came into being in a less purely rational manner than one would suppose. Then also the much *softer* social and political world about which boffins on technology often feel ill at ease sometimes prove to be much more influenced by technology than one would imagine possible at first glance. So it seems that there is enough reason to bring the natural and technological sciences and the social sciences much closer together. We should look at how the interaction between technology and society in fact takes place. From this we can learn lessons for our engagement with technology. A similar, empirical approach to scientific and technological development is found among the researchers and philosophers who are representative of what is sometimes called the empirical turn.

10.3 Constructivism: Latour

10.3.1 Symmetry

The French philosopher of science Bruno Latour as a scholar of social sciences has his own peculiar understanding of technology. As a cultural anthropologist Latour became aware of the fact that there is a significant basic presupposition to the cultural anthropological research that he did. All research is based on a fundamental borderline between his own culture and that of the tribes or nations investigated. Westerners with their scientific worldview set the norm for a normal, rational way of living, while the investigated people from other parts of the world are the people whose way of life is deviant and needs to be explained. Consequently the Western approach is regarded as 'normal' while the 'deviating', 'hard to explain' conduct of the non-Western people is seen as irrational (Latour, 1993: 16 et seq.). This presupposition is disputed by Latour. He raises the question whether it is not perhaps time to restore the symmetry. Should one not likewise explain the 'normal' and 'rational' conduct of Westerners? And is it not obvious then to begin with the segments of the Western world that most clearly express Western rationality: namely science and technology?

A second perspective of Latour in his studies of science and technology concerns scientific failures and successes. In the history of science the scientific failures are often interesting subjects for research. How is it possible that the particular scientists had it all wrong? On the other hand the successes need no explanation: for intelligence and rationality prove themselves. Formulated differently: success proves the rationality of the chosen solution. A theory has success because it is true. False points of departure and suppositions on the other hand automatically lead to failure. But is this really the case?

10.3.2 An example: Professors Buck and Goudsmit

In the 1990s a Dutch research group announced that they had made important progress in their research into the possibilities of fighting the AIDS-virus. Professor Buck from Eindhoven, a chemist, played a central role in this event. He claimed to have developed a method of 'sealing in' the AIDS-virus with the help of modified DNA so that it would no longer be able to work harm in the body of the carrier. Moreover the research of Professor Goudsmit (University of Amsterdam) proved that the DNA developed by Buck's research group actually did work in practice.

In the first instance the optimistic statements made by Buck on the Dutch television news on 12 April 1990 presented no problem. He confirmed that his discovery could within a few years mean an end to the hazard of AIDS (Hagendijk and Meeus, 1993). Initially hardly any critical voices were heard. Here and there an individual warned against too sanguine expectations since there is a long way from scientific research to full-scale production of a medicine. However, there were a number of concerned researchers close to Buck who doubted the soundness of the DNA produced by Buck and his team. Although Goudsmit initially remained solidly behind his colleague Buck, the situation changed drastically with a newscast on 30 August 1990. In this it was stated that the invention by Buck did not in the least constitute a breakthrough in the AIDS research and that the presence of the material produced by Buck could not be demonstrated in the sample that was tested (Hagendijk and Meeus, 1993: 404). Hereby Buck's position became untenable and it was not long before he resigned from the Technological University of Eindhoven. Afterwards hardly anything more was heard about the research which held so much promise for AIDS victims.

This is a typical example of a scientific failure. Yet the article was published in the authoritative journal *Science*. And many people were impressed by the results of his research. Following the dramatic developments several months after the publication of that article the research came to an end overnight and its basic assumptions aroused no further interest. Does this make Buck a bad scientist? Was his theory a bad theory? These are difficult questions to answer. Although Buck was at first presented in the press as a very reliable researcher of great repute, after the 30 August the image of him changed to one of a tyrannical leader of his institute, who had consistently presented the research to make it look better than it was. However, at issue are the images that were created. Who was the real Professor Buck behind the images created of him? After 30 August his colleague Goudsmit acquired the reputation of someone who had been misled by his colleague and who had naively shared his optimism. That, too, is an image and leaves unanswered the question how it was possible that he had come to positive results based on material supplied by Buck that seems not to have existed. Consequently the research by both professors lost all support.

The tragic example of Buck raises the question exactly how the success of scientific theories can be explained. Was Buck's problem that he had embraced an incorrect theory to gain fame? Or had he got hold of a fine theory but forfeited his chances of investigating this more closely by clumsy working methods and conflicts with his colleagues? Is it possible that he caused a good and useful theory to land on the scrap heap of history because he wanted to attain success too easily? Or was his explanatory model theoretically brilliant but empirically worthless? Or did he become a victim of the jealousy or malice of his colleagues?

According to Latour it is important that the success and failure of scientific research be explained symmetrically. It is not simply the case that a theory is successful because it is true or tallies with the facts, while other theories are rejected because they are contradicted by the facts. Rather, a theory is true because it has success. A theory to a great extent determines how a research community sees and explains phenomena. It constantly confirms such a community in their predictions. In this way theories (and technological solutions to specific problems) can become a *black box:* obvious truths which are no longer 'opened up' and no longer exposed to critical investigation, because doing so would cost the person too much, or be detrimental to their authority or reputation. For, once it is a matter of a *black box,* it becomes very difficult to come up with alternative explanations for the same phenomena.

10.3.3 Society as well as nature

According to Latour to explain the success of a certain theory, one should not only look at the 'nature' that is explained in the theory. One should also look at the 'society' in which this theory can become successful. Science is not only a logical rational manner of looking at reality but also a social practice. A professor who needs money for his/her research and does the round of the moneylenders uses more means than merely rational argumentation. What is more, many of the moneylenders cannot follow or evaluate her/his rational arguments. They go by the rumours they have heard about the specific professor ('It seems that s/he is brilliant in her/his field') or by the expectations the professor creates about her/his research ('I hold it for possible that with my discovery AIDS could be history within a few years'). Psychology and sociology, charm and sex appeal, politics and power, beautiful suits and smooth tongues: all these play their part in the development of science and technology and sometimes the intrinsic value of a theory cannot be distinguished from all the other factors that determine the success or failure of a certain theory or technology.

It is the supporters of the so-called Strong Programme who draw the most radical consequences from the above. In their opinion the success of theories has nothing to do with truth. 'Truth' is an effect of social and political relationships and not a feature of a theory. So, those theories are true which, for instance, serve the interest of those in power or which impart a sense of security to certain groups of people. On this point Latour dissociates himself from the Strong Programme. He states:

> Yes, the scientific facts are indeed constructed, but they cannot be reduced to the social dimension because this dimension is populated by objects mobilized to construct it. Yes, these objects are real but they look so much like social actors that they cannot be reduced to the reality 'out there' invented by the philosophers of science.
>
> *(Latour, 1993: 6)*

There always is the issue of 'coproduction' or of 'double construction'. The modified DNA of Buck is not only an empirically observable phenomenon, it is in the first place a theoretical construct. At the same time it is more than this, for the modified DNA was made with actual existing apparatus and carefully (or not so carefully) demonstrated. Both sides, concrete apparatus and objective measurements on the one side, and social and mental constructs on the other, are always present simultaneously. The result of the measurement shows that something which on the basis of theoretical formulation is supposed to exist, actually 'is' there. Simultaneously this measuring of the phenomenon is itself based on theoretical suppositions. 'Facts' and 'social constellations' in which these facts are demonstrated do not exist apart from one another.

According to Latour science and technology can neither be explained solely in terms of 'nature' (objective facts) nor solely in terms of 'society' (social relationships, political power). One could rather say it is a mixture of both factors, a *coproduction* of society and nature. A scientific theory or a technological solution always is a *construct.* However, these constructs do not only have a social origin. In the case of technologies this can be seen even better than with science. A technological artefact is a construct translated into synthetic material, metal or ceramic elements. These things work and play a part in the total network of actors which lead to the situation where a certain technology can assume a place in society.

10.3.4 Translations

The scientific and technological constructs studied by Latour therefore always have a *hard* (nature) and a *soft* (society) side. In a parable about a hotel key Latour explains how they are connected. A hotel owner has a constant problem with his guests forgetting to hand in their keys on leaving the hotel. His first solution is to put a notice at the reception desk saying: 'Please hand in your room key at reception.' This leads to a slight improvement but still too many guests forget to hand in their keys. He then gets the idea of printing at the bottom of the receipt a picture of the hotel key together with the sentence 'And I … I remain here!' This sentence reminds both the reception clerk and the hotel guest of the handing in of the key. With these measures the number of people involved (actors) has been increased. There now also is a printer that has to print the picture and text on each receipt. After introducing this measure there are, however, still guests who do not hand in their keys. The hotel owner invents the next solution: he gets a metal works company to attach a cast iron ball to each key. In order to realise this the hotel owner himself also has to take the necessary steps; for instance, he has to have his key cupboard fitted with heavier hooks on which to hang the keys. After this measure almost all hotel guests hand in their keys at departure. The number of actors or allies has been extended further, amongst others by the supplier of the cast iron balls.

This parable shows that someone who wants to reach a certain goal (in this case the hotel owner) will have to *translate* their interest into the interests of as many others as possible. The interest of the hotel owner to get back the key has been translated with the help of the cast iron ball into the interests of his guest to travel as light as possible. To make this translation he has set to work actors – human and non-human: the metal works company who manufactures the cast iron balls and the ball itself as a kind of moral weight in the pockets of the all too lazy hotel guests. It is precisely this last insight that makes Latour's view so unique: even non-human actors are a part of the social network of allies that create facts, make theories successful or allow technologies to succeed. Each component of the network can help but can also be fatal. DNA is not only a theoretical concept in the brains of scientists but also something that must give a sign of life in an experiment to convince numerous allies permanently of its existence. Not only should the forum of scientists believe in the claims of scientists about their part of reality, the instruments in the laboratory also have to cooperate.

10.3.5 The SCOT approach

In dealing with Latour we have looked in particular at his view of science. However, Latour gives just as much attention to technology. Therefore the consequences of the empirical turn are not limited to philosophy of science. It also is applicable to the field of research into technology and here it has led to a peculiar approach which has become generally known by the acronym SCOT, for *Social Construction of Technology*. This tendency is in some respects inspired by Latour, but also by the analytic tradition in philosophy of technology. The point of departure of this tendency is likewise that technological artefacts and systems, just like scientific knowledge, are social constructs. One of the founders of this tendency is the Dutchman Wiebe Bijker. He became known for his analysis of the history of the bicycle (Bijker, 1995). According to him a bicycle is exactly what its users consider it to be. In the first years of the bicycle they considered it a 'macho-machine'. It was a thing with which muscular young men could show off their skills in front of the girls. Hence they were quite satisfied with the great difference in size between the front and rear wheel. For because of this it was quite a feat to ride it without falling and that is exactly what one desires from a macho-machine. Later a bicycle was considered more as a means of transport, first particularly by women, but later by everybody. At that stage bicycles with wheels of equal size came on the market. So through the years a bicycle has always been what people wanted it to be. The appearance of the bicycle was adapted to this.

Besides Bijker one can also regard John Law, Donald MacKenzie, Trevor Pinch, Michel Callon and Latour himself as representatives of this approach. The latter wrote *Aramis, or the Love of Technology* (1996) about the Parisian metro system Aramis, which never came into being. It should have been a revolutionary new system in which travellers could enter into small compartments that would bring them almost directly to their destination and which would on the way be coupled and uncoupled again to other compartments together with which they would make a part of the journey. Latour points out how the engineers' plans were torpedoed because the parties involved – besides the engineers themselves, the mayor of Paris, the municipal council, and all kinds of companies – had widely diverging perceptions of what kind of transport system Aramis could become. Their different interests brought about that there was too little consensus on Aramis and thus all that remained were some prototypes of compartments that later stood deserted in the depot as visible evidence of the thwarting power of the social game. Callon formalised this insight in his co-called *actor-network theory* (Callon, 1991: 132–164).

With reference to case studies he demonstrates that technological developments always take place in networks of stakeholders. The stakeholders each have specific interests and certain powerful means of realising them at their disposal. These powerful means can be more or less effective. The interaction between these actors can be depicted as a combination of forces. In the middle there is the artefact or system that has to be developed. Each one of the actors attempts to push or draw this artefact in a particular direction. The direction depends on the interests of the individual actor. How great the force is depends on the effectivity of the powerful means of the actor. Thus one would be able to 'calculate' which direction the resulting force will have, what its size will be, and thus how the development of the artefact will work out further.

In the SCOT approach it is emphasised that technological developments have something conservative as a consequence of the social power game. Hence, technological innovations usually are made up of small steps. One could compare it with what Kuhn states about the growth of science. He wrote about the difference between '*normal science*' and a short period of revolution. One could also speak about a long period of '*normal engineering*' during which a certain technology is improved but not fundamentally changed, and a short period of revolution during which a fundamental shift in technology takes place. The historian Edward Constant has shown that the turbojet engine entered into aeronautics in a similar manner. In the SCOT approach people later on started speaking about technological routes and technological regimes. Both terms denote that during a certain period the application of a particular technology works like a deeply rutted path. At the time it is almost impossible to take a different route. The term 'regime' denotes that one is not dealing here with a 'loose' technology but with a close combination of technologies and infrastructures. One example is the electric engine in motor cars that has still not been introduced because the whole infrastructure is so geared towards combustion engines that it seems impossible to steer the development elsewhere.

10.3.6 Empiricism and normativity

Latour explains in an original and persuasive manner that the standard image of science is untenable. The supporters of the SCOT approach do the same for technology. The notion that pure science leads to pure knowledge does not correspond with reality. Science and technology are fully social practices in which next to logic and method, psychic, social, economic and other factors play a meaningful role.

The account given by Latour and the social constructivists at first sight seems plausible. However, one could ask whether social constructivism does not lay too much stress on the construed character of science and technology. Does this not render it too 'anti-realistic'? For is not science simply about the 'hard reality outside' while technology simply 'works'? The constructivists will counter this objection by saying that they do not in the least deny it, but that the way in which we discuss it is full of constructs. Nevertheless, this point remains a controversy between realists and constructivists. The realist will emphasise the real character of the technological artefacts in our environment. Thus a bicycle always remains a physical construct, consisting of wheels, cog wheels, crossbars, etc. These characteristic elements of the bicycle cannot be changed just because we get a different view of it. So, realists will claim that reality itself imposes much stricter restrictions on the success and failure of scientific research than the constructivists assert.

Another problem presented by Latour's view and by the SCOT approach is that their analysis of science, technology and society is strongly oriented towards the empirical reality. Within their perspective a normative analysis of technology is not really possible. However, this does not mean that Latour and the other supporters of the SCOT approach are not driven by ethical ideals. On the contrary. They pay particular attention to the political consequences for society of the development of science and technology.

10.4 Conclusion

Classical philosophers of technology look at technology in a general and critical way. In their eyes technology is much more than a means of controlling reality. Rather it is a particular way of looking at and interpreting reality. According to philosophers like Heidegger and Ellul modern technology almost entirely determines the way in which people experience life. Modern technology is the spectacles through which modern people look at reality. People are hardly conscious of the fact that they are wearing such spectacles, and because of this the development of technology seems to be autonomous. This autonomous development seems to be a process of growth in which technologies become increasingly clever and rational.

Recently, however, this general manner of thinking and speaking about technology has aroused growing resistance. There is no such thing as 'modern technology'. There are all kinds of different technologies and each one of these technologies has its own origin and history. The concrete analysis of scientific development processes renders knowledge of the many decisions taken in this kind of process. By means of this effective intervention is enabled. The development of technology is proved not to be an autonomous process, neither is it a development of unambiguous rationalisation. Many 'irrational' (soft) factors play a part. The 'soft' world of social reality and the 'hard' world of science and technology are not directly opposed but constantly infiltrate one another.

Although there is a great degree of consensus in current philosophy of technology about rejecting the general way of speaking about technology, in traditional philosophy of technology there are different opinions on the alternative way of doing it. The designation 'empirical turn' indicates several views. There is the approach of Latour and social constructivism in which social reality takes central position. There is also the SCOT approach, based more on analytical philosophy. Both approaches have the disadvantage – as the designation *empirical turn* indicates – of having to a great extent a descriptive character which makes it difficult to develop a more normative perspective.

Notes

1 By using the term *positivists* Schuurman does not directly refer to the philosophical trend of positivism. But there is an indirect reference when Schuurman states their view in the words: 'They direct their attention to the facts at hand' (Schuurman, 2009: 205).
2 The translated texts are published in Heidegger (1977).
3 Rudolf Augstein and Georg Wolff, '*Nur noch ein Gott kann uns retten*', in *Der Spiegel*, 31 May 1976: 193–219. The interview conducted on 23 September 1966 could only be published after Heidegger's death as he had stipulated.
4 In philosophy of technology there are two different ways of speaking about the *empirical* turn. In the more *analytically oriented* direction the term 'empirical turn' denotes the research into concrete techniques and technologies. In the more *socially oriented* direction 'empirical turn' denotes a comparable reversal in which the social processes of technological development take central position.

References

Augstein, R. and Wolff, G. (1976) 'Nur noch ein Gott kann uns retten', *Der Spiegel*, 31(May): 193–219.

Bijker, W.E. (1995) *Of Bicycles, Bakelites and Bulbs. Toward a Theory of Sociotechnical Change*, Cambridge, MA: MIT Press.

Callon, M. (1991) 'Techno-economic network and irreversibility', in Law, J. (ed.), *A Sociology of Monsters. Essays on Power, Technology and Domination*, London: Routledge, pp. 132–164.

Constant, E.W. (1980) *The Origin of the Turbojet Revolution*, Baltimore, MD: Johns Hopkins University Press.

Ellul, J. (1964) *The Technological Society*, New York: Knopf. Translation of *La technique ou l'enjeu du siècle* (1954).

Hagendijk, R.P. and Meeus, J. (1993) 'Blind faith: Fact, fiction and fraud in public controversy over science', *Public Understanding of Science*, 2(4): 391–415.

Heidegger, M. (1977) *The Question Concerning Technology – And Other Essays*, New York, London: Garland Publishing. Translation of amongst others *Die Technik und die Kehre* (1962).

Heidegger, M. (1996) *Being and Time*, New York: State University of New York Press. Translation of *Sein und Zeit* (1927).

Kuhn, T. (1962) *The Structure of Scientific Revolutions*, Chicago, IL: University of Chicago Press.

Latour, B. (1993) *We Never Have Been Modern*, Cambridge, MA: Harvard University Press. Translation of *Nous n'avons jamais été modernes* (1991).

Latour, B. (1996) *Aramis, or the Love of Technology*. Cambridge, MA: Harvard University Press. Translation of *Aramis ou L'amour des techniques* (1993).

Popper, K.R. (1959) *The Logic of Scientific Discovery*, London: Hutchinson. Translation of *Logik der Forschung* (1934).

Schuurman, E. (2009) *Technology and the Future. A Philosophical Challenge*, Grand Rapids, MI: Paideia Press. Translation of *Techniek en Toekomst. Confrontatie met wijsgerige beschouwingen* (1972).

Smits, M. (2001) 'Langdon Winner: Technology as a shadow constitution', in , H. Achterhuis (ed.), *American Philosophy of Technology: the Empirical Turn*, Bloomington: Indiana University Press, pp. 147–170.

Winner, L. (1977) *Autonomous Technology. Technics-out-of-Control as a Theme in Political Thought*, Cambridge, MA: MIT Press.

Winner, L. (1986) *The Whale and the Reactor. A Search for Limits in an Age of High Technology*, Chicago, IL: University of Chicago Press.

Winner, L. (1993) 'Upon opening the black box and finding it empty. Social constructivism and the philosophy of technology', *Science, Technology and Human Values*, 18: 362–378.

Further reading

Latour, B. (1987) *Science in Action. How to Follow Scientists and Engineers through Society*, Cambridge, MA: Harvard University Press.

Latour, B. (1991) 'The Berlin key or how to do words with things', in Graves-Brown, M.J., *Matter, Materiality and Modern Culture*, London: Routledge.

Verbeek, P.P. (2005) *What Things Do: Philosophical Reflections on Technology, Agency, and Design*, University Park, PA: The Pennsylvania State University Press.

JACQUES ELLUL

(1912–1994)

Jacques Ellul was born in Bordeaux in 1912 and died in 1994. Ellul studied in Bordeaux and Paris. During the First World War he was actively involved in the French Resistance. His well-known slogan is '*Think globally, act locally*'. With this slogan he inspired amongst others José Bové, the leader of the anti-globalists in France, who became known internationally in particular due to his actions against McDonalds and his resistance against genetic manipulation.

At the age of 19 Ellul was converted to Marxism and when he was 22 to Protestantism. These two trends to a great extent formed his frame of reference. On the one hand he was significantly inspired in his work by a Marxist-based analysis of society with a somewhat determinist character. On the other hand he developed, inspired by Karl Barth and Søren Kierkegaard, a decidedly existentialist life view in which the freedom of the individual takes central position. He also wrote a number of theological works. For the rest he did not spare Christianity. In this tradition, according to Ellul, the radical and revolutionary message of Christ is constantly hedged in by the powers of money, politics, religion, morale or culture. There are few philosophers who paint power and freedom in such shrill contrast to each other. Perhaps due to this his work has an immediacy that is hardly ever found in the work of current philosophers of technology.

Ellul's most prominent and influential book is *The Technological Society* (1964). The central theme is that in modern society technology has become autonomous. By this the natural living environment of the human being has been replaced by an artificial technological order. In two subsequent works Ellul continues his analysis. In *The Technological System* (1980) he claims that modern technology has grown into an interconnected and integrated whole of elements that can no longer be viewed separately. This system has eliminated the human being as an autonomous subject. In other words, human beings have themselves become a component of this system. The central message of *The Technological Bluff* (1990) is that too high expectations are being set for technology. Because of the massive dominance of technology in society modern people can only think about their problems in technological terms. Even when the problems have been caused by technology itself, the solution is immediately sought in technology.

Over against the dominance of technology Ellul sets an ethics of freedom that does not adjust and subject itself to the power of technology but on the contrary withdraws itself from this power by making no use of the means of technology. This ethics was substantially inspired by the nonviolent protest of Mahatma Gandhi and Martin Luther King. It is an ethics of '*non-power*' in which there is a conscious rejection of the methodical use of powerful means of exerting control and influence. Such a position can, however, not be held without concomitant tension and conflict because it breaks away from the dominant logic of technology in our society. It breaks through the restrictions imposed on people by the technological system and attempts to expose the belief in technology. In the final instance people are not at the mercy of the powers of technology without a will of their own. They can also withdraw from these.

Here follows a quotation from Ellul's work in which he defines what he understands by technology, and a quotation in which he explains the essence of his ethics of *non-power*:

> The term *technique*, as I use it, does not mean machines, technology, or this or that procedure for attaining an end. In our technological society, *technique* is the *totality of methods rationally arrived at and having absolute efficiency* (for a given stage of development) in *every field* of human activity. Its characteristics are new; the technique of the present has no common measure with that of the past.
>
> *(Ellul, 1964: xxv)*

> An ethics of non-power – the root of the affair – is obviously that human beings agree not to do everything they are able to do. Nevertheless, there is no more project, nor value, nor reason, nor divine law to oppose technique from the outside. It is thus necessary to examine technique from the inside and recognize the impossibility of living with it, indeed of just living, if one does not practice an ethics of non-power. This is the fundamental option. As long as people keep their minds oriented towards the spirit of power and the acquisition of power, towards an ever-increasing expansion (in production, consumption, etc.), nothing is possible. The issue is that we must search systematically and willingly for non-power, which of course does not mean accepting impotence (Non-power is far from being a synonym for impotence!), fate, passivity, etc. (But it is not this danger that lies in wait of us! On the contrary, since today 'destiny' equals 'more and more technique').
>
> *(Ellul in Ferré and Mitcham, 1989: 31)*

References

Ellul, J. (1964) *The Technological Society*, New York: Vintage Books.
Ellul, J. (1980) *The Technological System*, New York: Continuum.
Ellul, J. (1990) *The Technological Bluff*, Grand Rapids, MI: Eerdmans.
Ferré, F. and Mitcham, C. (eds) (1989) *Research in Philosophy and Technology*, Vol. 9, London: JAI Press.

11

CLASHING WORLDS

Globalisation and cultural diversity

Summary

In this chapter we will discuss the role of science and technology in the development of and the interaction between cultures and show that a balanced disclosure of meaning is only possible when good communication between cultures is brought about so that a process of differentiation, integration and individualisation of social practices can result. In practice it is found that the imperialism of Western society and the dominance of modern science and technology often lead to a disturbance of the relationships between different cultures. The result is *one* uniform and standardised monoculture. However, the process of globalisation also confronts us with new forms of plurality and diversity and adds a new dynamics to the interaction and interchange between cultures. Not only processes of power may lead to an unbalanced growth in cultural development. A warped or restricted view of life can also impede further development and unfolding of human existence.

In Chapter 10 the influence of technology on society was the central focus. Has modern technology developed into an autonomous factor determining the life and society of human beings? Or should we rather say that social processes influence technological development? In this chapter we will investigate an issue closely connected to this problem: how the development of modern technology changes the structure of society and how we have to interpret these processes of change. Some authors emphasise the integrating role of technology in society and warn against a worldwide monoculture of standardisation, increase in scale and love of gain. Others point out in particular the differentiating and pluralising effects of technology. By means of the new information and communication technology we come into contact with a diversity of cultures and lifestyles. We no longer need to subject ourselves to collective criteria and can choose our own particular individualistic lifestyle. The iPod is a good illustration of both homogenisation and pluralisation. On the one hand with this small apparatus we form part of a world-wide American culture of consumption; on the other hand we can draw up our own unique playlist and listen, for example, to local bands.

11.1 Cultural development as disclosure

We have described technology in terms of the 'disclosure' of meaning. People develop tools, instruments and methods to enable them to lead a meaningful existence. Disclosure can, however, also be understood as 'opening up oneself': new possibilities come into being, something that was once hidden can be caused to develop and unfurl. A new meaningful reality can be realised. Think for instance of a flower bud bursting open. Although in the bud the sepals, petals, stamens and pistils are present in rudimentary form, they only really start unfurling and showing up well when the bud opens. In a comparable way one can speak about the unfurling of a society or a culture. Traditional cultures have a closed character since no differentiation has taken place in practices and community relations. Giving children in marriage, economic production, political decision-making and handing down religious convictions all take place within one and the same local community. However, in modern cultures the different spheres of life are more differentiated and a difference is made between church and state, family relations and political relations, etc. In this way a diversity of expertise and responsibilities comes into being which should not be confused the one with the other. A policeman on duty for instance should be impartial and treat his son or daughter the same way as he does anybody else. Just so the state should reserve its judgment when it comes to matters of the church or religion.

As long as cultures lead an isolated existence with regard to other cultures or societies they cannot develop or unfurl fully either. Interchange with other cultures is the cause that all kinds of familial, economic and political relations develop relatively independently of the traditional communities. As a result of emigration and immigration citizenship and ethnic origin no longer correspond and as a result of trade relations a market which is to a great extent immune to the influence of the local community comes into being outside the traditional community. Since the process of cultural disclosure integrates several social functions in separate fields, a differentiation of relations becomes visible. State, market and school increasingly start developing according to their own nature which leads to a growing individualisation of practices and community relations. Each can focus completely on their own cause: science for the sake of science, art for the sake of art, and business for the sake of business. Formerly individual citizens were determined by the place they filled in local society. Now they can move around in a diversity of contexts and societal relationships. They are no longer locked up within one and the same traditional community.

The way we here discuss the development and disclosure of cultures is closely linked with biological theories about the development of an organism. However, this does not mean that we are also allowing ourselves to be led by a cultural evolutionism. Of great importance to our interpretation of culture as disclosure of meaning is a view of the good life with an understanding of the complex cohesion of the meaning of reality. In many traditional cultures the acts of human beings are subject to the powers and forces of nature. Because of this these cultures have a closed character and they can only develop and unfurl to a limited extent. Modern people, too, however, have the tendency to narrow their horizons and absolutise a certain aspect of reality. For instance, they reduce everything to physical or economic regularities. This absolutisation is coupled with much conflict and pain, because people stubbornly tend to deny the complexity of reality and thereby sell short their own existence as well as that of others.

Although the complex existential sides of the relation between science, technology and faith will be discussed more extensively in Chapter 14, it is important to say something about it here if we want to understand why science and technology have become such a driving force in the development of modern Western society. Science and technology have gone through an immense development in modern society and have hugely enlarged the action radius of human beings. People have instruments and possibilities of expanding the horizon of their thoughts, acts and judgments. They are no longer determined by the boundaries of the local community, but thanks to modern transport and means of communication they now move around in a global space with the result that the communication between cultures increases. However positive the role of science and technology may be in cultural development, a one-sided fixation on scientific-technological control can also cause numerous problems regarding the environment, animal welfare, national health and the circumstances of employees. People can cause great damage to human beings and nature when they suppose that the world is reparable. This can be called 'technicism': man's pretention to mould all of reality to his liking by means of technological-scientific control in a self-willed way, and thereby to solve all problems that do present themselves, and to guarantee material progress. The problems evoked by this technicism s/he attempts to overcome with the same or with new technological-scientific possibilities. But it is not the introduction of technological apparatus or the development of scientific knowledge as such that forms the problem, but the *belief* in technology and science as a more neutral and superior form of knowing. When people believe in technology, a kind of technologisation of the different community relations ensues. For instance in the agricultural sector where because of new methods of penning up, genetic manipulation, expansion and agricultural subsidies an industrial agriculture develops that can cause serious damage. Not only to plants and animals, but also to the scenic environment and society as a whole.

In the present societal order technicism and economism have formed a pact. As a consequence a tunnel society comes into being in which all blessings are expected to come from science, technology and economy. We first have to bring about good economic growth before we can really invest in the environment. Or we should economise on our wages as far as possible, to prevent running behind other countries in the near future. In this way every potential for development is made subservient to the ideal of bringing about a perfect society by means of science, technology and economy. The paradox of our society, however, is that we are promised a life of abundance, while in the meantime an impoverishment of life and society takes place and the ideal of the good life seems further removed than ever. The process of globalisation itself is not the problem, but the unbalanced growth resulting from a blind reliance on science, technology and economy. Cultural disclosure may not be identified with a materialistic and reductionist orientation. We must be aware that the way in which Westerners interpret the development and disclosure of cultural diversity might be prejudiced.

11.2 Preserving local cultures

The alterglobalists defend the meaning and value of local and traditional cultures against the dominance of modern Western science and technology. They are often called 'anti-globalists' but this name does not do justice to the input of the group. Alterglobalists do not reject globalisation as such. They are more concerned with the unbalanced growth of this process as a result of which natural and cultural diversity wane and a worldwide monoculture comes into being. They do not shun international politics and readily take part in worldwide organised manifestations and conferences.

11.2.1 Cultural and structural diversity

Alterglobalists think that science, technology and economy influence the process of globalisation too much. Because of this the diversity of cultures and community relations is lost. First, modern industrial society imposes its norms and values on other societies by means of the worldwide dominance of science, technology and economy. Examples of this are the World Bank and the International Monetary Fund (IMF). These bodies only give money to developing countries if they meet the standards of effectivity, efficiency and transparency, the typical features of the modern scientific-technologically directed market economy. Second, according to alterglobalists, science, technology and economy also dominate other social sectors. They all have to comply with a technological-economic rationality, which jeopardises the particularity of community relations. Both these forms of unbalanced growth are closely connected. They cannot be separated from one another, but they can be distinguished. In the first instance the cultural diversity is lost: the diversity of opinions on the good life. In the second instance the structural diversity is lost: that which gives to each practice its own peculiar nature and character.[1]

In *Conserving Cultures* (2004) Harry Redner states that the *cultural species* are threatened with extinction by globalisation on a scale at least as big as the natural species. This results in a monoculture in which there hardly is any room for diversity. Therefore an effort has to be made not only to conserve nature but also culture. If only for the reason that it would enable unique sources of knowledge and spirituality to be conserved. The preservation of local traditional cultures is urgent, according to Redner since these cultures can give better meaning to existence than modern Western culture and can give human beings a *sense of belonging*, a community feeling. For modern Western culture is in particular a *consumer culture* of brands, products and lifestyles. It is not a culture that can meet the basic psychological and existential needs of people. For that matter it does not mean that every culture is worth being preserved, or that cultures are static units. Just like natural species cultures have to adapt constantly to the ever-changing circumstances. Otherwise they become living fossils that belong in the museum.

The scientist and political activist Noreena Hertz also pays considerable attention to the threatened cultural diversity. In her book *I.O.U.* (2004) she especially emphasizes that the use of Western models for the financial-economic household is enforced by the IMF and the World Bank. In *The Silent Takeover. Global Capitalism and the Death of Democracy* (2001) she particularly goes into the threat to structural diversity, the diversity of the different social relationships and practices. With reference to a series of narratives she substantiates the statement that since the 1980s a hidden coup has been going on by which multinationals increasingly take over control from governments. It happens more and more that representatives of the population try to satisfy the markets. According to Hertz we have to re-orientate ourselves towards the interests of the state, both at a national and an international level. Politicians should no longer listen to managers and entrepreneurs but actively determine their own political agenda, thus offering a counterbalance to the market and creating the possibility of a more just social order. Jacques Ellul (Section 10.2.2) points out the more deep-seated cause behind this threat to structural diversity. In modern society the means have become more important than the goals we have set ourselves. Because of the primacy of instrumental rationality other forms of rationality are liable to be pushed aside. Where formerly people spontaneously developed various devices and technologies, nowadays they go about it in a planned manner and all actions are focused on the most functional and efficient method:

> It is no longer the best relative means which counts, as compared to other means also in use. The choice is less and less a subjective one among several means which are potentially applicable. It is really a question of finding the best means in the absolute sense, on the basis of numerical calculation.
>
> *(Ellul, 1964: 21)*

11.2.2 Preserving culture or disclosing cultures?

From the alterglobalists we can learn that in many non-Western societies there are sources of spirituality and knowledge available that can constitute a correction to the abstract and individualistic thought of modern society. In the same way that the extinction of plants and animals impoverishes nature, the dying out of cultural traditions leads to the impoverishment of human existence. Impoverishment sets in when science, technology and economy dominate the other spheres of society and the community is subjected to the norms of effectivity, efficiency and transparency. The result is homogenisation, standardisation and formalisation.

The alterglobalists focus particularly on the imperialistic character of Western culture and are very sensitive about any analysis of cultural development that takes Western presuppositions as a starting point. It is a matter of major concern to understand fully how well the modern differentiation and integration processes are established and to demand attention for the peculiar character of many indigenous cultures. For instance, why should the specific social differentiation that has taken place in Western societies be taken as the norm for every type of society? When we make ourselves dependent on modern science and technology we deny that very valuable forms of knowledge have also originated in traditional societies. The alterglobalists point out that the syllabi in schools and universities in non-Western societies force many students into a position where they have to forget about the knowledge and skills they learned in their homes. They have to assume a system of universal knowledge with definite Western presuppositions. This often leads to a process of disruption and alienation causing the indigenous population to regard their own culture as inferior with the result that they can no longer be proud of their own tradition and heritage (Norberg-Hodge, 1996). The modern scientific-technological worldview and the dominance of modern science and technology seem to have become so all-determinative that other cultures can hardly resist them and there is scarcely such a thing as balanced development and disclosure of cultures.

When we look a little more systematically at the stance of Redner and other alterglobalists we see that they are constantly working with contrasts used to mark the difference between 'the' local traditional culture and 'the' modern culture:

- The first distinction is the difference between local and universal knowledge. Indigenous knowledge is local: knowledge of the immediate environment handed down by way of customs, rituals and narratives and mostly having an implicit character. This is *knowledge by acquaintance,* knowledge we have because we are familiar with things surrounding us. Modern scientific knowledge, on the other hand, is universal: abstract knowledge learnt at school and that is generally valid and applicable. It is *knowledge by description,* knowledge compiled in symbols, texts and models giving us information about reality. One could compare these two forms of knowledge with, on the one hand, people who can easily orientate themselves in a certain environment because they have lived there for a long time and, on the other hand, someone who orientates themself in the same environment with the help of a map.[2]
- A second distinction is that between a mythical and a demythologised worldview. In the mythical worldview much emphasis is laid on the spiritual character of nature, which results in a particular way of thinking and acting. Such a worldview is expressed for instance in traditional medicinal practices. In the modern demythologised worldview, on the contrary, there is hardly any room for anything sacred. Another difference concerns time. In the traditional

concept of time every day of the year and every moment of the day have their own religious value and meaning. People go with the rhythm of time. In the modern worldview, however, time is seen as something that is empty and that people have to fill with meaning themselves, for instance with the aid of a diary.

- A third distinction concerns the difference between community and individual. Traditional cultures are often oriented towards a community, wider social relationships within which people take up a fixed position. In modern culture, on the contrary, separate individuals have to bring about relations themselves and deliberately do their own networking. The development of modern society is regarded by the famous sociologist Ferdinand Tönnies as a development from *Gemeinschaft* to *Gesellschaft,* from a society held together by natural ties (family, tribal connections, ethnicity) to a society based on social communication between individuals, expressed amongst other things in modern market relations and the noncommittal way in which people make contact with one another.

A problematic side of the analyses of alterglobalists is that they often approach culture in an essentialist way as if cultures can be compared with biological species which indeed do develop and adapt themselves, but hardly have any mutual interaction. Thus they often suggest a stark contrast between modern Western cultures and other more traditional cultures. Traditional knowledge and abstract scientific technological knowledge should not, however, be played off against one another as if they are at odds and in competition. We could much rather say that the scientific-technological knowledge is always embedded in traditional knowledge or naïve experience. It can deepen this knowledge and if necessary correct it, but never replace it. Therefore the suggestion that people in modern society dominated by science and technology no longer make use of traditional sources of spirituality is undeserved. We also have to shun the suggestion that the knowledge gained from experience in traditional societies is more valuable than the knowledge gained from experience by people in a highly industrialised society. However, a one-sided fixation with science and technology in the process of globalisation can indeed relegate other sources of knowledge, rationality and spirituality to the background.

A similar remark can be made when contrasting individual and community. People who have worked in a third world country on development projects know that thinking in terms of a community is not exclusively wholesome. It can kill personal initiative and lead to various unhealthy kinds of nepotism (which for that matter also occur in Western countries more often than one would suppose). Earlier on we stated that an important norm for cultural disclosure of meaning is the individualisation of practices and community relations. By that we did not mean that each individual should be seen as an independent and atomic unit who should be able to enter freely into relationships with other individuals. On the contrary, it is important that the different relationships should each come into their own according to their nature. In addition we can be concerned with relations between individuals, like trade relations or club connections, but also with relations having a definite community character, as family contexts or national communities. It is only when the different relations develop parallel to one another and in interaction with one another that the lives and communities of people can unfold properly. Without inter-individual relations people would be left at the mercy of the community but without community relations one cannot develop lasting loyalties and people lose their roots and their identity.

The fact that our opinions on how cultures are in relationships can also have practical consequences, becomes obvious when for instance we listen to the present intensifying clamour for cultural rights, supported also by the United Nations.[3] Will Kymlicka, himself not an alterglobalist but a well-known Canadian champion for the rights of Indians and Inuits, says in his book *Multicultural Citizenship. A Liberal Theory of Minority Rights* (1995) that for people to build a meaningful existence they must have access to their own cultural sources and traditions. Therefore their language should be recognised as an official language and they should be given the opportunity to establish their own institutions like schools. It also is conceivable that there may be groups who maintain their own forms of family law and criminal law. However, others argue that cultural identity in itself could not be a reason for claiming rights for specific groups. In jurisprudence arguments like 'it's a vital part of our culture' or 'we have been doing this for a long time' carry little weight (Barry, 2001: 254). Besides, rights for specific groups could easily confirm internal hierarchies and thus weaken the position of, for instance, women even further. When we can no longer have recourse to a universal criterion we slide into cultural relativism. Kymlicka for that matter is also conscious of this. In spite of his open mind regarding other cultures he keeps on making an explicit distinction between practices which can still or can no longer be tolerated within a modern liberal social order.

11.3 The cosmopolitan agenda

Cosmopolitans like Manuel Castells and Ulrich Beck give an interpretation to science, technology and globalisation which in many respects is at odds with that of the alterglobalists. An important difference is that they emphasise that new technologies do not lead to homogenisation, levelling and standardisation but on the contrary to making things more flexible, and to pluralisation and hybridisation. They also support a policy that, different from the alterglobalists, openly seeks alliance with recent technological and social developments. Their stance is that with the rise and continued effect of information and communication technology, genetic manipulation and nanotechnology a transition is taking place from an industrial society to a post-industrial society. While industrial society is characterised by clear-cut distinctions and more or less stable forms of organisation, post-industrial society is typically a society of constant innovation where one can hardly think in terms of fixed frameworks.

11.3.1 Typifying a post-industrial society

The term 'post-industrial society' was introduced in 1973 by the sociologist Daniel Bell. Other sociologists use a different terminology. Alvin Toffler speaks about the 'third wave', Castells about the 'information age' and the 'network society', Anthony Giddens about the 'late modern age' and Beck about 'risk society' and 'second modernity'. The transition from the industrial society to the post-industrial society is often called 'post-modernisation'. The stance taken by the above-mentioned sociologists shows a definite connection with the views of post-modern philosophers like Jean-François Lyotard, Richard Rorty and Jacques Derrida. From a sociological viewpoint we could call the philosophy of post-modernism typifying of the mind-set in post-industrial society. However, seen from a philosophical angle, the above-mentioned sociologists are still working within a modernist frame of reference. They still depart from a grand narrative in which history is described as an evolution from an agrarian society via an industrial society to a post-industrial society. In this they differ from post-modern philosophers who only think in terms of small narratives and who distrust beforehand any reconstruction of the past in terms of progress.

The numerous interpretations of post-industrial society are an indication that describing post-industrial society unambiguously is not a simple matter. Many authors do indeed describe more or less the same phenomena but they often interpret them differently or emphasise different things. However, we can develop some form of systematics when we differentiate the levels at which their analyses are made. The sociologists of post-industrial society make statements at three levels: (1) the level of technological foundation, (2) the level of social structure and (3) the level of cultural diversity.[4]

The first level is the one of scientific-technological foundation. In *The Rise of the Network Society* (2000) Castells investigates information technology, biotechnology and nanotechnology. These separate technologies are closely interlaced in post-industrial society: they use the same paradigm of encoding and decoding. What they have in common is a manipulation of symbols. In this they are not merely concerned with the distribution and exchange of information (that already took place in industrial society: just think of the telephone or the mass media), but with the editing and manipulation of data flow to enable a continuous reorganisation of (social) processes and structures. In industrial society production was organised and coordinated with the aid of the conveyor belt according to a fixed scheme. Each action was carried out in a certain spot and in a certain time span and integrated with other actions via the system of the conveyor belt. In post-industrial society such processes are by far not as fixed as formerly. Automisation and ICT facilitate continuous adaptation and reorganisation of processes which result in a huge diversification of products. Where Henry Ford at the beginning of the previous century said that anyone could get his own model T Ford as long as it was black, nowadays the consumer can choose a motor car to suit them from a multitude of brands, colours and accessories.

Changes in technological foundations are closely connected with changes at the level of social structure. Social processes are not exclusively the product of technological changes, nor are technological changes exclusively determined by changes in the social structure. A characteristic of the new social structure is what Castells calls its *network logic*. Organisations no longer are closed units characterised by a relatively static hierarchy in which the various components are joined into a greater whole and are coordinated from a central managing organ. Horizontal networks now develop between the different decentralised components of the organisation and across the borders of the organisation. These networks are liable to continuous change which means that the employees have to be flexible in their adjustment to new circumstances all the time. Where loyalty and faithful service formerly determined the quality of an employee, employees are now evaluated for the extent to which they can cope with dynamics and can make swift changes in jobs and environment. One example of *network logic* is Business Process Re-engineering. One of the basic assumptions of this approach is that the triad of time, place and action is broken by information and communication technology. Different activities (production, marketing, management, etc.) need no longer take place in the same place and in direct succession in time, but can be connected with one another in a flexible way.

Beck gives an analysis of the network structure of post-industrial society which tallies closely with Castells' analyses. An important symbol of industrialised society, he says, was the national state that maintained its own demarcated space in which market and society could be developed unhindered without interference from outside. The expansion of information networks and transnational capital flow, however, resulted in a removal of borders of the national domain so that market, politics and society had to position themselves anew with relation to one another which meant that the state became one of many players within the global domain (Beck, 2002: 16–17). Amongst other things this meant that governments are no longer able to guarantee their own people a secure existence, job opportunities and safety. The state which took care of the people and pursued a fair distribution of property, is replaced by a high-risk society in which threats and risks are minimised as much as possible and reduced to socially acceptable proportions. These risks are brought on by modern scientific-technological development itself. They often have a global character and occur unexpectedly now here, then there. The nature of these risks can moreover be either ecological (nuclear disaster, greenhouse effect), social (labour insecurity, increase in the number of divorces), or political-religious (fundamentalism, terrorism).

While the elite in the different world cities can take double-quick advantage of new developments via electronic media, there comes into being in the slums of Rio de Janeiro or Cape Town, the ghettos of the USA and the outskirts of Paris a new social subclass which has no access to the network and is cut off from global capital flow. There also comes into being a huge group of employees who can at any moment lose their jobs and with it often their security of income. In principle these people are free to build their own future, but in fact they are exposed to the whims of the market (Bauman, 2005: 23). As an example we can cite the clothing industry in former homelands in South Africa. With the termination of Apartheid the trade embargos were removed and the markets stood open to international trade. The direct consequence was that the clothing industry in the homelands now had to compete with the much cheaper factories and businesses in Asia. It proved that this was a battle that could not be won and many companies chose to revert to niche markets and act as liaisons between the companies in Asia and the clothing industry in southern Africa. Many companies were also bought by Asian entrepreneurs who saw the possibilities of setting up new industries in the former homelands. The consequences for the poor, relatively unskilled inhabitants of the region go without saying. When something in world markets changes or a Chinese firm many thousands of kilometres away takes a specific decision, it can have direct consequences for their income and ways of sustaining life while they hardly have means to influence this.

Finally, at the level of cultural diversity there was a fragmentation and pluralisation of identities. The former plurality of worldviews in which the great stories of liberalism, socialism and Christianity stood over against one another is replaced by a plurality of lifestyles, in which each individual gives form to their own identity, choosing from the diversity of views and convictions presented by diverse media. These lifestyles are no longer nurtured by a consciousness of forming part of one and the same national state. An individualisation and a globalisation ensue by virtue of which human beings feel themselves ever less connected with others by a collective identity. Life becomes a *do-it-yourself biography*. Instead of subjecting themselves to the standard of a collective identity, individuals constantly have to take decisions that determine their identity and construe their own life story from the diversity of mutually contradictory opinions and convictions. Where formerly misfortune and adversity could be ascribed to fate, social position or the will of God, nowadays the finger is pointed at judgements made by the individual themself. One is what one chooses and that means that one is also wholly responsible for the mix one makes of one's own life. Castells talks about a transition from *Weltanschauung* (worldview) to *Selbstanschauung* (self-view) (Castells, 2004: 421). In a *Weltanschauung* or worldview the focus is on a common culture or collective identity. The *Selbstanschauung* on the other hand revolves purely around one's own personal identity. In other words, a kind of Copernican turn occurs: instead of society being the centre around which the different individuals circle, the centre position is now taken by the individual who gives form to his/her own identity from a mix of roles and life views.

11.3.2 Variation of cultures or cultural variety?

Up to now little has been said about the cosmopolitan agenda of sociologists like Castells and Beck. What has become clear, however, is that the quest for a common cultural identity is not something they are likely to support. According to Beck, as a result of the growing interaction between cultures we can only erect a new political power if we free ourselves from categories derived from the national state. We must detach ourselves from nationalistic presuppositions and open up a 'cosmopolitan space' for political actions by means of a new critical theory. A stronger emphasis on domestic employment security and the participation of one's own citizens only serve to deflect attention from the real contrast between poor and rich in the world that is between North and South. What is more, the idea of a national state upholds a distinction between inside and outside and between us and them which denies that a number of nationalities and cultures can exist in one country and that these could have been dispersed over different states. It does not mean, for that matter, that a cosmopolitan should not feel themself attached to a specific local context. Citizens can also have shifting loyalties. One can consider oneself a Dutchman as well as an Antillean or a Moroccan. The main issue is *die Anerkennung der Andersheit der kulturellen Anderen* (the recognition of the otherness of culturally different people) and this is not readily compatible with a fixation on collective identities (Beck, 2002: 71–74, 408–414).

Castells, too, holds that it is important that the awareness of one's own cultural identity should not evoke reactionary powers, but be coupled with an awareness of the global space in which one lives. It is detectable in his work that he sympathises with progressive movements like the ecology movement and the women's movement which indeed often have local roots but still have a cosmopolitan agenda. However, Castells' descriptions are far from neutral. For instance, he feels little for what he terms reactionary movements that cling to certain religious convictions or that put the interests of their own national state first. One example which Castells quotes in a positive sense is that of the nationalists in Catalonia (Castells, 2004: 45–54, 332–337). On the one hand they attempt to preserve their own cultural identity and also pursue a measure of self-government; on the other hand, they form all kinds of alliances at regional, national and transnational level. From this it is proved that nation and state need no longer automatically coincide. Sometimes the Catalonian, sometimes the Spanish or European identity can prevail. The state then acquires the character of a network. It no longer is a homogenous and closed organisational whole with an unequivocal centre. The difference between alter-globalists and cosmopolitans is in this respect best described with reference to the distinction made by Zygmunt Bauman between a *variety of cultures* and *cultural variety* (Bauman, 2002). In the first instance a culture is primarily regarded as a complete whole, clearly distinguished from other cultures, which has to preserve its particularity. In the second instance cultural diversity is expressed more in a diversity of lifestyles offered on the world market from which one can make up one's own repertoire. Instead of, as the alterglobalists do, focusing on a *variety of cultures* and thereby limiting the individual choice, cosmopolitans are looking for ways of maximising the number of cultural forms of expression in the world market.

The advantage of Castells' and Beck's analyses is that they show that technological development and homogenisation need not necessarily keep up with each other. This is a significant difference from the alterglobalists who constantly stress that technology has a levelling and homogenising effect. Sytse Strijbos points out that the neutralising division of functions (see Section 5.4) need not automatically lead to standardisation but can also be utilised to bring about a pluralising of products. The diverse neutral components can be integrated with one another in multiple ways, and definitely so when automatisation and ICT enable a flexible integration of functions (Strijbos, 1997). In this way it becomes possible to operate in a more client-oriented way and to coordinate products with the wishes of the consumer (see Section 7.7). Another advantage is that the cosmopolitans emphasise the importance of inter-individual relations and show that a state can only operate within a network of differently qualified relationships. The modern ideal of the national state mistakenly supposes that the state is elevated above the other relationships and that it can place in a framework and manage these other relationships in a meaningful way. Thus it denies that citizens also enter into numerous relationships which withdraw from the sphere of government, like trade relations, cultural identifications and religious loyalties. The cosmopolitans depart from a dynamic concept of culture and completely avoid thinking in terms of relatively closed cultures. This links up closely with our view that communication between cultures is essential to reach disclosure of cultural diversity.

However, it also is important to pay attention to the one-sided fixation of the cosmopolitans with plurality, individuality and network-forming. Strijbos indeed is positive about the pluralisation brought about by new production technologies but he also shows that the cultural variety offered via media and market is enormously one-sided. Finally the cosmopolitans do not debate the technological worldview. They depart from a plurality of lifestyles but conceal the fact that this way of thinking presupposes a modern secularised frame of giving meaning which contradicts frames of reference developed in other cultural and religious traditions. The view that faith is an individual life project where one can commute between various repertoires, ignores the nature of and the dedication to convictions that appeal to tradition, revelation or some other authority. As long as this asymmetry regarding other cultural and religious traditions is not recognised, there will never be good communication between cultures. There is the danger that other cultures are forced into the mould of the postmodern lifestyle. They become part of a *consumer culture* and can only increase the variation on the culinary and tourist market if they adopt the Western lifestyle. In this respect alterglobalists are right that globalisation is accompanied by levelling, standardisation and an imperialism of modern Western culture. The cosmopolitans, moreover, hardly leave any room for thinking in terms of borders so that all experience of collective identity is wiped out and the existence of community relationships is denied. It is correct that the state forms only one relationship in a network of relationships but it also is a relationship which keeps together citizens in a specific way and effects mutual solidarity. The state marks out its own territory and determines what is permissible within its own borders and what is not. When borders become porous and people forge relationships across borders, it still does not mean that borders are superfluous. Whatever one's opinion on free commerce and the opening up of borders, one would have to admit that the South African government, for example, has a special responsibility towards the employees in the clothing industry in the homelands (Section 11.3.1).

11.4 Power, faith and cultural imperialism

Above we have seen that scientific knowledge and technological artefacts are important engines behind the interaction between cultures and that in this way they also contribute substantially to the process of cultural disclosure. But we also saw that science and technology disturb the balance of power between cultures by the dominance of Western knowledge and skills in education or by enforcing a *consumer market* according to the logic of lifestyle pluralism. For a long time it was presumed that science and technology were neutral means. Cooperation in development was mainly seen as *technology transfer,* transferring technology from the North to the South. Nowadays we are much more aware that the introduction of Western science and technology brings with it a whole system of Western values. Wolfgang Sachs (1999: 12–16) says in *Planet Dialectics. Explorations in Environment and Development* that s/he who brings a mixer or another electric apparatus on the market in a traditional society, does not only introduce a technological aid but also makes the population dependent on a global network of suppliers and energy producers. Moreover, a modern way of preparing food could easily destroy the local eating habits or throw into confusion traditional family relations. For instance, since less time is needed for cooking, women can get other social roles. At the conclusion of this chapter we pause to consider the question whether the dominance of Western values in the process of globalisation is caused by their inherent superiority or by political power. Has modern scientific-technological development become dominant simply because Europe was more powerful or did modern science and technology give a better picture of reality so that Europeans were better able to make progress?

11.4.1 Postcolonial studies of science and technology

A first trend in which much consideration was given to the relation between science, technology and power is the *postcolonial science and technology studies.* One of the points of departure of this trend is that cultures are not unchangeable and sealed wholes but networks of texts, artefacts, institutions and practices. According to them classical cultural anthropology departs too much from the immutability of cultures. This results in the incorrect presentation that in many non-Western societies time has stood still, while a historic development was taking place in the West which led to a higher level of civilisation. The population of the world is divided into groups that have a history and groups without a history. In order to avoid such a simplistic representation of matters it is vital to document the complicated relations between the rise of the modern practice of science and the expansion of European powers. In doing this we depart from the supposition we have already met in the work of Latour that science and society are going through a co-evolution (see Section 10.4). The scientific-technological development is the product of social relationships and structures but these social relationships and structures are also the product of the scientific-technological development.

The most prominent representative of the *postcolonial science and technology studies* is Sandra Harding. In her book *Is Science Multicultural? Postcolonialisms, Feminisms, and Epistemologies* (1998) she makes a connection between the domain of post-colonial studies and the domain of science, technology and society. With this she has laid open a completely new field of research and also gathered a following. In *Is Science Multicultural?* Harding first rejects an internalist explanation for the rise of modern science practice. In such a version it is assumed that about the time of the voyages of discovery something special happened in Western thought which enabled Europeans to traverse the whole world and subject continents. But when one tries to fathom what was so special behind the scientific-technological revolution in Europe of the fifteenth century, one finds hardly any plausible explanations according to Harding. We have seen earlier that the Europeans made good use of knowledge and skills that they had acquired in interaction with other civilisations. Moreover, the colonisation of other parts of the world was linked with the destruction of cultural traditions of a high quality which contained much knowledge and skills. The success of the European scientific-technological project occurred as a consequence of the slightly superior development of the Europeans (fire-arms) and because the Indians in North America were wiped out to a great extent by the illnesses brought in by the Europeans and to which the Indians had no resistance.

Nor does an externalist explanation of history suffice according to Harding. In this account it is presumed that the scientific-technological revolution can be explained by looking at a certain political-economic constellation. However valuable such a version of history may be, it cannot explain the peculiar character and the social effect of the scientific-technological dynamics. In such an approach there is too much emphasis on the idea that scientific-technological knowledge is utilised as a passive instrument to realise social and political goals. But in this view there is no recognition of the fact that scientists and technologists themselves actively develop knowledge and skills and so also influence the political-social constellation. We not only have to describe all external factors, but also to investigate how a peculiar, specific form of power is created within technological science. The social dependence that we find nowadays in the medical and agrarian industry was already present at the time of the West European voyages of discovery. Cartography, mathematics, mechanics and astronomy received a lot of attention because they facilitated faster movement in an unknown space and thereby constituted an advantage in building up a great colonial empire.

The great stress laid on the relation between science, technology and power in *postcolonial science and technology studies* has also evoked strong resistance. The problem of such analyses is that the whole of human existence is reduced to a struggle for power in which any claim to truth is decried as a covert seizure of power. Sandra Harding, for instance, sees no fundamental difference between science on the one hand and folk religion, pseudo-science and witchcraft on the other hand. All these cases, according to her, are systematic efforts to acquire knowledge. The difference between Western science and other non-Western systems of knowledge is that the former has become dominant and has elevated the peculiar local context, the peculiar specific research questions, the peculiar value orientations and the peculiar way of organising science to the absolute norm. Thus it has ousted other systems of knowledge to the periphery on a worldwide scale. Harding proposes that this hegemony of Western science be broken by bringing in as many perspectives and claims of knowledge as possible from elsewhere. According to her no single viewpoint can be fully objective but we can make manifest the prejudices of Western science from other perspectives and thus reach a higher degree of objectivity. Harding does not recognise any exclusive and absolute claims to the truth. The difference between scientific claims and political and religious claims, according to her, is that the first kind of claim is always open to revision while the second kind always appeals to a certain authority (Harding, 1998: 143–145). The problem is, however, that with this she is herself laying a claim to an absolute truth, namely that every claim to the truth is based on a covert seizure of power and should therefore be excluded from the scientific discussion. And she scarcely avoids making such a claim. However, should she give up the pretention that science is different from other forms of knowledge, the jeopardy of relativism becomes great. How else can one maintain that a system of knowledge in which as many perspectives and opinions as possible have been collected is more objective and thus comes nearer to the truth?

11.4.2 The neoconservative contra-revolution

In reaction to the latent relativism there is a turn in the forming of theories on science, technology and power that we can typify as a neoconservative contra-revolution. In contrast to the social constructivism of *postcolonial science and technology studies* neoconservative philosophers once more point out the importance of moral formation. People who in their formative years acquired a strong work ethic are also stimulated to develop their abilities as far as possible. But if they grow up in a cultural environment in which a work ethic is not that well developed it can lead to an attitude of resignation by which people do not do justice to themselves or their environment. In neoconservative opinion the view that the gap between poor and rich, developed and underdeveloped countries is the product of a long history of conquest, extortion, colonialism and neo-colonialism needs to be debated. According to David Landes in his article 'Culture makes almost all the difference' (Harrison and Huntington, 2000: 2–13), such 'dependence theories' have a negative effect on the efforts and morale of human beings. All too easily the blame for one's own misery is laid on someone else, resulting in a loss of social and economic energy. Therefore we should not take too much notice of the reproach that we leave the poor too much to fend for themselves. When someone has low self-esteem because as a child s/he had some or other horrible experience, do we not send him/her to a psychiatrist and point out to him/her that they have a responsibility to take action him/herself? According to neoconservative philosophers history has taught us that genuine solutions to the problem of poverty have to come from inside. Foreign aid can indeed be necessary for a while but eventually people have to take their fate in their own hands and bring about development and progress themselves.

The views of the neo-conservatives seem to tally to a great extent with our views on culture as disclosure. Religious convictions, as we have seen, can open people's eyes to a certain reality, but can also make people blind to the reality around them. An animistic culture with strong emphasis on community relations, in which the spirits of the ancestors are also worshipped, for instance, offers little room and opportunity for an open attitude towards the future, the unfurling of science and technology and the development of a democratic constitution. Thus it impedes to no small extent the further unfurling and development of social life. However, the same can be said about Western consumption culture in which efficiency and technological progress have been elevated to the most important criterion and the organisation of society is almost entirely dictated by the market mechanism.

There are, however, also sufficient arguments against the neoconservative approach from which it transpires that the *postcolonial science and technology studies* are relatively right. Cultural disclosure is still led from the aspect of faith, that is, from a certain view of the good life. In this the neoconservative approach is right. But at the same time this cultural development also has to be initiated by the actions of human beings. It is therefore founded on the form-giving or power aspect. And this is exactly what the *postcolonial science and technology studies* point out. This means that not only religious factors determine the way in which cultures are developed and interact with each other, but also that factors of power have an important role. If we denied this we would have to support a Hegelian view of history in which the highest form of culture will finally prevail. But then we would be denying the tragedy of history. For it is not so that good always prevails. We do an injustice to the situation in many developing countries should we state that the most important problem of, for instance, African or Latin American societies is that they lack good intelligence and labour spirit. To interpret the situation properly we would at least have to recognize that there are real differences in power between Western and non-Western societies which often have made themselves felt for generations.

The combination of power and ideology thus proves to be hard to fathom. It therefore is of the utmost importance not to form too swift a judgment, whether it is of the imperialism of Western culture, or of the supposed impotence and passivity of other cultures. Cultural traditions should not lock themselves up in their own existence but keep an open mind towards the world and dare to enter into confrontation with other traditions. Even though Western science and technology inevitably bear the traces of Western culture, it does not preclude them from disclosing something of the structure of reality for which people in different cultural settings have been blind up to now. Conversely, however, they have no patent on the truth either. They could be corrected by the knowledge that, for instance, traditional communities have discovered about herbs and treatment techniques. Whoever regards Western knowledge systems as the universal norm takes them far too much as absolute. On the other hand, whoever writes off every claim to universality as ethnocentric has not understood that the fact that these knowledge systems developed in a certain social and cultural setting does not automatically cause them to lose their truth and validity.

11.5 Conclusion

In this chapter we have investigated how science and technology help to make possible cultural disclosure. When the process of cultural disclosure goes off well, differentiation, integration and individualisation of societal relationships occur. By this better justice can be done to the multiformity of human existence. Now it becomes clear why designing new technological artefacts and organisational structures has become so complex. Scientific and technological practices are incorporated into an intricate interplay with social practices. It is of great social importance that engineers are aware of this complexity. A one-sided emphasis on the effectivity and efficiency of technological designs in the long run produces only one uniform and standardised social structure. Disclosure of possibilities is jeopardised by this. When, on the other hand, engineers become aware of the plurality of social relationships and the diversity of cultural lifestyles, and do not see technology purely as a means to control reality, they are all the more capable of disclosing new forms of life and communities in a creative way by means of science and technology. In order to get a better idea of how the development of new artefacts and organisational structures can both enrich and impoverish life, in Chapter 12 we will investigate more closely the way in which technology intervenes in human actions.

Notes

1 The concepts 'cultural diversity' and 'structural diversity' are connected with the concepts 'regulative side' and 'constitutive side' introduced earlier (see Chapter 9).
2 The coupled concept *knowledge by acquaintance/knowledge by description* has been derived from Borgmann (1999: 14–15).
3 See for instance the Human Development Report 2004 titled *Cultural Liberty in Today's Diverse World*, published by the United Nations Development Program (UNDP) which can be downloaded via http://hdr.undp.org.
4 Just as in Section 11.2.1 we here make a distinction between a structural and a cultural dimension, connected with the constitutive and regulative sides of practices respectively. However, we now add a foundational dimension to this. This foundational dimension is concerned in particular with the technological form-giving of practices and thus with the foundational rules which apply to these practices (see Section 9.5).

References

Barry, B. (2001) *Culture and Equality. An Egalitarian Critique of Multiculturalism*, Cambridge, MA: Harvard University Press.

Bauman, Z. (2002) 'Cultural variety or variety of cultures', in Malesevic, S. and Haugaard, M. (eds), *Making Sense of Collectivity. Ethnicity, Nationalism and Globalisation*, London: Pluto, pp. 167–180.

Bauman, Z. (2005) *Liquid Life*, Cambridge: Polity Press.

Beck, U. (1992) *Risk Society. Towards a New Modernity*, London: Sage Publications.

Beck, U. (2002) *Macht und Gegenmacht im globalen Zeitalter. Neue weltpolitische Ökonomie*, Frankfurt am Main: Suhrkamp Verlag.

Borgmann, A. (1999) *Holding on to Reality. The Nature of Information at the Turn of the Millennium*, Chicago, IL: University of Chicago Press.

Castells, M. (2000) *The Rise of the Network Society*, 2nd edn, Malden, MA: Blackwell Publishing.

Castells, M. (2004) *The Power of Identity*, 2nd edn, Malden, MA: Blackwell Publishing.

Ellul, J. (1964) *The Technological Society*, New York: Vintage Books.

Harding, S. (1998) *Is Science Multicultural? Postcolonialisms, Feminisms, and Epistemologies*, Bloomington: Indiana University Press.

Harrison, L.E. and Huntington, S.P. (eds) (2000) *Culture Matters. How Values Shape Human Progress*, New York: Basic Books.

Hertz, N. (2001) *The Silent Takeover. Global Capitalism and the Death of Democracy*, London: Arrow Books.

Hertz, N. (2005) *I.O.U. The Debt Threat and Why We Must Defuse It*, London: Harper Collins Publishers.

Kymlicka, W. (1995) *Multicultural Citizenship. A Liberal Theory of Minority Rights*, Oxford: Clarendon Press.

Norberg-Hodge, H. (1996) 'The pressure to modernize and globalize' in Mander, J. and Goldsmith, E. (eds), *The Case Against the Global Economy. And for a Turn Towards the Local*, San Francisco, CA: Sierra Club Books, pp. 33–46.

Redner, H. (2004) *Conserving Cultures. Technology, Globalization and the Future of Local Cultures*, Lanham, MD: Rowman and Littlefield Publishers.

Sachs, W. (1999) *Planet Dialectics. Explorations in Environment and Development*, London: Zed Books.

Strijbos, S. (1997) 'The paradox of uniformity and plurality in technological society', *Technology in Society*, 19(2): 177–194.

Further reading

Carrithers, M. (1992) *Why Humans Have Cultures. Explaining Anthropology and Social Diversity*, Oxford: Oxford University Press.

Feenberg, A. (1995) *Alternative Modernity. The Technical Turn in Philosophy and Social Theory*, Berkeley: University of California Press.

Goodman, A. (2004) *The Exception to the Rulers. Exposing Oily Politicians, War Profiteers, and the Media that Love Them*, New York: Hyperion.

Goudzwaard, B. (1979) *Capitalism and Progress*, Grand Rapids, MI: Eerdmans. Translation of *Kapitalisme en vooruitgang* (1976).

Gray, J. (1998) *False Dawn. The Delusions of Global Capitalism*, London: Granta.

Harding, S. (1991) *Whose Science? Whose Knowledge? Thinking from Women's Lives*, New York: Cornell University Press.

Klein, N. (2002) *No Logo. No Space, No Choice, No Jobs*, New York: Picador.

Lyon, D. (1999) *Postmodernity*, 2nd edn, Minneapolis: University of Minnesota Press.

Stiglitz, J. (2002) *Globalization and its Discontents*, New York: W.W. Norton & Company.

ALBERT BORGMANN

(1937)

PORTRAIT 11 Albert Borgmann

Albert Borgmann was born in Freiburg in Germany in 1937. At the University of Freiburg he studied under the well-known philosopher Martin Heidegger. Under the influence of the latter's lectures on language and identity he decided to study further in German and English literature. To improve his English he went to America by boat. As a poor student he had to do dishes while on board to pay for the voyage. In America he nevertheless decided to study philosophy and when back in Germany he finished his doctorate degree in 1963 at the University of Munich. Thereafter he became a lecturer at the DePaul University in Chicago. In 1970 he was appointed professor at the University of Montana. However shortly after taking up this position he became interested in philosophy of technology.

In the book *Technology and the Character of Contemporary Life* (1984) he first wrote down his ideas on technology for which he would become famous. The influence of Heidegger cannot be missed. He takes over from Heidegger the idea that technology lessens our experience of reality. Heidegger claims that as a result of technology we only see reality as something that has to be worked over to be of any value. Consequently we lose sight of the value of reality itself. Borgmann elaborates on this idea and says that as a result of technology we have a much less intensive relation with reality than formerly. Due to technology we hardly need to exert ourselves to actualise all kinds of comfort. Instead of going to the forest to chop wood for a fire, we merely turn up the thermostat of the heater. Borgmann calls this '*disengagement*'. His term '*device paradigm*' indicates that we take it for granted that technological devices fulfil their purpose. However, he is less pessimistic than Heidegger who claimed that only a 'god' would be able to save us (and seeing that he did not believe in a god this expression was meant to be quite fatalistic). According to Borgmann the 'focal activities' offer a way out. These are activities that once more involve us in reality (for instance, we have to go back to the forest to fetch wood). The term 'focus' here has the same meaning as when speaking of a lens which on the one hand draws rays of light together and on the other hand causes them to fan out. In this way a focal activity brings together all kinds of aspects of life and simultaneously shines forth to all of life. Borgmann, however, realised that our whole life cannot consist of such activities and therefore pleads for a 'part economy' in which focal activities are not nipped in the bud by economic preconditions. As a (Roman Catholic) Christian in his book *Power Failure* he mentions church services as a focal activity.

Later on in *Holding on to Reality* (1999) Borgmann wrote more specifically about information technology. In *Crossing the Postmodern Divide* (1992) he wrote about the influence of postmodernism on technology. In both books he expresses the criticism that technology has too great an influence on our lives. We close with some citations from *Technology and the Character of Contemporary Life:*

> To bring the distinctiveness of availability into relief we must turn to the distinction between things and devices. A thing, in the sense in which I want to use the word here, is inseparable from its context and its world, namely engagement. The experience of a thing is always and also a bodily and social

> engagement with the thing's world. In calling forth a manifold engagement, a thing necessarily provides more than one commodity. Thus a stove used to furnish more than mere warmth. It was a focus, a hearth, a place that gathered the work and leisure of a family and gave the house a centre... A device such as a central heating plant procures mere warmth and disburdens us of all other elements. These are taken over by the machinery of the device. The machinery makes no demands on our skill, strength or attention, and it is less demanding the less it makes its presence felt.
>
> *(Borgmann, 1984: 41–42)*

> The great meal of the day, be it at noon or in the evening, is a focal event par excellence. It gathers the scattered family around the table. And on the table it gathers the most delectable things nature has brought forth. But it also recollects and presents a tradition, the immemorial experiences of the race in identifying and cultivating edible plants, in domesticating and butchering animals; it brings into focus closer relations of national or regional customs, and more intimate traditions still of family recipes and dishes... In a Big Mac the sequence of courses has been compacted into one object and the discipline of table manners has been reduced to grabbing and eating.
>
> *(Borgmann, 1984: 204–205)*

References

Borgmann, A. (1984) *Technology and the Character of Contemporary Life: A Philosophical Inquiry*, Chicago, IL: University of Chicago Press.

Borgmann, A. (1992) *Crossing the Postmodern Divide*, London: University of Chicago Press.

Borgmann, A. (1999) *Holding on to Reality*, Chicago, IL: University of Chicago Press.

Borgmann, A. (2003) *Power Failure: Christianity in the Culture of Technology*, Grand Rapids, MI: Brazos Press.

CASE STUDY III

Network enabled military operations

In this case study[1] the usefulness of the concept of normative practices will be illustrated by an example from the world of military operations. We will see how information technology can create connections between practices that cause confusion of norms with – in this case – dramatic consequences. Instead of simple 'whodunnit' stories, the normative practices concept provides insights into complexity of interacting practices and tensions between norms that may arise. The case study was headline news in its time, and still today it is worth reading because of the importance of avoiding these kind of situations.

1 Technology and military practice

Technology has always played a role in military practice. In the course of time all kinds of weapons have been developed. As the striking power of these weapons increased, more and more ethical concern was expressed about what is morally acceptable and what is not. Particularly the atomic bomb that ended the Second World War, but also the nuclear weapons during the Cold War, and still today, are a cause of concern and ethical debate. Weapons are, however, only one type of technological support in military practice. There is an increasing amount of technologies that are used for surveying, communicating, transportation, etc. Information and communication technology (ICT) has become a substantial part of the military supportive technology. These technologies largely escape from the ethical debates, as most people tend to see them as neutral instruments. The case that will be described here, will show that this is not an adequate view on these technologies. These technologies have an influence on us. They invite us to act, even in certain directions. In that sense, they are by no means neutral and it is necessary to assess their impact on us, if they are to fulfil a proper role.

2 The Kunduz airstrike

Military operations have become more and more networked by information technologies. 'Network Enabled Operations' (NEOs) such operations are called. All actors involved in such an operation are now in contact through sophisticated ICT. Military staff in a camp can, as it were, look over a pilot's shoulder and watch the screen in his aircraft. This seems like an improvement compared to the former situation in which the pilot was pretty much on his own when he had to decide whether or not to drop a bomb in case of a suspicious situation on the ground. The accident that happened in the province of Kunduz in Afghanistan in September 2009 shows that ICT networked military operations pose new challenges to pilots, soldiers and their superiors.

In August 2009, the acting German colonel in a camp for the International Security Assistance Force (ISAF) stabilisation mission got news that two fuel trucks had been stolen by Taliban. This caused a direct threat for a nearby camp to be attacked and therefore the Colonel ordered for a survey of the area from the air. An American long-range bomber flew over the area at that time and the German colonel could follow what happened on the ground by watching the pilot's screen by means of a remotely operated video-enhanced receiver (ROVER) device. He

FIGURE C3.1 One of the soldiers involved in the Afghanistan mission working with the network technology

Thanks to soldier Dennis for granting us the right to use the picture.

spotted what he interpreted as some trucks on the ground with people walking back and forth between the vehicles. Two American F-15 fighter jets took over, as the long-range bomber was involved in a different assignment already. The German colonel had asked his Afghan informant about the possibility of civilians being present in that area, and the answer had been: very unlikely. The Colonel's conclusion was therefore that the people on the ground were Taliban transferring fuel from one truck to the other in preparation for an attack on the nearby German camp. He ordered the pilots to attack. The American pilots, however, did not have a good feeling about this, as there was still a risk that what they saw on the screen were civilians. In their practice, the pilots had the right to make their own decisions in case they felt that there was any uncertainty about what was underneath. They proposed to fly over on a low altitude in order to chase away the people, use a remotely operated drone to do so, or ask confirmation from a higher-ranked American Combined Air Operations centre in Qatar. The German colonel, however, urged them to take immediate action, given the threat the situation caused in his view. The pilots then dropped two 550 pound bombs and … killed approximately 90 civilians. Later it appeared that these people had tried to get going two trucks that had become bogged down in a riverbed by removing fuel to make them lighter.

3 Analysis of the Kunduz case

Of course the press were quick to blame the German colonel under whose authority the strike had been committed. However, it can be questioned if the colonel's decision was not the result of a situation in which the networking technology played an ambiguous role. The concept of normative practices, as described in Chapter 9, can help to get a more sophisticated analysis and appreciation of what happened in this sad incident.

In fact, we are dealing with several sub-practices within the military practice here. The two most important ones for our case are the pilot's and the colonel's sub-practices. These clearly have distinct structures and directions. In the structure the responsibilities of the various military functions are defined. The colonel has certain rights and responsibilities and the pilots have different rights and responsibilities. In the colonel's sub-practice's structure it is laid down that he can give orders to pilots. In the pilot sub-practice it is laid down that he can refuse an order for bombing if he does not have a good feeling about the possibility that it can be civilians below him. Such a norm is part of what in military terms is called 'rules of engagement': when to attack and when not to. The directions of the colonel's and the pilot's sub-practices also differ. In the colonel's sub-practice, concern for his own people and taking out possible threats is a primary value. In the pilot's sub-practice, however, a concern for the people below him is a primary concern. That is why there is such a good fit between his sub-practice's structure and function, because the right to refuse a bombing order in case the pilot is not sure if it is the enemy or civilians below him (part of the structure) fits well with his primary concern, namely the safety of civilians (part of the direction). One may question whether or not it is desirable that two sub-practices within one military operation have different directions. On the one hand, there are two distinct values in such an operation: protection of the soldiers and prevention of civil casualties. The idea behind the operation is that the two are compatible and therefore different sub-practices are given different concerns. But the case study shows that they can be in conflict and in that case, there is an inherent tension between the sub-practices. This tension was enhanced when the colonel more or less forced the pilot to drop the bomb, based on the concern that was caused by what he had seen on the pilot's screen, but which he had not been trained to interpret. Besides that, his focus on taking out possible threats to his own people (part of his sub-practice's direction) made him vulnerable for what is called the 'predator view effect': if one is focused on identifying threats, one is more inclined to see things as possible threats. Due to the fact that the ICT network gave him direct access to information that before only came to him through the pilot's interpretation of the screen – which was based on his training to interpret this device – he now felt entitled to interfere with the pilot's rights and responsibilities. Thus we see how the technology's presence caused a change in the relation between sub-practices, thereby disturbing an existing harmony between structure and direction. In Chapter 12 more is discussed regarding the 'hermeneutic' relation between reality and what we see in a technologically mediated observation.

What also played a role here is the effect of distance. There are two kinds of distances involved here: the geographical distance and the psychological distance. There was a difference in geographical distance between the pilot and the colonel. The pilot flew directly above the people he was about to bomb. The colonel, however, was in a camp away from that area. Psychological studies have shown that greater geographical distances generally cause greater psychological distances. The pilot feels much more 'connected' to the people below him than the colonel and therefore has more hesitation to push the bombing button than the colonel, for whom this all happens at great geographical distance. This, in fact, is another reason why the pilot's right to refuse a bombing order in case of absence of a good feeling fits so well with his concern about the people below him. This concern is 'natural' for him, given the small geographical distance between himself and those people. For the colonel it would be much more difficult to base his judgement on a good feeling, as he did not feel so much 'connected' to the people that were to be attacked. This psychological distance also plays an important role in the discussions on military drones. Drones are also controlled from great distance and therefore bring along the effect of psychological distance. Based on studies on psychological distance (although there are some contradictory outcomes), one could expect that distance makes it easier for people controlling a drone to attack than for soldiers and pilots that are in the direct vicinity of the people they attack.

4 The role of context

Does the context, the third property of a practice (see Chapter 9) also play a role in this? Yes, because military practices are country-dependent. German military practices (either pilots' or colonels') are defined in the context of German policy, and likewise American military practices in the context of American policy. The 'rules of engagement', for instance, are different for American military than for German. The consequences of this for international missions, such as the ISAF missions in Afghanistan, can be illustrated with another case from the same mission as the one in which the Kunduz strike took place. A Dutch reconnaissance soldier on the ground had to gather information about a certain area in which a reconnaissance mission had to be carried out. An American colleague flew in a Kiowa helicopter in the same area and offered assistance in checking out the local villages. The Dutch soldier gladly accepted the offer and after a while had contact again with the American pilot, who told him that he had spotted fighting men in a circle of people. His guess was that they were Taliban. 'Shall I take them out?', was his question to the Dutch soldier. Fortunately, in this case the decision was not made on the spot and the Dutch soldier checked out himself to discover that these people were civilians celebrating the end of Ramadan. The American pilot could ask the question about 'taking out' (killing) the people he had spotted, because the American rules of engagement are such that a pilot can attack on his own initiative. This is not the case in the Dutch rules of engagement and that is why the Dutch soldier had to make a more careful investigation and report back first. For both practices these rules of engagement (part of the structure of the practice) fit well with their practice's direction. In the American military practice, taking out possible threats immediately is what in this practice is seen as a primary value. In the Dutch practice, there is more like a 'winning hearts' (of locals, not of Taliban of course) kind of attitude towards stabilisation missions. This means harmony within each of the American and Dutch pilot practices, but frictions between the practices when they become entangled in an international mission, particularly when the presence of technology creates intense and frequent contacts. That is particularly the case because their relation to the overall mission is what we called in Chapter 5, a part–whole relation (not an encaptic enlacement). That means without the embedding in the 'umbrella' practice (the ISAF mission), their presence would lose all sense. They are not like the canteen in the basis that could equally well be the canteen of a factory or a school. That sub-practice could function in a different 'umbrella' practice without any problem. But the American and Dutch pilot's and soldier's presence in Afghanistan fully determines the purpose of their actions. This makes the connection the more intense and the more easily problems can emerge when two sub-practices (the American pilot's practice and the Dutch soldier's practice) have to work together in the 'umbrella' practice of the ISAF mission. The differences between practices of different nationalities also played a role in the Kunduz case. In the overall ISAF norms, reduction of civil deaths had recently been added to the directional norms in the ISAF practice, but the German military command had just before deleted certain norms from the structure of their practice (the rules of engagement) that were aimed at prohibiting civil casualties. As a consequence, there were hidden inconsistencies between the direction of the ISAF practice and the structure of the German sub-practice in ISAF.

5 Moral decision making in military training

What can be done to prevent these kind of events from happening? One way is to go back to the old non-networked situation. It is unlikely that this will be seen as a preferable alternative, given the investments that have been made to create the possibility of networked operations. In the likely case that the technology remains in place, what measures can be taken? As we have seen, theoretically the practices have been defined in such a way that there is harmony between structure and direction. The problem was that the colonel and the pilot, stimulated by the technology, did not hold to the norms as defined in their practices. This is due to the fact that this consequence of the networked situation was unforeseen and both colonel and pilots were not prepared for it. That can be prevented when in military training more attention is paid to the impact of technologies like the networked operations technology on moral decision making. Pilots have to be trained to make moral decisions that are even more complex in a networked situation than in a non-networked situation. In the countries that are involved in international military operations, generally moral decision making is part of military training. The introduction of new technologies, such as Network Enabled Operations, should have consequences for the ethical dimension in the training of people in a military practice.

Note

1 This text is based on research done by Christine Boshuijzen-van Burken (2014).

Reference

Boshuijzen-van Burken, C. (2014) *Moral Decision Making in Network Enabled Operations*, thesis, Eindhoven, the Netherlands.

12

HOMO TECHNICUS

From device to cyborg

Summary

In this chapter we demonstrate that technology is a human activity. Technological artefacts are extensions of our human capacities but also a shift in the way we observe, as the philosophers Ernst Kapp, Arnold Gehlen and Don Ihde have shown. To the question how far this extension may go, different answers are given. Lewis Mumford warns against the dominating influence of technology on the life of human beings. Donna Haraway sketches postmodern man mainly as a cyborg, a hybrid of organism and machine. Ray Kurzweil pleads for a radical integration of man and technology. According to him it will lead to infinite possibilities. Subsequently we go into the various human motives leading to technology. Finally we emphasise the human responsibility for technological activity.

'Have you seen my glasses?' 'You're wearing them.' A well-known joke. Sometimes we become so used to a technological device that we no longer realise we are using it. It is as if it has become a part of us. Spectacles are meant as a small artificial extension to our bodies. And the longer we use them the more we begin to consider them as a part of ourselves. The technological device, the extension of one's body, can even mean the difference between life and death as in the case of a pacemaker. Even then in the long run one may forget that originally it was not part of the body. Until one day you pass through security at the airport and the alarm goes off although you have completely emptied your pockets. And you are reminded of the presence of that apparatus in your body, that piece of metal between the organic tissues. It is not altogether natural that it is there.

12.1 Technology as an extension of human beings

In 1877 Ernst Kapp's book *Grundlinien einer Philosophie der Technik* (*The Basics of a Philosophy of Technology*) was published. It can perhaps be called the first real publication on the philosophy of technology. In this book he states that technology has to be seen primarily as a projection of our human organs and limbs in an artefact. A pair of scissors and an axe are projections of our hands, a camera is a projection of our eyes, and the telegraph cable a projection of our nervous system. Kapp saw this as the logical next step in the process of evolution by which human beings came about. A human being needed the organ projections when his own organ no longer worked properly, or when he wanted to expand the capacity of that organ. In this way the lens in the spectacles is meant to complement the functioning of the lens in the eye. Likewise the hammer was made to enhance the striking power of our hands. And by means of these organ projections human beings could get to know themselves better. By projecting organs and limbs they could objectively observe their nature and in this way learn more about who they were. Technology thus plays an important role in the growing self-consciousness of human beings. That growing self-consciousness was seen by Kapp as a significant driving force behind history. In this he was inspired in particular by the German philosopher Georg Wilhelm Friedrich Hegel. What human beings learn by this organ projection is that they constantly feel the need for changing the environment. Evidently this is inherent in being human. A human being is a *homo technicus*.

Some things should be detracted from Kapp's line of thought. His idea of technology as an organ projection looks more like an a priori argument than the result of a careful investigation of reality. It is true that Kapp illustrates his statements with various examples but it would not be difficult either to think of examples of technological artefacts that cannot be explained as organ projections. Explaining a motor car as an extension of our feet is really a feeble description. And what about a match? Is it an extension of a finger because its tip is sadly not flammable?

12.2 Technology as a means of survival

The German philosopher Arnold Gehlen said that human beings need technology to survive. Gehlen made a significant contribution to the development of philosophical anthropology. His book *Der Mensch. Seine Natur und seine Stellung in der Welt* (Man. His nature and place in the world) (1940) has become a classic with a tenth edition in 1974. The core of this analysis is that a human being is a *Mängelwesen*, a handicapped or deficient being, who cannot survive in a natural environment since s/he has no specialised organs and instincts. Therefore by an intelligent adaptation to their environment human beings themselves have to create the conditions for sustaining themselves, for instance by using fire and weapons. Thus Gehlen characterises technology as devices and skills developed by human beings in order to maintain themselves since they lack certain instincts and are defenceless beings. In Gehlen's view there is a fundamental difference between animals and human beings. An animal can just be what it is. An animal is bound and secure in its *Umwelt* (milieu). He calls this 'determined'. But humans stand open to the world, *weltoffen*. They have not been 'determined' but lead their own life. They are acting beings with a complete structure or *Gestalt* of their own. And where human beings act culture ensues.

This activity originates in the nature of human beings, their deficiency. They are imperfect beings. Gehlen says that animals have a 'natural' head start on human beings. They have inherent instincts enabling them to react with absolute certainty to the various situations in their own specific environment. They also have been equipped with a natural protection against the influences of the weather, they have natural organs for catching and devouring prey, and they have bodies that enable them to flee easily. Humans have none of these. In order to survive they are compelled to create conditions for their own existence. In this way culture comes into being.

Thus Gehlen attempts to understand culture-forming activity by human beings as resulting from their biological conditions. He sees culture as a network of institutions comprising a collection of behaviour patterns. Since a human being is a *Mängelwesen*, a deficient being, s/he cannot 'just' maintain her/himself in a natural environment but has to rework nature into a form in which s/he can maintain her/himself. This form Gehlen calls culture or *zweite Natur* (second nature) (Gehlen, 1940: 38). Not only technological artefacts like weapons, tools and dwellings come under this second nature, but also social structures like marriage, families and community. Only by institutionalising their activities do human beings overcome their deficiencies and become less vulnerable. Institutions therefore are crystallised patterns of behaviour. Man as a deficient being is best expressed in the figure of Prometheus: by constant forethought (the literal meaning of his name) he has to succeed in maintaining his position as a half-god (therefore also a kind of *Mängelwesen*) among the gods (Gehlen, 1961: 48). Gehlen emphasises that institutions relieve man from the pressure of taking decisions. They also function as pointers giving direction to the human being inundated by a mass of impressions and stimuli. A human being standing open in the world cannot do without institutions. Although institutions prod human beings in a certain direction and to a great extent determine their thoughts, they also give human beings the energy for creative and productive work (Gehlen, 1961: 71–72).

The nature and character of technology can likewise only be comprehended by adopting the idea that a human being is a deficient being. In his book *Die Seele im technischen Zeitalter. Sozialpsychologische Probleme in der industriellen Gesellschaft* (The soul in the technological age. Social-psychological problems in industrial society) (1957) Gehlen further expounds this view. He argues that only technology enables human beings to remain standing in the struggle for existence. So in his view a human being is a *homo technicus*. Gehlen differentiates between three kinds of technologies: complementing technologies, strengthening technologies and disburdening technologies. Weapons clearly are technologies complementing human beings: they do something the human being's organs cannot do. Another complementing technology is the heating technology in heaters and ovens. The hammer, the microscope and the telephone clearly are technologies that strengthen the organs of human beings, they take the natural skills of a human being to a higher plane. Finally, a wagon is an example of a technology that disburdens: dragging burdens is made redundant. Gehlen also uses this categorisation to characterise modern technological artefacts. So, for instance, in an aeroplane all three technologies are present: it gives man wings to fly (complement), supersedes all organic flying achievements (strengthening) and saves exertion in moving forward over enormous distances (disburdening). Man as an intellectual being is not bound to 'organic' examples from nature. Inventions often have no parallels in nature. For instance, Gehlen draws attention to the invention of the wheel. Such an invention he calls 'abstract'. The world of technology is a *nature artificielle* having its own dynamics (Gehlen, 1957: 9). During the course of history materials of organic origin (wood, leather, hemp) are replaced by synthetic materials (steel, plastic, electricity). Simultaneously the powers of organs (arms, legs) are replaced by inorganic powers (wind, oil, gas).

Gehlen emphasises that during the course of history technology has undergone a qualitative change. Earlier technologies like stone tools and bows are very different from contemporary technology. The development from the old to the new technology, however, cannot be characterised as a transition from tools to machines. We can only understand this development if we also consider the structural changes in other important sectors of our culture. In particular Gehlen means the development of science and the growth of the industrial system in the seventeenth and eighteenth centuries. He speaks about the *superstructure* of science, technology and industry. Gehlen stresses that we can only comprehend the tempestuous development of technology against the backdrop of the natural sciences and the capitalistic production system. This development also has a downside: human beings have to adapt to the technological patterns of the superstructure. The methods of science, technology and industry are applied in the non-technical sectors of society and they have a huge influence: not only does the superstructure have an influence on interpersonal relations but also on our way of interpreting, thinking and reflecting. The consequences are enormous. By adapting to the superstructure man loses sight of reality. According to Gehlen this crisis is not of a religious character but can be called a total crisis: the entire interpretation of the world has become suspect. The fundamental needs of human beings can no longer be fulfilled: the technological culture cannot stabilise the (social) life of human beings. This is because the industrial culture is characterised by 'creative destruction'. According to Gehlen we can only break the power of the superstructure by setting limits to the beginning (the wish to know) and the end (the wish to consume) of the development of the superstructure. Abstinence (ascesis) could be the signal for a new era (Gehlen, 1957: 54).

Gehlen's view is clearly related to Kapp's. Kapp also is of the opinion that human beings have to do something to complement the shortcomings of the human body. When dealing with Kapp we questioned whether one could adequately describe all forms of technology as organ projection. Now dealing with Gehlen we could ask whether it really is true that technology originates solely in the fact that we are deficient beings. We will return to this at the end of this chapter. At this stage we can already state that we are not so well off if we really need as much technology as we have produced. For technology is ubiquitous. We can almost speak about a technological environment. Was this necessary only to survive? It seems unlikely. We have to ask whether all of technology really has a positive effect on the quality of life. One of the first philosophers of technology who dealt with this question in detail was Lewis Mumford.

12.3 Technology in the framework of human culture

The historian and philosopher Lewis Mumford has a different approach from Gehlen's. According to him technological inventions are always preceded by cultural change. This means that we can only understand the role of technology in Western society if we focus on ideological and social changes. Also it means that we may not limit our research to the question why a certain technology was developed, but should give consideration to the question why the technology was applied on a grand scale. Technology in itself is no new phenomenon in history. What is new is the fact that technologies are being 'projected' and 'embodied' in organisations that affect all aspects of our existence. The development of culture, according to Mumford, gives a better insight into the nature of human beings than the development of tools. Contrary to what Gehlen says, he is of the opinion that we should not understand a human being as *homo faber* (a fabricating human being) but as *homo sapiens* (a thinking human being). It is 'thinking' man who opened up the way to the development of the machine in Western culture.

In *Technics and Civilization* (1934) Mumford deals with the relationship between culture and technology. He wonders why technology plays such an important part in Western civilisation. In the past other cultures also reached such a high level of technological perfection without being influenced by the methods of technology to the same extent that we are. So, for instance, the Chinese, Arabs and Greeks had the clock, printing press, water mill, compass and loom. Yet they developed no 'machine'. Mumford asks himself how it came about that the machine could take possession of European society. It is linked to a change in worldview. He makes the statement that *the* industrial revolution of the eighteenth century was in fact the result of a development going back over a much longer period. According to him the machine 'thundered' over our civilisation in three overlapping waves. First came the wave of eotechnology, lasting from the tenth to the middle of the eighteenth century. In this period horses, water and wind were the main sources of energy, and wood and glass the most important materials. After that came the phase of paleotechnology, lasting from the middle of the eighteenth century to the second half of the nineteenth century. Coal was the main energy source and iron the most important material. The last wave is that of neotechnology that started about the second half of the nineteenth century. Oil and electricity now are the main sources of energy while steel and other alloys are the most prominent materials.

Mumford places the beginning of the new technological order in the tenth century. From that time the categories time and space changed fundamentally and the road was paved for the machine to advance. According to Mumford the origin of these changes lies in the fixed timetables of life in the monasteries. Time there developed into a chronological time. Time became a measurable quantity – hours, minutes, seconds. According to Mumford the clock and not the steam engine is the key to understanding the modern technological era. The clock prepared human beings for the 'regular collective beat and rhythm of the machine' (Mumford, 1967: 14). For, so he writes, not only is the clock a means for measuring time; it also is an instrument to control the behaviour of human beings and attune them to one another. In a comparable way space got a different meaning: space as a hierarchy of religious values was replaced by space as a system of quantities (length, width, height). For instance, Mumford criticises the advancing infrastructure for motor cars in the USA because it devours so much space that it forms an obstacle to the other ways of transport like walking and cycling. Step by step the new views on time and space started influencing every aspect of life. Thus space was created by the breakthroughs of technology.

In *The Myth of the Machine, I. Technics and Human Development* (1967) Mumford sketches the contours of his anthropology. In his prologue he makes the statement that contemporary theories overestimate the role of tools and machines in the development of human beings. If we see man as *homo faber* we overlook the most important period of human history, namely the period during which the development of dreams, rituals, language, social interaction and social organisations held a prominent place. Mumford emphasises that these cultural activities precede and surpass the making of tools. Human beings had to get to know and give form to their inner life. This achievement he judges to be of greater significance than the knowledge of and the form-giving to the exterior world. For in order to develop culture human beings have to have control of all their natural functions, including their excretory organs, rising emotions, adulterous sexual activities and terrifying dreams. On the other hand, the use of tools only requires control of hands, muscles and eyes. The view that cultural activities precede the development of tools is mirrored in the order of his book: the first hundred pages deal mainly with language and dreams and only then does he get to tools. In Mumford's way of thinking, people are beings that in the first place have to develop and control themselves before being able to fashion the world. In the later chapters of *Technics and Human Development* he elaborates on the concept 'megamachine'. It is an invisible structure consisting of living but rigid human parts, each one of which has a special function, role or duty. By means of this, grand ventures can be realised jointly. An example of this is the construction of the pyramids. For this new forms of organisation and communication were needed, like the invention of writing. The enormous size of this megamachine can be seen in an inscription on a sceptre of the Narmer dynasty which mentions more than 120,000 prisoners, 400,000 oxen and 1,422,000 goats. For the development of such a megamachine the mechanical powers first had to be 'socialised'. In other words, cultural development precedes technology.

In the second part of *The Myth of the Machine* (1971), bearing the subtitle *The Pentagon of Power*, Mumford sketches how the secret of the megamachine was rediscovered in the sixteenth century and in the span of less than two centuries was transformed into a new and improved model. The new catchwords are power, speed, movement, mass production, precision, uniformity and especially control. The modern megamachine surpasses its earler self in both the positive and the negative sense. Where the destructive powers of the modern megamachine are concerned, Mumford sketches a very dark picture. Our culture has become unbalanced because technology dominates the other aspects of civilisation. The result is that culture produces 'warped and unbalanced minds'. The modern megamachine requires a system of total control. Some descriptions by Mumford forcefully recall the pessimistic views of Jacques Ellul (see Chapter 10). However, Ellul depicts human beings as passive victims of technological development while Mumford regards human beings as acting cultural beings who have the ability to resist the technological system. The supremacy of the megamachine can be shattered if human beings decide to abandon the 'myth' of the machine.

12.4 Infinite possibilities?

Mumford's optimism is a continuation of the increased possibilities to put into practice Kapp's idea on projection by means of prostheses. For currently there are quite a few components of the human body that can be replaced with prostheses. Just think of false teeth, artificial arms, legs, kidneys, lungs and even artificial hearts. We can equip a person with technological artefacts as happened with impressive results in the *Six Million Dollar Man* (a television series from the 1970s). He could for instance run extremely fast, hit right through a piece of metal with his bare hand and hear two people whispering at a distance of a kilometre. The film makers professed that this had no effect on his self-image. But what would happen should his brain also be replaced by a prosthesis? Would that not lead to alienation? In science fiction films this step is usually not taken. Not much was left of Robocop Alex Murphy either when he was reconstructed but his brains were still his own. It seems that for the makers of science fiction films there is a limit. However, it remains to be seen whether one would retain one's individuality if all parts of one's body – except one's brain – were replaced by technological artefacts. Some people are of the opinion that the whole body forms part of being human, not only the brain as the 'carrier' of the human personality. Therefore the question is where the boundary is and whether there actually is a boundary.

The advance of information technology, biotechnology, nanotechnology and cognitive sciences more and more confronts us with questions concerning the difference between human being and machine, material and spirit, life and death. Can artificial intelligence match the human spirit? Are the boundaries between species still fixed in the light of the possibilities of genetic manipulation and animal organ transplants? And by the use of nanotechnology and information technology the possibilities of building technology into the human body escalate enormously. There are two scholars who occupied themselves specifically with the difference between man and machine. The first is Donna Haraway who in 1985 wrote the much-discussed essay 'Manifesto for cyborgs: science, technology, and socialist feminism in the 1980s'.[1] She makes the statement that today's world is populated by cyborgs, hybrid forms in which organisms and machines have been fused. The second is Ray Kurzweil. He regards the convergence of nanotechnology, biotechnology, information technology and the cognitive sciences as a prospect for escaping from human finiteness and developing a form of intelligence that will surpass by far our current imaginative powers in its speed and extensiveness.

On reading the first pages of 'Manifesto for cyborgs' it becomes clear that Haraway has a political intention with the text. She does not want to broach the relationship between human beings and technology, or between organism and machine in an abstract theoretical discussion, but demonstrates that these kinds of differences originate in the essentialist way of thinking that is characteristic of the modern scientific-technological worldview. In modern ontology the human being, that is, 'man', is pictured as the lord and master of nature. He has to make manageable this nature, generally identified with the feminine, and subject it to his goals. In reply to this essentialism Haraway writes a myth about cyborgs. In her criticism the distinction between fact and fiction falls away: to Haraway it is not important whether these cyborgs only are figures in science fiction films or have a concrete existence. Actually it is enough that a different view is taken of reality, so that the interlinking between man and nature, organism and machine becomes perceptible. Are not many people in fact already cyborgs by virtue of their glasses, watch, cellular phone (and in the case of the aged their hearing aids and dentures)? It seems obvious that one would answer this question in the negative, but at the same time we have to state that it is no easy matter to define where the boundary lies between a human being with various kinds of aids and a cyborg.

An example of a cyborg frequently quoted by Haraway is the *Oncomouse*™. This is a mouse tribe that carries information on the human breast cancer gene in its genetic material. Since the company Dupont de Nemours has the patent rights on this 'product' the discussion is whether we may indeed regard a living being as an invention. For is the Oncomouse in the first instance just a mouse bred in a special way like many of our pets, or is it really a product of scientific technological intervention whereby something completely new was brought about? According to Haraway what becomes clear with the example of the *Oncomouse*™ and other cyborgs is that many boundaries in our scientific-technological culture are constructions rather than fixed data which we cannot pass by. A first boundary crossed by cyborgs, according to Haraway, is the boundary between human beings and animals. Just like the centaurs from Greek mythology, the sphinx from Ancient Egypt or the mermaids from our own local legends and narratives, the *Oncomouse*™ consists of a configuration of animal and human elements. The second is the boundary between organism and machine. On the one hand computers and robots are beginning to look more and more like human beings, while on the other hand organic functions like the eyes, the heart and hearing are ever more frequently supported by apparatus and prostheses. The third boundary, between the material and the immaterial, is a subcategory of the second. Here Haraway thinks particularly of the development of nanotechnology by means of which apparatus are increasingly hidden to the eye. A microchip, for instance, can be fitted subcutaneously by means of an injection.

According to Haraway it is characteristic of cyborgs that they have no original nature and no fixed identity either. They are hybrid and heterogeneous constructions that overthrow all existing order. At this point in particular Haraway's political agenda becomes clear. Socialist feminism, according to Haraway, all too frequently was led by established typologies and often identified the essence of 'the' woman with the natural, care of the family or the private sphere. However, modern science and technology break out of these kinds of dichotomies. With ICT technology penetrating into the living room and modern market economy, care-taking is increasingly becoming a component of the formal labour scene. This means that science and technology penetrate further and further into the private sphere but also that what is private is being incorporated into what Haraway calls the *integrated circuit*, the network of modern high-tech society. This has its consequences for the feminist agenda. Just like cyborgs women have a hybrid identity. On the one hand they are a part of the highly modern technological society but on the other hand they also still represent the 'other', that which is not a part of the system. Hereby they have at their disposal the potential to break through the dominant male discourse and create confusion by their hybrid identity. The cyborg politics that Haraway champions resists the totalitarian character of science, technology and society, and simultaneously attempts to assume responsibility for and take up position within the power network of our high-tech society. Rather than withdrawing into a safe domain outside the sphere of influence of science and technology, Haraway wants to participate fully in the existence and give a voice to the weak.

The technologist Ray Kurzweil also strongly emphasises rupturing the boundaries between human being and animal, organism and mechanism, and the material and the immaterial. He sees NBIC convergence (the convergence of nanotechnology, biotechnology, information technology and cognitive sciences) as an impulse to expand the action radius of human beings and by this break through the boundaries of human existence which are set by nature. Kurzweil is often regarded as belonging to transhumanism, a movement that attempts improving human nature to such an extent that it will give rise to forms of intelligence surpassing the human being and obtaining an immortal character. At this point Kurzweil's discourse is at odds with that of Haraway. He believes in total control of reality by means of science and technology. Breaking through the boundaries between species, according to him, leads to our getting more control of existence and so to the possibility of radically doing away with being situated and the finiteness of human beings on which Haraway places so much emphasis.

Ray Kurzweil is of the opinion that the world from science fiction films like *The Matrix* will become reality in the next three to four decades. He forecasts that then there will no longer be a distinct difference between human and artificial intelligence (Kurzweil in Yeffeth, 2003: 234). In *The Age of Spiritual Machines* (2000) Kurzweil, on the basis of an analysis of the developments in the computer industry, concludes that we are at the moment living through exponential technological growth. He expects that this exponential growth will be accelerated further when the current *integrated circuits* technology has come to the end of its potential and is replaced by three-dimensional technologies. To give an impression of this growth he compares the *computing power* of a PC with that of animals and people. At this moment the computing power of a PC lies somewhere between an insect and a mouse. By about the year 2020, however, it will be comparable with the brain of a human being and in 2050 with the brains of a billion people! Kurzweil is aware of the fact that *computing power* does not automatically produce (human) intelligence. Yet according to him there are a large number of technologies in development that will realise a breakthrough. Of great importance is the so-called technology of the *reverse-engineering* of the brain. In this technology the human brain functions as a model for intelligent machines. Thus biologically inspired models will lead to 'genetic algorithms' and 'neural networks'. Another trail-blazing technology is that of implants based on so-called *nanobots*. These are miniscule robots (on a nano scale) which can multiply themselves. They can be designed in such a way that they can take care of communication between neurons and electronics (neuro-transistor). Kurzweil is of the opinion that the development of these nanobots will lead to an expansion of human intelligence and an enormous escalation of our skills.

Kurzweil concisely sums up his message in the title of his latest book, *The Singularity is Near. When Humans Transcend Biology* (2005). According to him we are approaching a unique period ('*singularity*') in which the speed of technological developments will be so great that human life will be irreversibly changed. We will exceed the limitations of our biological bodies and brains and there will no longer be any difference between human being and machine.

Philosophically speaking, Kurzweil forms part of the modern version of the so-called *Strong Artificial Intelligence*. He sees human intelligence and emotions as emerging characteristics of complex systems. 'Emerging' means to say that these are characteristics that rise from or are inherent to complex systems. Douglas Hofstadter and Daniel Dennett are well-known representatives of this line of thought. In *The Singularity is Near* Kurzweil enters into a detailed discussion with John Searle who became known for his 'Chinese Room Experiment' by means of which he attempted to demonstrate that intelligence is a feature of biological systems and not of physical systems. In this discussion the final issue is whether the concepts at higher level (intelligence, will, emotion) are fully reducible to physical phenomena. In the view of Kurzweil and Dennett that is the case. Searle opposes such a materialistic view.

Both Haraway and Kurzweil and the transhumanists are on the same line as Kapp. They see technology as a logical extension of a human being, just as human as man himself. No fear of the cyborg, and no fear of *The Matrix*. In Kurzweil's work we distinctly discern a materialistic view of human beings forming the basis of his optimism. However, a dualistic view of the human being could likewise lead to such optimism. For if a human being could be divided into a physical and a mental part, replacing parts of the body would have no consequences for the spirit. But it is questionable whether materialism and dualism provide satisfactory views of our humanity.

12.5 Technology as an intermediary in our sense of reality

The scholars up to now draw a picture in which technology is something we cannot escape. However, there also are philosophers of technology who hold a more moderate view in which technology does not necessarily have to become one with us. One such is Don Ihde, an American philosopher of technology, who wrote much about the intermediary role of technology in our sense of reality. Ihde was inspired by Heidegger (see Chapter 10) but does not accept everything he says without questioning. From Heidegger he took the idea that human beings inhabit a world with which they are in contact through technology. Heidegger was very pessimistic about the intermediary role of technology because, according to him, technology causes us to see our social environment solely as something that still has to be cultivated. The intrinsic value of reality completely eludes us. Ihde, on the contrary, is much more optimistic. According to him the intermediary position of technology enhances our sense of reality. He illustrates this by describing four different ways in which technology supports our observation (Ihde, 1990).

Ihde derived the first directly from Heidegger. For Heidegger described what happens when we handle a hammer. As long as the hammer lies on the table we realise that it is separated from us. However, when a skilled carpenter takes up the hammer and drives a nail into the wall with it, the hammer becomes one with his hand. While concentrating on hammering s/he no longer realises that s/he has a hammer in her/his hand. The hammer only re-enters her/his conscious thought when for instance the head comes loose from the handle. According to Ihde this is but one example of what he calls the '*embodiment*' relation between a human being and technology. The same happens in the example of the spectacles with which we began this chapter. They become one with our eyes to such an extent that we no longer realise that they are on our noses.

There is a second possibility, a relation that Ihde calls '*hermeneutic*' ('interpreting'). Now the technological device does not become one with ourselves but with the environment. Take for instance an operator in a power station. S/he sees her/his control panels as one with the whole power station. When s/he looks at a display giving the temperature in the combustion chamber s/he sees the flames before her/him in the numbers on the display. We find another example in one of the *The Matrix* films. Someone is looking at the screen where the symbols of the Matrix simulation programme flick by and suddenly calls out that he sees a shapely blonde. Due to his experience with the programme he sees the simulated

Matrix-world in the symbols on the screen. The operator and the computer specialist interpret what they see on the display or the screen. Therefore Ihde speaks about a hermeneutic relation. Technology stands between ourselves and the environment in such a way that we need translation skills to 'read' the device in order to know what is happening. We have to be conscious of this so as not to misjudge ourselves. A person looking at a photo of the universe who does not realise that the colours represent temperature, would think that there are red, green and yellow heavenly bodies to be seen. It is therefore important not to lose sight of the intermediary role of technology.

The third possibility Ihde calls the '*alterity*' relation. In this case technology changes the observed reality to some extent (hence 'alterity', derived from 'to alter'). This relation is found for instance in a science fiction film or a western. We then see the reality in which we live not through a technological device; we only see the technology itself. Another example of this is playing a video game. Here, too, there is no reality behind the technology; there is only the reality of the technological device.

The fourth relation Ihde mentions, is the '*background*' relation. Technology sometimes presents sounds, smells or colours that influence our observation of reality only in the background. We are no longer conscious of the humming of the fridge. What happens here is comparable to the '*embodiment*' relation: when the fridge stops working we suddenly hear the silence. Yet there is a difference. The humming of the fridge is no device for noticing more accurately the sounds in the kitchen. It is merely an addition in the background.

Ihde is mostly positive about the intermediating role of technology. But he does warn about misconceptions that can occur when we are not conscious of the role of technology, as may occur for instance in the hermeneutic relation. Ihde here opposes Heidegger for whom technology had an especially impoverishing effect on our picture of reality. In this Heidegger is not alone. The philosopher Hubert Dreyfus also remarked on a diminishing effect of technology but for a different reason than Heidegger. Dreyfus wrote mostly about computers and the internet. In contrast to Kurzweil he sees fundamental differences between computers and people when it comes to intelligence. According to him computers will never be able to do what people do, since they have no human spirit. Neither can computers replace the bodily aspect of human beings. Dreyfus sees it as an impoverishment should this happen. One example is the Internet. For via the Internet we can communicate with one another in ever more sophisticated ways. We can also 'visit' other places all over the world. By means of images and sounds we get an impression of what the places look like. We can even build complete virtual worlds in which we can wander around as virtual personalities. However, Dreyfus points out that in all this, impressive as it may be, something essential is missing: we are not bodily involved (Dreyfus, 2001). We do communicate with one another, but we are not physically in each other's presence, we cannot shake hands or embrace. We travel through the whole world, but only with our eyes, ears and thoughts. We cannot feel the sand of the Caribbean shore between our toes or smell the scent of the tropical rain forest. For education in particular Internet is an impairment. According to Dreyfus nothing beats being physically together as teacher and learner in a space, however useful conveying information via telematic instruction may be.

Dreyfus warns against the confusion between our reality and the virtual worlds we can create with the aid of computers and Internet. Sherry Turkle has remarked that virtual worlds are a means to assume different personalities and to experience what it is like to be such a personality. But you can also shed that personality with equal ease. Still, that is deceptive, for one cannot shed one's real (non-virtual) personality. A sense of responsibility is definitely not promoted by virtual worlds. Turkle emphasises that we should realise constantly the difference between the real and the virtual world and that the sophistication of technology sometimes causes us to lose sight of this. When a rape was committed in a virtual world, the perpetrator shrugged it off with the argument that it was not 'real'. But it proved that for the victim it was different. She definitely felt it as a far-reaching event. The borderline between virtual and real then begins to blur. Mistakenly, says Dreyfus, for the virtual can never be as complete as the real. So according to Dreyfus no real rape has occurred (which does not go to say that on the whole no crime was committed). This issue is very topical again as a result of virtual worlds like Second Life. The fact that the problem indicated by Turkle and others is relevant transpires from the fact that the Dutch organisation for Technology Assessment, the Rathenau Institute, is researching the question how legislation should be adapted for Second Life-like situations.

The critique expressed by philosopher Albert Borgmann on the intermediary role of technology is also a continuation of Heidegger. The so-called '*device paradigm*' forms the core of his philosophy of technology. It means that human beings to a degree lose contact with reality by means of the devices (Borgmann, 1984). Borgmann calls it: '*disengagement*'. Formerly, people had to go into the forest to chop wood for the hearth. This resulted in an intensive relation with reality. Nowadays they turn up the thermostat. The effect is the same, but the contact with reality has been much diminished. People lose their insight into what exactly is happening. They are merely after 'commodity' and are not interested in the question as to how the 'device' produces that commodity. What Borgmann finds the worst is that the commodity becomes so important that we lose interest in more essential matters. Together with Hannah Arendt he wonders whether technology currently serves a real purpose or whether it has become the purpose in itself. In this Borgmann is entirely in line with Heidegger to whom technology also meant a diminished view of reality. Borgmann's remedy is to look for what he calls 'focal activities'. For instance, chopping wood yourself or cooking a meal (not buying a microwave meal). Involvement in the liturgy of a church service Borgmann also mentions as an example of a focal activity. Technology can definitely be involved in this. A Gothic cathedral for instance can involve us in a most intensive manner in a higher reality. But we will then have to have a much more conscious interaction with technology. In this context Borgmann refers to the book *Zen and the Art of Motorcycle Maintenance* by Robert Pirsig. The profound conversations between father and son on their journey by motorcycle are partly evoked by their relation with the motorcycle. Therefore technology as such need not be an obstacle to an intensive relation with reality. Borgmann realises that the whole of daily living cannot consist of focal activities. That would completely turn the economy upside down. But he does plead for making room for focal activities in the economy that is focused on 'commodities'.

Criticising Borgmann's view is not difficult. He himself admitted that focal activities could never fill all of life. It also can be debated whether it would remedy the problem that Borgmann indicated. Besides, one could wonder where exactly Borgmann is heading. Sometimes he suggests that the exertion of chopping wood has value in itself while it actually is nothing more than a means. Elsewhere, however, he argues that we should not focus on means but on ends. Then heating by means of self-chopped wood or by central heating would be all the same, for the effect is the same. One could also ask whether every technology promotes 'disengagement'. Has not the internet with its many new possibilities for communication enormously intensified the relations between people? Yet Borgmann definitely has a point. For instance, he refers to the way in which a restaurant sign like McDonalds leads to a culinary and cultural unity sausage (or rather a unity hamburger). Our technological activity may indeed not lead to a diminished experience of reality. Whether this says it all, remains to be seen, however.

12.6 Determinism or freedom?

Thus the influence of technology on our humanity and on culture and society can be judged in different ways. It makes a considerable difference what place we allot to technology. But do we have a choice? Have we as human beings not been 'sentenced' to using technology? It seems as if some of the views discussed above imply this. Gehlen for instance, states that technology is inevitable for survival. In the work of Kurzweil we sense the conviction that the development of technology will go on because of its great advantages, even if some people resist it. Both these authors therefore doubt the possibility of a human decision. However, one could ask the question whether this is a good view of technology. Is not technology in the final instance the result of decisions by human beings?

What actually are the motives for developing technology? This question was raised towards the end of the 1960s and the beginning of the 1970s in all kinds of movements trying to expose technological domination. We could even call it an entire *anti*culture. People like Theodore Roszak and Charles Reich protested in their publications against the supremacy of reason-driven technology. According to them technology only served for dominating and exploiting nature. Instead of this they propose that feeling and fantasy should regain power. Often this ideal was coupled with a kind of romancing of nature. But why the natural was better than the artificial was never spelt out clearly. Eventually the anticulture was relegated to the background again. This happened because engineers were capable of solving the indicated problems to some extent. For when it became clear that technology had detrimental effects on the environment they started developing more 'sustainable' technologies. These technologies partially unnerved the critique of the anti-movement. Moreover, the anticulture cannot offer a good alternative to a responsible development of technology. It reminds one of the Luddites in England who, when steam technology was being developed, protested against this 'progress' by vandalising steam engines. Their only alternative was to maintain manual labour. So eventually they lost. To a great extent this is what happened to the anticulture. An upshot of this we currently find in *New-Age* thinking. In this, too, we see a protest against rationalised technology without a ready alternative.

To others technological progress was so obvious that discussion was superfluous. Marxists saw technology as the very means for bringing about the essential social revolution. System technologists and philosophers saw in cybernetics the obvious means to enable control of social developments. Well-known figures in this context are Norbert Wiener and Ervin Laszlo. 'Control' is an important word in their terminology. To capitalists again, technology was the adopted means of generating income and profit. In short, the anticulture raised a question that to many people had ceased to be a question. Why technology? For progress. And what is progress? Sophisticated technological control of course.

The philosopher of technology Egbert Schuurman (2009) however, pointed out that the anticulture was indeed making a valid point. The negative consequences of technology that become ever more distinct, should be a reason for engineers, politicians and consumers to think more critically about the motives for our technological activity. Following the philosopher Hans Jonas, Schuurman uses the concept 'responsibility'. As human beings we always have to be able to give account of our technological activity. To many this would mainly mean accountability towards human beings, for instance towards the government or clients. To Schuurman, however, it also means accountability to God. This emphasis on responsibility links up very well with the approach of this book. Above we argued that human existence has a responsive character. Human beings want to make their existence as meaningful as possible and give account of their deeds. Science and technology can be means towards discovering this meaning (see Chapter 2).

However, what often obstructs the acknowledgement of human responsibility is the ever more dominant character of technology. Technology is supposedly essential to our survival and therefore we need not think about the issue whether we want to launch it or not. Arguments against are no longer relevant. Before considering ethics and the everyday responsibility of human beings in the next chapter, it is important to look critically at the idea of the inevitable supremacy of technology. It also touches on the theme of this chapter. Are artefacts and apparatus merely extensions of our bodies or is there more to it and are human beings increasingly subjected to technology by the development of biotechnology, information technology, communication technology and nanotechnology?

Something that strikes one is that both Haraway and Kurzweil fail to escape a paradoxical tension between technological determinism and human freedom. On the one hand Haraway departs from a materialistic worldview in which no distinction can be made between human being and machine, while on the other hand her discourse is very political and appeals to a specifically human freedom of activity and responsibility. In the work of Kurzweil, on the contrary, the problem is focused mainly on the question whether improvement of human nature means that human beings get more control of their existence and can govern their own evolution, or that they are merely objects of manipulation, a phase in the process of evolution to a higher form of intelligence. Kurzweil concludes the first and states that with the new possibilities human freedom of acting as well as responsibility are expanding. Other transhumanists, however, emphasise that in the new phase of evolution people inevitably lose control over the developments.

Henk Geertsema (2006) who adheres to the same tradition as Schuurman and who gives much consideration in his work to the responsive structure of human existence shows in a plausible way that the paradoxical tension between determinism and freedom can be solved by thinking less absolutely about knowledge. As knowing subjects human beings do not take up position *over against* reality but move around *within* this reality where they attempt to build an existence and link up connections. As human subjects we can only have and develop knowledge when there also are objects which allow themselves to be known. And with our knowledge and skills we can only give form to objects if these objects allow themselves to be formed. Translated to technology, this means that our interaction with apparatus and artefacts does not imply a confrontation between subject and object but on the contrary a relation between them. The human being as the subject takes the initiative and takes up the relevant tools. This does not alter the fact that the structure of the tools or the material that is being processed has a role in determining what is technologically possible or not. The tools and the material therefore also give form to the relation but always in a passive way (see Chapter 4). Even though apparatus work in an ever more independent way it does not mean that they then also gain a certain consciousness and can be called to account as responsible beings. Geertsema (2006: 294) says one can only speak of an autonomous working of machines in the same way that one says physical and psychic processes in the body are autonomous because they are hidden from the conscious observation of human beings. This is a thought that links up well with the phenomenology of Dreyfus and Ihde who emphasise that in a symbiotic relation between human being and machine human beings always remain the initiating party.

Geertsema's view also assists us in reflecting more accurately on boundaries and delimitation. When we state that human beings are involved in a web of relations with other beings or things we no longer need to maintain that there is an absolute boundary between human being and animal, organism and machine, the material and the immaterial. So then, it is not at all strange when physical or technological structures are implanted into the human body where they influence or take over certain bodily processes. When we analyse the body from an abstract physical perspective we can indeed discover many similarities between the human body and apparatus and can even describe the body as an intricate machine. It becomes problematic when we forget that we are herein dealing with an abstraction and that the body is not only a physical structure but also a living organism with certain processes of consciousness. It therefore still makes sense to distinguish between human being and animal, organism and machine, the material and the immaterial, but that does not exclude every possibility of a relation between the different poles. We should not separate the poles from each other, but distinguish between them.

12.7 Conclusion

This chapter is about the relation between human being and technology. In the initial phase of philosophy of technology, in particular in the work of the German Ernst Kapp, technology was seen as an extension of the human body. In the continuation of this view the philosopher Arnold Gehlen emphasised that the human being is a deficient being who needs technology for survival. The historian Lewis Mumford, on the other hand, regards a human being as an acting cultural being who during the course of history has opened up the way to the development and application of technology. According to him, new technologies are 'embodied' in organisational structures that attempt to exercise complete control over all human activities. However, Mumford holds the opinion that human beings have the ability to shake off the controlling power of technology.

The rapid developments in technology tempt many modern day technologists and philosophers into breaking through the boundaries between human being and animal, organism and mechanism, the material and the immaterial by their thinking. A *merger* of human being and technology could, according to Kurzweil, hold infinite prospects, among which is immortality. A more moderate view is found in the philosophers Ihde and Borgmann who state that the intermediary role of technology can lead to a diminished experience of reality.

The price for the utopian visions of Kurzweil and others is a complete surrender to technology: a Faustian transaction whereby immortality is obtained at the price of human freedom. It is these kinds of utopian visions – where the issue of the reality of their nature is less relevant – that compel us to raise the issue of the motive for our technological activity. We postulate that the relation between technology and humanity may only be raised within the perspective of human responsibility. This opens the eyes to a network of qualitative and normative relations in which human beings relate to other human beings, animals, plants and things.

Note

1 We use the 1991 version.

References

Borgmann, A. (1984) *Technology and the Character of Contemporary Life: A Philosophical Inquiry*, Chicago, IL: Chicago University Press.

Dreyfus, H.L. (2001) *On the Internet*, London: Routledge.
Ihde, D. (1990) *Technology and the Lifeworld: From Garden to Earth*, Bloomington: Indiana University Press.
Geertsema, H.G. (2006) 'Cyborg: Myth or Reality?', *Zygon*, 41(2): 289–328.
Gehlen, A. (1940) *Der Mensch. Seine Natur und seine Stellung in der Welt*, Berlin: Junker und Dünnhaupt. Translated as *Man, his Nature and Place in the World*.
Gehlen, A. (1957) *Die Seele im technischen Zeitalter. Sozialpsychologische Probleme in der industriellen Gesellschaft*, Hamburg: Rowohlt. Translated as *Man in the Age of Technology*.
Gehlen, A. (1961) *Anthropologische Forschung*, Reinbek bei Hamburg: Rowohlt.
Haraway, D. (1991) 'A Cyborg Manifesto: Science, Technology, and Socialist-Feminism in the Late Twentieth Century', in Haraway, D., *Simians, Cyborgs and Women: The Reinvention of Nature*, New York: Routledge.
Kapp, E. (1877) *Grundlinien einer Philosophie der Technik. Zur Entstehungsgeschichte der Kultur aus neuen Gesichtspunkten*, Braunschweig: Verlag George Westermann.
Kurzweil, R. (2000) *The Age of Spiritual Machines*, New York: Penguin.
Kurzweil, R. (2005) *The Singularity is Near. When Humans Transcend Biology*, New York: Viking Press.
Mumford, L. (1934) *Technics and Civilization*, New York: Harcourt Brace.
Mumford, L. (1967) *The Myth of the Machine, I. Technics and Human Development*, New York: Harcourt Brace.
Mumford, L. (1971) *The Pentagon of Power, II. Technics and Human Development*, New York: Harcourt Brace.
Pirsig, R. (1974) *Zen and the Art of Motorcycle Maintainance*, New York: William Morrow & Company.
Schuurman, E. (2009) *Technology and the Future: A Philosophical Challenge*, Grand Rapids, MI: Reformational Publishing Project and Paideia Press.
Yeffeth, G. (ed.) (2003) *Taking the Red Pill. Science, Philosophy and Religion in The Matrix*, Dallas, TX: Benbella Books.

Further reading

Roszak, T. (1969) *The Making of a Counter Culture. Reflections on the Technocratic Society and its Youthful Opposition*, New York: Anchor Books.
Verbeek, P.-P. (2005) *What Things Do. Philosophical Reflections on Technology, Agency and Design*, University Park, PA: Penn State University Press.

DON IHDE

(1934)

PORTRAIT 12 Don Ihde

Don Ihde was born in 1934. He studied philosophy at the University of Kansas and theology at the Andover Newton Theological School with the well-known theologian Paul Tillich. He worked at Boston University and Southern Illinois University for some years and since 1969 he has been Distinguished Professor of Philosophy at the State University of New York (SUNY). He is the Chief Editor of an important series of books, the *Indiana Series in the Philosophy of Technology*.

Don Ihde is one of the doyens of current philosophy of technology. He is still very active in the discipline. In 1979 he published his first book on philosophy of technology, *Technics and Praxis*, which showed his philosophical orientation towards Heidegger and Husserl. His next book, *Technology and the Lifeworld* was published in 1990. Here he worked out for the first time his ideas on the role of technology in our observation of reality referring to empirical examples. The book forms his 'empirical turn' that we find in many current philosophers of technology (see also for instance the portrait of Andrew Feenberg). In 1991 *Instrumental Realism* came out which in a certain sense is the philosophy of science counterpart to the more philosophy of technology oriented *Technology and the Lifeworld*. In 2002 Ihde published *Bodies in Technology* in which he gives special consideration to the role of corporality during our observations. He is also concerned here with observations of our own bodies (for instance by means of medical imaging techniques).

Ihde's interest in philosophy of technology came about especially through his study of the role of scientific instruments. Such instruments play an intermediating part in observations. This role remained the most important focus of his attention in all his work on philosophy of technology. This work is in the tradition of the phenomenology which was developed by, amongst others, Edmund Husserl. Heidegger, too, wrote about the role of technology in our observation and experience of reality but often in a very critical vein. To him technology in particular meant a reduction of our experience of reality. Instead of the full, gorgeous reality we merely see things from which something still has to be made. Ihde joins up with this in a sense when he emphasises that technology gives us a framed version of reality. This is what literally happens when we look out through a window frame: we then see only that part of reality that can be seen through the window. Technology therefore does not play a neutral role in our observation. It always changes somewhat the way in which we see and experience things (see also Chapter 12). To Heidegger it was mainly a negative change, a narrowing down, but Ihde is much more positive. Technology can, according to him, enhance our observation of reality. For instance, our realisation of the finiteness of the earth was enhanced when astronauts could take photos of the earth on which it could be seen as a beautiful coloured ball in a vast dark environment. Technology always plays its intermediating role in a cultural context, which also helps to determine the way it functions. Ihde illustrates this amongst other things with reference to an example about observation when navigating. He demonstrates that the instruments used by Columbus were useless to the inhabitants of the islands that he discovered because these instruments did not fit into their cultural context.

Let us look at a quotation from the work of Ihde:

If the role of perception, especially at the microperceptual order of things, does not immediately seem to fit the concerns of philosophy of science, the opposite is the case with respect to the philosophy of technology. The materiality of body carries with it, particularly with regard to artifact use, certain rather immediate implications. Merleau-Ponty offers three examples in the discussion of bodily spatiality: 'A woman may, without any calculation, keep a safe distance between the feather in her hat and things which might break it off. She feels where the feather is just as we feel where our hand is. If I am in the habit of driving a car, I enter a narrow opening and see that I can "get through" without comparing the width of the opening with that of the wings, just as I go through a doorway without checking the width of the doorway against that of my body.' In Polanyi's parlance, these phenomena would be examples of *tacit knowledge,* since they are a kind of 'know how' without explicit conceptual judgment attached to them. But they are more – they are examples of how artifacts (technologies) may be used or experienced in use. They are examples of what I call *embodiment relations.* Such relations are existential (bodily-sensory), but they implicate how we utilize technologies and how such use transforms what it is we experience through such technologies.

(Ihde, 1991: 29–30)

References

Ihde, D. (1979) *Technics and Practice*, Dordrecht: Reidel.
Ihde, D. (1990) *Technology and the Lifeworld*, Bloomington: Indiana University Press.
Ihde, D. (1991) *Instrumental Realism*, Bloomington: Indiana University Press.
Ihde, D. (2002) *Bodies in Technology*, Minneapolis: University of Minnesota Press.

13

'GOOD' TECHNOLOGY?

Normative artefacts and the web of responsibilities

Summary

In this chapter we will be dealing with a number of ethical theories, each one of which sheds light on one specific aspect of human conduct. We will be discussing in succession deontology, consequentialism and virtue ethics. Deontology emphasises in particular the constitutive side of operational practice, consequentialism emphasises the contextual side and virtue ethics the regulative side. As a result of modern technology the radius of action of people expands, and thus human accountability is extended. In our opinion engineers and managers are not powerless when facing scientific-technological development. The technological-organisational form-giving of social practices can also support actors in making sound choices. Moreover, the differentiation of social practices facilitates a better staggering of responsibilities.

Do technology and ethics really go together? Is it useful to develop a separate ethics of technology? Does ethics have any significance for engineers who simply want to solve concrete problems or for managers who have to run a factory? Is ethics not rather something for politicians and policy makers? We start with two statements which illustrate the way people in our culture think about the relation between technology and ethics.

> As an information scientist I do not bother with ethics. It is not my field of study. I have to see that I deliver a good computer system that functions properly. What happens further with the product is the user's responsibility.

> I'm not in favour of genetic engineering in agriculture myself, but if we do not do it, it will be done by someone else. Besides, we cannot afford to lag behind.

It is striking that in these statements technology is treated as an autonomous phenomenon which can be seen separate from human accountability, or which has its own scientific-technological logic which one cannot influence. And if ethical reflection were possible, it would be a reflection that comes not from technology itself, but one from outside that lays down limits to technology. In the statement cited first this emerges in the supposition that technology has no normativity in itself. It is not designing technology that demands ethical reflection but its application in a concrete context. Therefore technology here is no more than an instrument that can be utilised in a good or a bad manner. The second statement, on the contrary, stresses the autonomy of technological development. You may very well ask moral questions about science or the use of technology but it has little use as science and technology nevertheless run their own course and as human beings we have very little influence on that.

Both arguments are plausible. It is indeed true that a distinction should be made between the design phase and the utilisation phase (see Chapter 7): the designer cannot be held responsible for everything the consumer does with the product. In this respect it is justified when, for instance, a producer of weapons states that they are not responsible for the wars waged with the weapons they have developed. Yet this is not all there is to it. For even in the design of weapons numerous moral decisions are taken: requirements for the safety of the weapon or for the range and nature of the damage it can cause. May one, for instance, develop cluster bombs when many clever weapon systems can be developed that restrict the number of civil victims? The designer is not always responsible for what is done with the weapons they design, but they may be held responsible for the choices they make while in the process of designing. They can choose whether to make use of certain possibilities for development or not.

It cannot be denied either that numerous technologies are developed, like information and communication technology, genetic engineering and nanotechnology, each one of which exhibits its own dynamics and by reason of that is difficult to manage. It has also been proved that technology and economy are so closely interwoven that they form a power to which nobody is equal (see Chapter 11). This does not mean, however, that engineers and managers are not responsible for the choices they make. The above-mentioned technologies are ethically ambivalent and the people involved can take crucial decisions in many areas. It is therefore good to be aware that the room for action by engineers and managers is limited and that they can therefore not be held accountable for everything that happens to their product. Still, in their own sphere of activity they have a degree of latitude which carries with it a specific responsibility.

13.1 Ethical theory and practice

Contemporary practice of ethics often begins with a theoretical reflection on abstract moral principles and subsequently evaluates the actions of human beings in practice. This method can be queried in diverse ways. First we will discuss the three most common ethical theories: deontology, consequentialism and virtue ethics. We will make these three different ethical approaches concrete by an example from technological practice. We will use a case study written down by Simone van der Burg and Anke van Gorp (2005) on the design of trailers. One of the dangers for pedestrians, cyclists and the drivers of motor cars when colliding with a lorry is that they land between the wheels of the trailer and are decapitated. To prevent this kind of situation the Dutch government has drawn up a number of safety regulations. Lorries must now have a mirror that shows the blind spot and have to install beams between the pivotal axle and the wheels of the trailer. It would be even safer to cover completely the underside of the trailer, both in front, behind and on the sides. Practice has shown, however, that designers hardly have time for extra safety precautions. In what follows, we will see that the different ethical perspectives continually throw light on other aspects of this case.

13.1.1 Deontology

In deontology one investigates whether actions comply with moral rules and principles ('deon' means 'that which ought to be done'). Deontology investigates whether people do their duties and act with respect for moral law. The motivation of an activity or its consequences are less important. In the case of the safety of lorries and trailers deontology would primarily think of rules of conduct laid down by government to protect the interests of other users of the road and thus to take care that the freedom of conduct of the transport company or the lorry driver does not infringe on the freedom of other users of the road: the compulsory use of a mirror for the blind spot or the prescribed beams to the front of the trailer.

The German philosopher Immanuel Kant (1724–1804) did the best known and trend-setting elaboration of deontology. Kant states that a human being is a citizen of two worlds: the world of nature and the world of freedom. In the world of nature everything takes place according to laws of causality in which one event follows from the other. In the world of freedom a human being is capable of taking decisions and thus altering the course of history. As long as we are led by our nature, needs, wishes and urges, we as human beings remain heteronomous. This means that we are at the mercy of alien powers. But if we act with freedom and allow our conduct to be determined by our will and not by our primary urges, then we are autonomous. According to Kant we should not allow ourselves to be ruled by powers from outside, but by the moral law of reason which is present in every human being.

In his ethical theory Kant looked for *one* universal principle from which all other moral laws, rules and principles can be deduced. For autonomy demands that we do not allow ourselves to be led by circumstances but by a law that is generally valid for all and applies to everyone in every situation. Kant formulates his universal rule of categorical imperative as follows: 'Act only on that maxim whereby thou canst at the same time will that it should become a universal law' (Kant, 2005: 39). This means that an activity can only be rationally justified when it can also be carried out by everybody else in a different situation without thereby jeopardising a third party; or, 'do as you would have others do to you'. Moreover, Kant considers it important that the freedom and rationality of others are not denied, for in this way one would detract from their human dignity. This so-called principle of reciprocity is considered by Kant not to be contradictory to the above-mentioned principle of universality. On the contrary, according to him both principles eventually lead to the same conclusion. If one formulates the categorical imperative in terms of the principle of reciprocity one arrives at the following principle: 'So act as to treat humanity, whether in thine own person or in that of any other, in every case as an end withal, never as a means only' (Kant, 2005: 46). What Kant wants to express with this principle in particular is that we should have respect for others, just as they should have respect for us. It is not permissible to use another person as an instrument for one's own goals, because we ourselves would be robbed of our humanity if other people treated us as an instrument.

During the course of history diverse objections have been raised against deontology. One of the objections is that such an ethics offers no solution to situations in which two principles clash which can both be universalised. In everyday practice, however, we often have to choose between two evils, for instance between the safety of the traffic and the extra cost to be incurred to adapt lorries and trailers. The outcome of the decision is always closely dependent on the specific situation. And with this we arrive at a second objection against deontology, namely that it does not take into consideration the consequences or the undesired side effects that the action may have in a specific situation. Deontology abstracts too much of the specific situation and regards the action as an entity in itself that only has to comply with the generally valid rules. For instance, it could happen that in a country with a much used and congested road system weighing up the safety and cost efficiency could be approached differently from the same consideration in a country with an extensive road system on which traffic is much lighter.

13.1.2 Consequentialism

Consequentialism is directed towards the consequences of an activity in a certain situation. According to consequentialism actions should not be evaluated by themselves but always within a given situation or context in which goals are pursued. One and the same conduct could be the right thing in one situation while in another situation it could be wrong. By means of this consequentialism highlights an issue that was underemphasised in deontology: differences in circumstances and context. Approached from the angle of consequentialism the designer of the trailer in one situation would have to make a different choice from their choice in another situation. If for instance the client imposed harsh restrictions and the competition was deadly, one could not expect much from the designer's inventiveness. However, it is entirely different if the designer is guaranteed a purchase and thus can him/herself also set certain requirements because s/he need not be so apprehensive of damaging his/her position in the market. Under the denominator *Technology Assessment* diverse instruments have been developed which can make consequentialism concrete. For instance, methods and technologies have been developed for measuring the effects of decisions, making prognoses and formulating requirements for technological designs in a democratic manner (see Box 13.1).

BOX 13.1 FRAMEWORK

Technology Assessment

Technology Assessment is a field of study in which the social effects of technological activity are investigated and a responsible embedding of technology within the context of nature and society is pursued (Sollie and Düwall, 2009). Over time methods and technologies have been developed in this relatively new field of study. Some are focused especially on investigating new technological developments, others more on assessing the social consequences and environmental effects, and still others on considering social interests and perspectives in a balanced way. Generally speaking, Technology Assessment is comprised of two activities, namely to investigate new technological developments and to assess the effects (impact) of these technological developments on society and environment. The objective is to influence the technological developments in the desired direction. If this is not possible then the negative effects have to be anticipated and provisions have to be taken to absorb these effects. The primary concern is not with the direct and intended effects but with the indirect or higher order effects which are mostly unintentional too.

We can distinguish three periods in the history of Technology Assessment. The first covers the period from directly after the Second World War up to the 1960s. This was the time of reconstruction during which the whole of society was characterised by a belief in progress, national safety and prestige. Science and technology were then mainly seen as a driving force behind social development. Moreover it was generally accepted that fundamental research would in itself lead to new useful applications. From the 1960s and 1980s a more critical period followed. Society became increasingly conscious of the negative ecological and social effects of new technology. Therefore ways were explored to direct scientific-technological development and keep it in check. There also was an endeavour to develop more nature-friendly and society-friendly alternatives. The idea was that science and technology should no longer develop autonomously but had to be directed in a democratic way. The last period is the period from the 1980s up to today. Under the influence of the economic recession the operation of the markets is increasingly regarded as very important and government enterprises are swiftly privatised. Science and technology are now mainly seen as strategic means to strengthen the competitive position and to reap the fruits of globalisation. The scientific-technological development is not rejected altogether as in the 1960s and 1970s but is not applauded without criticism either as it was in the preceding period. Within the existing order space is created to intervene in a strategic manner and alter the direction.

The most well-known and trend-setting approach in consequentialism is utilitarianism. Utilitarianism raises the question whether an activity is useful or serves a moral purpose ('utilis' means 'useful'). Often utilitarianism is presented as a somewhat crude form of ethics, an ethics that is only after efficiency and general usefulness. However, when we look at the rise of utilitarianism it can be doubted whether this reproach is justified. The founders of utilitarianism at the time were important social innovators who pursued a more humane society. Jeremy Bentham (1748–1832) with his theory, for instance, wanted to develop a judicial system that was not based on retribution but on laying down punitive measures that on the one hand would sufficiently deter and on the other would leave open enough possibilities for rehabilitation. John Stuart Mill (1773–1836) was a great advocate for an open dialogue in which everyone would be able to form their own personal opinions and convictions and express these freely.

Although the principle of utility already played a prominent role in the works of Francis Hutcheson and David Hume, Jeremy Bentham is generally viewed as the father of utilitarianism. According to him an activity is useful when it leads to the greatest happiness for the greatest number of people. For Bentham happiness finally means the presence of 'pleasure' and the absence of 'pain'. In this he stands in the tradition of hedonism that elevates pleasure to the highest moral value and allots a subordinate position to other values. Bentham eventually also wants to quantify the highest form of pleasure. For this he developed the 'hedonist calculus' in which costs and benefits, pleasure and pain could be weighed up and the most balanced choice be sought. The money standard can be used as a general criterion. It is not surprising that his theory does well in a society where trade and consumption is increasing rapidly under the influence of the industrial revolution.

John Stuart Mill elaborated on Bentham's work. According to him one cannot merely describe pleasure in quantitative terms, one should also take into account the quality of the pleasure. Intellectual desires like reading poetry or studying philosophy, according to him, have a higher value than bodily or beastly desires like eating, lounging or sexual pleasure. It is better to be a dissatisfied Socrates than a satisfied pig. To Mill it is important that people can fully develop their individual talents according to their own insight. Everybody should be free to pursue their own pleasure in their own manner. We may not take away or obstruct one another's pleasure. The principle of freedom is therefore restricted by the principle that one person should not harm another.

Consequentialism has also been criticised over time. A much-heard objection is that it could easily lead to an ethics in which there is an over-emphasis on the goals to be realised and too little stress on the internal motivation and moral duty either to act or to refrain from acting. Thus we have seen that the designers of the trailer did adhere to the compulsory safety regulations but took no extra trouble to promote the safety of other road users. Safety is seen mainly as the responsibility of government and the users. Sometimes the emphasis on goal-oriented activity can extend to where the end justifies the means and for instance certain fundamental rights are infringed to keep the peace or to bring about greater wealth. Another objection against the approach of goal ethics is that it is always hard to calculate the exact effects of a certain measure. One never knows where the development of a certain technology might lead. And even if a reasonable assessment of the effects were possible, one still has to find a normative principle to determine what is permissible and what is not. For instance, how do we weigh up the driver's safety against the safety of other road users? It is highly dubious whether a criterion like 'the highest pleasure to the greatest number of people' would offer enough to hold on to.

13.1.3 Virtue ethics

Virtue ethics finally concentrates on the question of how one becomes a person of good morals. Formulated in contemporary terms it is the integrity of the acting person with which we are here concerned. For instance, when engineers who are designing a trailer neglect to pay sufficient attention to the safety precautions we could ask ourselves how they can give account of this. For designers not only want to be good professionals but also morally good people and that means that they have an accountability that transcends the borders of their practice. Every group of professionals works with unwritten rules about what is and is not morally permissible. Sometimes these rules are also formalised in a code of conduct, as for medical practitioners, barristers and the military. Also professional engineering organisations often have their own code of conduct.

Virtue ethics goes back to the work of Aristotle (384–322 BC) and had much influence in the Middle Ages. Modern thought has always rebelled against virtue ethics. However, in the last few decades it has been enjoying renewed interest. *After Virtue* (1981) by Alasdair MacIntyre has played an important part in this. According to Aristotle a person has a good character when s/he has certain virtues and moreover knows how to use these virtues in a good and balanced way. Examples of virtues are courage, temperance, justice, truthfulness, friendliness and a sense of humour. But in forming an excellent character it still is imperative to hold the golden mean and beware of extremes. Courage, for instance, supposes a measure of nerve, but there is the danger of becoming reckless. To find the best dose and optimal balance people have to rely on the practical wisdom of their common sense. So this can contribute to forming a moral character and also play a part in appreciating the good life. Upbringing and education are of the utmost importance here. Ethics should be focused on a state of *eudaimonia,* a state of ultimate happiness in which people can fully reach their destination. However, people can never generate this good life on their own, they only find this in interaction with others. Therefore, by definition the good life is a life in communities.

Virtue ethics is not focused on the realisation of duties or goals but on forming a virtuous life that has value in itself. MacIntyre elaborates on this difference by making a distinction between goods that are internal to a certain practice and goods that are external (see Chapter 9). Following Aristotle, he states that virtues are always formed within practices. For instance, one can only become a good football player if one becomes a member of a football association to bring the playing of football into practice. In addition it also is important that a good context is created in which virtues can be developed. With reference to football one could say that there should be a situation of *fair play*. In current top football where money plays a great role there certainly is the risk that the only thing that matters is the end result of the match, the number of goals scored. However, the score is an external good to be distinctly differentiated from the internal good of playing the game itself. One can lose a match and still feel satisfied with the way one played. The major concern then is what is most important in football: the quality of the game itself. In addition it is significant to realise that practices according to MacIntyre's view never can exist all by themselves. A specific practice can help people to develop certain virtues, but a virtue only really becomes a virtue when it pervades one's whole life. A person cannot be courageous in one context and cowardly in another, or honest in one context while telling lies in another. The virtues that are developed in various practices should get a position within an unambiguous life story, and the different life stories should again be embedded in a common tradition.

A first objection against virtue ethics, however, is that it seems difficult to fit it into the modern fragmented way of life. As modern people we function in a diversity of relationships and contexts and in the one context a different attitude and position is expected of us than in another context. This also is expressed in the dilemma of the engineers who have to design a trailer. As parents, friends or neighbours they may be highly concerned about the safety of their loved ones, but in their professional life, with its pressure, urgency and restrictions, they may see no way of realising this value on the shop floor. They are compelled to take decisions that clash with decisions they would take in other spheres of life. Another objection is that virtue ethics has a strongly elitist character. In fact one can only become a virtuous person if one grows up in a good harmonious environment, with parents who know how to bring up their children, teachers who know how to discipline their pupils, and friends who keep us on the right track. It is sometimes said that one has to have a certain measure of '*moral luck*'. Only citizens who had enough wealth and free time to reflect and who had access to the polis, could develop into virtuous people in the time of Aristotle. Likewise a boy in Barcelona or Munich whose parents can buy him football boots, has a much greater chance of excelling in football than another boy of the same age living in Cameroon.

13.2 Acting responsibly in practice

We have discussed three common ethical theories. Deontology puts special emphasis on the structure of activities themselves, consequentialism is geared towards the consequences of the action in a specific context, and virtue ethics finally concentrates on the actor, looking at the question of how people give direction to their conduct and how they can develop into virtuous people. When we look at the three different ethical theories it seems that a clear connection can be made with the concept of practice as used in this book (see Chapter 9). Deontology focuses mainly on the constitutive side of practice: which rules and normative points of departure apply to a practice and how can the normative points of departure best be realised? Consequentialism looks at the contextual side of operational practice. The concern here is the result and the effect that actions can have. Finally virtue ethics is focused on the regulative side of operational practice, i.e. on the way in which people, groups and communities express their intentions, motives, views and convictions. These different perspectives are summarized in Table 13.1.

In this book we start from the assumption that each theoretical reflection and therefore also the forming of ethical theory isolates a certain aspect from the complex cohesion of meaning in which people function. A consequence of our approach is that primacy is given to practice since we presume that practical experience precedes theoretical thought. This is diametrically opposed to the current view of ethics as a theoretical foundation for good decisions. The sequence is exactly the other way around: the forming of ethical theory and reflection takes further the moral intuition that people already have in their daily lives. Theoretical reflection can actually help to sharpen our practical intuition and if necessary criticise it, but there are no independent moral rules and principles that we first have to find by means of theoretical reflection and subsequently apply in a specific context.

However, the idea that the forming of ethical theory abstracts from the complex coherence of meaning has more consequences. Often a distinction is mistakenly made between a factual and a normative reality, where the former is regarded as a given, present world while the latter reality is seen as something that should be disclosed by ethical reflection. The different operational practices, however, actually exist by the grace of the fact that they are intrinsically normative by nature. Therefore during the last few decades much criticism has been levelled at the rigid distinction between fact and norm, *is* and *ought.* Thus Jürgen Habermas, for example, holds that we cannot comprehend the factual development of justice and morality without accepting that the opinions and convictions on which they are based have intrinsic normative power. The other way around Pierre Bourdieu argues that the development of justice and morality can well be explained sociologically and that an important driving force behind this development is that people are attempting to acquire social status by pretending to be altruistic. The traditionally accepted division between the social sciences and ethics, regarding the

TABLE 13.1 Overview of different ethical perspectives in relation to the practice model

	Perspective of action	*Dimension of the action*	*The side of practice*
Deontology	Activity	Structure	Constitutive side
Consequentialism	Situation	Context	Contextual side
Virtue ethics	Actor	Direction	Regulative side

former as occupying itself with the factual structures of society and the latter with what the norms for society should be, has been proved to be no longer tenable. Every forming of social theory is normative by nature and every practice of ethics should also be concerned with the factual situation. It is striking that after an empirical turn in philosophy of science and an empirical turn in the philosophy of technology nowadays one also hears about an empirical turn in ethics.

A third and last consequence of the turn to practice is that the moral dimension can no longer be isolated from the other aspects of human existence. It may be useful in itself to develop an ethics in the narrow sense of the word in which reality is investigated from a specific ethical or moral perspective. For instance, it is important to make an explicit distinction between the moral aspect and the juridical. In this way one could show that a moral appeal can indeed be made to the designers from the previous section to raise the safety of the trailer, but that juridical sanctions cannot automatically be attached to this. It can also be good to distinguish the moral aspect from the aspect of faith. It is true that to a great extent religion and philosophy of life do direct ethical judgement, which does not mean, however, that moral intuitions can be derived directly from religious convictions. A commandment like 'Thou shalt not kill' for instance is endorsed by almost everyone even though people differ in their opinions on the origin of that commandment. Yet we cannot avoid anchoring such a constricted ethics in an ethics in the broad sense of the word that investigates human activity in the diversity of normative contexts and relations (for instance lingual, social or aesthetic).

13.3 Modern technology and the extension of the human operational radius

We return to the questions raised at the beginning of the chapter. Of course no engineer or manager will deny that s/he is responsible for the choices s/he makes, neither will s/he say that it does not matter how s/he acts. So why the resolute statement that engineers and managers cannot be held responsible for the social effects of their products and that they cannot influence technological development? The answer to this question, according to us, is that in modern scientific-technological development the accountability of people becomes so huge that it easily exceeds human limits. It seems that science and technology are leaving human beings behind. The only possibility left is to adapt passively to technology or merely direct oneself to components. Moreover, it happens more and more that people work in extensive associations where they only have control of one small link in the entire process. Thus they can hardly be held accountable for the process as a whole.

The pressure of modern technology on human responsibility is not something new. Where formerly circumstances, however, were experienced as a given, reality is now seen as something constructed. Thinking in terms of threats, danger and fate is replaced by thinking in terms of risks (Giddens, 1991). Illnesses and disasters formerly outside our sphere of influence, have now for instance become part of our scientific experience and technological skill. The fact that we can calculate the chance of a seawall or a dam wall breaking makes us partly responsible for potential disastrous floods, should we as hydraulic engineers and as policy makers neglect to take effective precautionary measures. Also engineers who develop a new material for aeroplanes are responsible for defects in this material. If one thinks of the consequences of a plane crash, one realises how heavy this burden really is.

The expansion of the human operational radius is closely connected with the specific character of modern technology. In contrast to more traditional forms of technology it is characterised by an independent functioning of its tools. A hammer still has to be gripped in the hand but technological apparatus like the clock, the steam engine or the computer system function as relatively independent entities. You only have to press a button and the specific action is performed. This directly affects the practice of the engineer who becomes more and more dependent on apparatus and who finds it increasingly difficult to oversee the effects of his/her actions. Moreover, modern technology is characterised by a symbiosis of science and technology (therefore nowadays one often uses the term *technoscience*). In this way a systematic exploitation of nature becomes possible and an enormous expansion takes place. The philosopher Hans Jonas in particular lays much emphasis on this last aspect. He even states that a complete new ethics is needed. All earlier forms of ethics had as a common presumption that the nature of human action always is a given. But as currently under the influence of science and technology the radius of action of people increases infinitely, and with it the nature of human action also changes, we cannot evade developing new forms of ethics (Jonas, 1984: 1).

We can explain Jonas' view by using an example from his work. He begins his book *The Imperative of Responsibility* with the image of a farmer from antiquity cultivating his land. The effect of his intervention in nature remains limited and has little consequence for the earth. So no special commandment is needed to protect nature. Neither need one worry about what is happening elsewhere. In such a situation it is enough for people to have the rule that they should love their neighbour as themselves. Nowadays the situation has changed drastically. An accident with a nuclear reactor can inflict damage on a large area and radioactive waste can constitute a risk to the health of people for millions of years to come. Likewise the effects of genetic manipulation and other forms of biotechnology can often not be calculated. Global warming and the hole in the ozone layer moreover demonstrate that human beings are capable of destroying the earth and with it their own living environment. Thus human beings have become co-responsible for the future. So nowadays our duty is to take care of the survival of the earth. Traditional ethics, in which the relationship of one human being to another takes central position, no longer suffices for this. We should gather knowledge on the possible effects of our actions and go in search of an ethics in which caring for the earth takes a central position. Instead of Kant's categorical imperative we now have to devise an imperative for actions that is focused on the future world: 'Conduct yourself in such a way that the effects of your conduct are always in agreement with the continued existence of dignified human life' (Jonas, 1984: 11).

One could wonder whether Jonas' statement that the nature of human actions has changed and his emphasis on the necessity of a new ethics are not exaggerated. Is it really true that with the extension of their operational radius human beings also get another kind of accountability? Or is their responsibility merely extended? Is the plea by Jonas to focus ethics on the continued existence of dignified human life not better understood as a new interpretation of the commandment to love the neighbour as oneself? One could extend the Kantian demand for universalising one's actions in such a way that it also entails the fate of future generations. However, that does not detract from the fact that the implications of our conduct and the uncertainty about its effects set before us new far-reaching moral choices.

13.4 The normativity of technology

That under the influence of modern technology the radius of action of people is enlarged thus also carries with it an extension of human accountability. Jonas doubts whether we modern people can rise to this extended responsibility. According to Jonas we live in a moral vacuum since science and technology have demythologised our worldview, thereby causing our moral sources to run dry. Nature is regarded as something that in itself has no intrinsic value, but to which human beings should give meaning (Jonas, 1984: 23–24). Concerning the short-term effects of risks in our direct sphere of life, the fear of a disaster may get us to take care of our environment. But what about care for future generations or people from other parts of the world? Only reverence for the sacred, that is, the awareness that one cannot just do to nature as one likes, can still help, according to Jonas. As modern people we hardly know what that means. Modern people can hardly fulfil their extended responsibility. It is therefore not surprising that scientific-technological development is coupled with an increasing resignation and passivity regarding science and technology.

We have seen in the introduction to this chapter that the first reason why engineers and managers think they are not capable of taking responsibility for the results of their actions is that they regard science and technology as an independent reality of its own, detached from the morality they have as human beings. As experts they are merely doing their work and trying to do it technologically as best as they can. They do not feel that they are responsible for what is done with the product further along the line. Their technological-scientific actions have nothing to do with their subjective opinions and motives. This argument, however, is mistaken. Engineers and managers cannot appeal to the neutrality of science and technology as an alibi for evading their responsibility. Technological development practice is far less neutral and also much less subject to the logic of science and technology than is generally supposed. Engineers and managers are indeed responsible for the choices they make. People should never allow themselves to be imprisoned in a scientific-technological system but should learn to stand in a relationship to technology where they retain their freedom. However, this is only possible if we also consciously let go of the idea of having control over everything. We should have the courage to shrug off technology so that we can extricate ourselves from its power. We should not do everything that is technologically possible, nor adapt to the laws of technological society, nor allow technology to determine the horizon of our thinking (Ellul in Ferré and Mitcham, 1989: 23–36).

The ethics of freedom puts much emphasis on the accountability of the designer and the manager when it comes to technological development. It is, however, a question whether this ethics is not asking too much of engineers. Hans Achterhuis who links up with Bonhoeffer states that it is abnormal to make every moment of our lives into an existential moment in which we have to take a radical decision. This would make life unbearable (Dorrestijn and Verbeek, 2013). The ethics of freedom denies that technological artefacts and systems also have a morality of their own that gives direction to the user's conduct. Thus it is blind to the possibility that designers can also give form to artefacts and systems in such a way that they stimulate certain types of responsible conduct or, to the contrary, frustrate such conduct. We can indeed appeal to the responsibility of each consumer, but if the structures work against it, then sometimes it is asking the impossible from the consumer. Technological and organisational structures can, however, also be given form in such a way that consumers receive a little push in the right direction. Referring to this, Achterhuis speaks of a 'moralising of apparatus' (Dorrestijn and Verbeek, 2013). For instance motor cars that send out a signal when the driver or passengers have forgotten to fasten their seat belts. The norm of safety is thus 'internalised' in the apparatus. Therefore we can be certain that everyone will use their safety belts. Managers can actively promote this kind of development by discussing with designers the desirability of specific forms of moralising and possibly including these in codes, procedures and instructions for designs. In this way designers are moreover stimulated to design in a responsible manner.

Speaking about 'moralising' apparatus can easily lead to misunderstanding. It is not true that apparatus and technological systems are neutral in principle and that afterwards a certain norm has to be built into these apparatus and systems. As if the morality has to be brought into the technology from the outside like a *Fremdkörper* (foreign body). Technological artefacts and systems have always – even when the designer is not aware of it – been given form according to a certain morality. The philosopher of technology Andrew Feenberg says that every apparatus or system has a technological code in which social relationships and values are reflected. So, for instance, the size of tools and apparatus in the workshop or the height of the conveyor belt in the factory have been tailored to the size and the height of adults (Feenberg, 1999: 87–89). The social meaning of this can only be grasped when we are aware of the fact that in Western societies until late in the nineteenth century it was normal for children to work in the factories. So the fact that nowadays the size of tools and apparatus have been tailored to full-grown bodies says something about the division of labour in our society, in the same way that the size of the radu says something about the fact that only young adult men in Guinea-Bissau and Senegal are allowed to handle it (see Section 2.3). The interesting thing about the view of Feenberg is that as a philosopher who is critical of society he actively attempts to link up with social constructivists like Bruno Latour (see Chapter 10) and Donna Haraway (see Chapter 12). He does not regard technology as a *black box*, as something to which limits have to be imposed from the outside. On the contrary, he attempts to use the insights of the social constructivists in his own critical theory. This renders a fascinating and productive approach in which he pursues ways of bringing about a more democratic form-giving for industrial practice (Feenberg, 2001: 74–88).

The analyses by Achterhuis and Feenberg link up well with our own statement that every practice has its own normative structure. Inherent to this is the supposition that each technological-organisational system has to be tuned to the practice within which it functions. It should cooperate in the realisation of the normative nucleus of this practice which comes to the fore in particular in the qualifying function (Strijbos, 2003). So, for instance, the layout of a traffic and transport system should promote accessibility between people and counteract traffic-jams. Simultaneously justice should also be done to the realisation of norms like liveability and durability. This can be brought about for instance by developing information systems that guide the traffic or by discouraging certain forms of road use by means of pay-as-you-drive. However, it is essential that the technological-organisational structures support the operations in the various practices and not dominate them. Systems for guiding traffic may not become a straitjacket, but should actually promote the flow and the further disclosure of systems for guiding traffic and transport systems. Such an ideal actually goes further than a democratisation of the technological and institutional apparatus which Feenberg propagates. Even if all parties involved shared the power and had a say in the form-giving of the technological-organisational structure it still is no guarantee that this would bring about a more durable and just society. There also are normative principles which thrust themselves on the design, apart from what society wants. Should a traffic and transport system function on a permanent basis and meet the goals with which it was set up, it would have to give account of values like accessibility, safety and liveability, whether there is democratic consensus about it or not. Otherwise the system, in the long run, would destroy itself.

13.5 A web of responsibilities

We have seen in the introduction that a second reason why engineers and managers do not consider themselves responsible for the results of their actions is that they form only one small link in a much greater process. As individual actors they are not able to direct the development of science and technology as a whole. The argument therefore is that their actions are to a great extent dependent on what other subjects do. They are compelled to take part in the rat race whether they want to or not. If you do not keep up with the new developments you quickly land on the side-line. Therefore a ban on stem cell research or cloning, for instance, seen from this angle makes little sense, because then others get the opportunity to run with the new scientific technological possibilities and get a head start. This feeling of powerlessness originates in particular because scientific-technological development inevitably brings with it specialisation. Individuals have less and less influence on the complex whole. They have to work more and more in a team and are dependent for what they do on what is happening elsewhere in society and in their profession. With reference to this Ulrich Beck speaks about an 'organised irresponsibility'. The technological system is constructed in such a way that everyone takes responsibility only for his own small contribution. There no longer is anybody who takes responsibility for the whole process (Beck, 1995: 58–69).

In this increased specialisation we should on the one hand avoid conveniently assuming an attitude of resignation, but on the other hand we should not impose all responsibility on the individual engineer or manager either. For the increased specialisation not only brings with it an increased individual responsibility but also a better decentralisation and staggering of responsibilities. A laboratory, company or university, for instance, is organised on different levels. Not only are choices made at workshop level, but also and in particular at the level of higher management. Every organisation, moreover, again forms part of a network of institutions which often have a global character. This does not relieve individuals from their responsibility but it can lead to a more relaxed attitude. The burden of scientific-technological development need not rest on individual shoulders altogether. People are responsible for what they can do in their own situation, within their own manoeuvring space, not for what they could have done if they had all the power in their hands. It also is important to realise that the actions of individual engineers always take place in a web of relations and contexts, each of which has its own particular normative structure. This means that the accountability of the engineer regarding their profession is different from the accountability that the manager or the politician has. All these actors have a specific responsibility which fits their occupation. On the one hand this lays down the limits within which they can and may act and on the other hand it says what can or cannot be expected from them in their particular profession.

The fact that engineers have a specific responsibility regarding the product they are developing means that we may not only expect from them to be knowledgeable in their profession, or to know 'how' things should be made, but also that they should have a sense of their responsibility for the question of the 'why' of technology, or the public interest that is served by their product (Winner in Durbin, 1990: 53–64). They should not only search for possibilities of serving society as best they can with their products, but should also make a specific contribution to the forming of public opinion. Because they are experts they often can get a comprehensive view of the consequences a certain technological development could have and they can help other citizens to form a better informed opinion. Still, engineers cannot take the responsibility for everything that happens in design practice. The way in which technological artefacts and systems are designed and given form also is a public matter. Not only should the most effective solution in a technological sense be sought, but also a product that contributes to the quality of life in the most comprehensive sense of the word. For this negotiations and debate with other actors are needed. In the case of the lorry trailer from Section 13.1 it means, for instance, that engineers should also actively involve lorry drivers, cyclists and policy makers in the safety precautions they are trying to develop. In this way the development of the new product can become a joint responsibility.

Engineers are mainly focused on delivering professional work, but they also take into consideration the interests of the company, the client and the public in a wide sense. Managers, on the other hand, should take special note of the institutional conditions under which the design practice and the production process can function optimally, so that the product can also be marketed successfully. While engineers, just like football players, should in the first instance be focused on delivering an achievement of good content, managers, whether they run a factory or a football club, should mainly focus on creating a situation in which the professionals can do their work well and can deliver the maximum performance. Here we are thinking not only of the formulation of policy and the management of the stakeholders but also of the payment of salaries and the execution of the administration. This means that managers have a responsibility which cannot be reduced to the responsibility of the other actors and which also transcends merely making a profit or reorganising organisational structures. In the last few years there has even originated an entirely new field regarding this: *Corporate Social Responsibility*. In this field numerous methods and other aids have been developed for managers to help them fulfil their responsibility better, not only regarding the financial achievements of the company but also where staff policy, and both the social and natural environment are concerned. In short, managers should pay attention to '*people*', '*planet*' and '*profit*' (Elkington, 1997). At the same time managers have a restricted responsibility. The primary task of managers is to create the preconditions for a professional practice to function properly. In doing this they need not take over the responsibility of the professionals. Managers make room for the responsibilities of the different stakeholders and where needed they can 'remind' them of these. In any case they should ask their own employees to give account of the extent to which they engage in dialogue with interested parties and do their designing in a responsible manner.

Besides engineers and managers there also are other actors. Government for instance has a significant role in taking care that no product is detrimental to the consumer or the environment. Its most important instrument to effect this is legislation (for instance laws on the quality and safety of products, on the environmental impact of the production process or environmentally friendly means of disposal at the end of the life cycle). But government may also give subsidies promoting the design or purchase of new products. Another important actor is the citizen who can exert influence in various ways not only as a voter but also as a consumer. One of the most important ways is 'voting by practical choices': consumers no longer buy a certain product because it uses too much energy or contains too many harmful substances. Still another actor is the environmental movement. Greenpeace succeeded, justly or unjustly, in organising consumers in such a way that Shell did not sink the oil platform *Brent Spar* but had it dismantled somewhere in a Norwegian harbour. These three different actors can be called the defenders of public interest. Just like the referees at the football matches they see to it that the activities of the professionals and managers also serve a broad social purpose. However, the responsibility of government, citizens and interest groups is limited. The time when everyone believed that citizens and interest groups could steer the entire national society via government is definitely past (see Chapter 11). The state is an important actor in the worldwide network structure, but has to operate in interaction with other parties in the global market at all times.

13.6 Conclusion

In this chapter we have discussed three different kinds of ethics: deontology, consequentialism and virtue ethics. We demonstrated that each one of them looks at the practice from a specific perspective and can in this way make a useful contribution to the reflection on the accountability of engineer and manager. Simultaneously we have, however, stated emphatically that practice precedes theory and that theoretical reflection always constitutes an abstraction of the practice. With a turn towards practice it becomes possible in ethics to give attention to the whole spectrum of human conduct and the normativity inherent to it. This is important because only in this way can we get a view of the fact that the scientific-technological development has extended the radius of action of people exponentially which makes it very hard for human beings to fulfil their responsibility. In order not to succumb to the burden of this extended responsibility we have dealt with two rules of thumb which can give engineers and managers a grip on the task they have in society. The first rule of thumb deals with the way in which normativity is objectified in technological structures and the second with the way in which the acting subject has been taken up in a web of normatively qualified contexts and practices. It is all too easy to evade one's own accountability, but it is vital to know that one has only a limited range and capacity and that one cannot do the impossible.

References

Beck, U. (1995) *Ecological Politics in an Age of Risk*, Cambridge: Polity Press.

Burg, S. van der and Gorp, A. Van (2005) 'Understanding moral responsibility in the design of trailers', *Science and Engineering Ethics*, 11(2): 235–256.

Dorrestijn, S. and Verbeek, P.-P. (2013) 'Technology, wellbeing, and freedom: the legacy of utopian design', *International Journal of Design* 7(3): 45–56.

Durbin, P.T. (1990) *Broad and Narrow Interpretations of Philosophy of Technology*, Dordrecht: Kluwer Academic Publishers.

Elkington, J. (1997) *Cannibals with Forks. The Triple Bottom Line of 21st Century Business*, Oxford: Capstone Publishing.

Feenberg, A. (1999) *Questioning Technology*, London: Routledge.

Feenberg, A. (2001) *Transforming Technology. A Critical Theory Revisited*, Oxford: Oxford University Press.
Ferré, F. and Mitcham, C. (eds) (1989) *Research in Philosophy and Technology*, Vol. 9, London: JAI Press.
Giddens, A. (1991) *Modernity and Self-identity. Self and Society in the Late Modern Age*, Stanford, CA: Stanford University Press.
Jonas, H. (1984) *The Imperative of Responsibility. In Search of an Ethics for the Technological Age*, Chicago, IL: University of Chicago Press.
Kant, I. (2005) *Fundamental Principles of the Metaphysics of Ethics*, tr. T.K. Abbott, Mineola, NY: Dover Publications.
MacIntyre, A. (1981) *After Virtue. A Study in Moral Theory*, London: Duckworth.
Sollie, P. and Düwall, M. (eds) (2009) *Evaluating New Technologies: Methodological Problems for the Ethical Assessment of Technology Developments*, Dordrecht: Springer.
Strijbos, S. (2003) 'Systems thinking and the disclosure of a technological society', *Systems Research and Behavioral Science*, 20: 119–131.

Further reading

Bauman, Z. (1993) *Postmodern Ethics*, Cambridge, MA: Basil Blackwell.
Davis, M. (1998) *Thinking Like an Engineer. Studies in the Ethics of a Profession*, Oxford: Oxford University Press.
Jonas, H. (1973) 'Technology and responsibility. Reflections on the new tasks of ethics', *Social Research*, 40(1): 31–54.
Jonas, H. (1976) 'Responsibility today. The ethics of an endangered future', *Social Research*, 43(1): 77–97.
Kaptein, M. and Wempe, J. (2002) *The Balanced Company. A Theory of Corporate Integrity*, Oxford: Oxford University Press.
Kitcher, P. (2001) *Science, Truth and Democracy*, Oxford: Oxford University Press.
Lyon, D. (2006) *Theorizing Surveillance: The Panopticon and Beyond*, Devon: Willan Publishing.
MacIntyre, A. (1998) *A Short History of Ethics. A History of Moral Philosophy from the Homeric Age to the Twentieth Century*, 2nd edn, London: Routledge.
Radder, H. (1996) *In and About the World. Philosophical Studies of Science and Technology*, Albany: State University of New York Press.
Solomon, R.C. (1993) *Ethics and Excellence*, New York: Oxford University Press.
Solomon, R.C. and Higgins, K.M. (1996) *A Short History of Philosophy*, Oxford: Oxford University Press.
Strijbos, S. and Basden, A. (eds) (2006) *In Search of an Integrative Vision for Technology. Interdisciplinary Studies in Information Systems*, New York: Springer Verlag.

EGBERT SCHUURMAN

(1937)

PORTRAIT 13 Egbert Schuurman

Egbert Schuurman was born in 1937 in Borger and studied civil technology at the Technological University Delft and Philosophy at the VU University in Amsterdam. In 1972 he was awarded his PhD on the thesis *Techniek en toekomst. Confrontatie met wijsgerige beschouwingen* (published as *Technology and the Future: A Philosophical Challenge* in 2009). In the same year he became an extraordinary professor in Reformational Philosophy at the Technological University in Eindhoven. Later there followed extraordinary professorships in philosophy at the Technological University of Delft (1974) and the Agricultural University of Wageningen (1986). Since 1983 Schuurman is a member of Parliament, first representing the Reformational Party Federation and later the ChristenUnie (ChristianUnion). He was one of the founders of the Prof. dr. G.A. Lindeboom Institute in Ede and the Institute for Cultural Ethics in Amersfoort. For some time he was the chairman of both these institutes. On 15 May 2002 Schuurman officially retired as professor, delivering his valedictory lecture on *Bevrijding van het technische wereldbeeld. Uitdaging tot een andere ethiek* (Liberation from the technological worldview. The challenge of a different ethics).

Amongst other works Schuurman has written *Technologie: middel of moloch?* (Technology: means or idol?) (1977), *Filosofie van de technische wetenschappen* (Philosophy of the technological sciences) (1990) and *Geloven in wetenschap en techniek. Hoop voor de toekomst* (1998) (translated as *Faith and Hope in Technology* (2003)). In 1994 Schuurman received an honorary doctorate from the North West University in South Africa. In 1995 he also received the Tempelton Award.

An ever-recurring theme in his work is the technological messianism by which modern Western culture is driven. The technisation of the different domains in society is expressed in a pursuit for technological perfection, during which human activity is constantly measured by norms of effectivity and efficiency. This technological messianism, according to Schuurman, is a secularised version of what in the Christian faith is called the Kingdom of God. Instead of expecting their salvation from God, people want to organise the total reality to their preference in a self-willed way. In spite of his criticism on the faith in technology Schuurman wants no part of the tendencies which reject the development of science and technology as such. What is needed is an ethics of responsibility in which everyone has to state on which motives, principles, norms, criteria and goals s/he acts and contributes her/his part to scientific–technological development. Technology should be made serviceable to the maintenance and form-giving of a diversity of forms of life. By connecting cultural development to an integral framework of norms a meaningful perspective on the future can be disclosed.

> In the spectrum variations of thought about our technological culture we can clearly discern two extremes as well as options between these extremes. At the one extreme we have the technicists who consistently pursue and put into practice everywhere the scientific-technical idea of control; at the other end of the spectrum we encounter the naturalists who condemn and even destroy every science and technology. A dream and a nightmare stand opposed to each

other [...] The one absolutisation, that of science and technology, stands in opposition to the other, that of freedom, spirituality, and nature. Our culture is dominated and torn by that polarizing tension.

(Schuurman, 2003: 147–148)

A liberating perspective opens up when it is understood that the specific meaning of technology ought to be led by the normativity of the various post-cultural aspects of our reality, namely, the lingual, social, economic, aesthetic, juridical, ethical, and pistical aspects. Belief is placed last in this sequence not because the disclosure of meaning is rounded off by belief – on the contrary, belief is precisely the most necessary condition for disclosure – but because it is the belief that the utter insufficiency of technological development is made clear.

(Schuurman, 2009: 416).

The biblical view of responsible cultural development evokes a different image: that of creation as a garden to be cared for and kept by human beings. Primarily in that image is the conviction that everything has been given, that all things are from, through, and to God and that they therefore possess a value or character of their own. That intrinsic worth should be respected according to this view before we even begin to deal with these things in a scientific-technical manner. Every human activity ought to begin with careful contact and respectful treatment. Creation as a whole and all creatures within it need to be approached in accordance with their own nature – or else life will vanish. That does not mean the deification of nature, for created phenomena have no independence in relation to God. On the contrary, it means recognition of the Creators' endless care to which we humans must respond appropriately.

(Schuurman, 2003: 211–212)

References

Schuurman, E. (1977) *Techniek: middel of moloch?*, Kampen: Kok.

Schuurman, E. (1990) *Filosofie van de technische wetenschappen*, Leiden: Meullenhoff.

Schuurman, E. (2002) *Bevrijding van het technische wereldbeeld. Uitdaging tot een andere ethiek*, Delft.

Schuurman, E. (2003) *Faith and Hope in Technology*, Toronto: Clements Publishing. Translation of *Geloven in wetenschap en techniek. Hoop voor de toekomst* (1998).

Schuurman, E. (2009) *Technology and the Future. A Philosophical Challenge*, Grand Rapids, MI: Paideia Press. Translation of *Techniek en Toekomst. Confrontatie met wijsgerige beschouwingen* (1972).

CASE STUDY IV

Innovation in health care

In Western countries health care faces enormous challenges. Demographic changes – the grey wave – transform the face of the health care system, rising costs make strong demands on economies, the rise of individualism and neoliberal thought undermines the idea of solidarity, and the dominance of the medical-technological approach clashes with social and spiritual needs of patients. In addition, it appears an uphill battle to convince governments to invest in prevention and to persuade citizens to adopt healthier lifestyles. It is widely recognised that radical changes are required to develop a sustainable health care system. It is believed that innovation will be key to realise such a change.

Western society is shaped by innovations and will be shaped further by future innovations. The extent to which an innovation becomes a success or a failure depends heavily on the way the whole innovation process is managed and the (organisational) culture in which it takes place. Recently, the 'secrets' of innovations and breakthroughs were investigated by Henk van den Breemen et al. (2014) and compiled in the book *Breakthrough: from Innovation to Impact*. Innovations and breakthroughs in the field of business, geopolitics, philanthropy economics, media, technology, agriculture, logistics and healthcare were explored. This case study is mainly based on the chapters 'Health care innovation in practice' and 'Mayo Clinic in the 21st century: reinventing a healthcare icon' (Breemen et al. 2014: 291–328).

1 Nature of the health care practice

In this case study we will focus on the health care system in the USA: the most expensive system in the whole world. Michael Porter and Elizabeth Olmsted Teisberg argue in *Redefining Health Care* for a radical re-orientation of health care policy of America. They conclude that the 'combination of high costs, unsatisfactory quality, and limited access to health care has created anxiety and frustration for all participants' (Porter and Teisberg, 2006: 1). They state that health care does not appear to be a 'normal market' because 'competition does not reward the best providers, nor do weaker providers go out of business' and they therefore conclude that the market mechanism fails in this segment of society (Porter and Teisberg, 2006: 3). Porter and Teisberg believe that the problems of health care only can be 'cured' by realigning 'competition with *value for patients*', where value for patients is defined as 'the health outcome per dollar of cost expended' (Porter and Teisberg, 2006: 4).

The problems that Porter and Teisberg identify have been addressed by several American authors. Shannon Brownlee estimates in her book *Overtreated: Why Too Much Medicine is Making us Sicker and Poorer* (2007) that Americans spend between one-fifth and one-third of their health care dollars on care that does nothing to improve their health. Gilbert Welch in *Overdiagnosed: Making People Sick in the Pursuit of Health* (2011) shows that paradigm changes in health care induce over-diagnosis, resulting in worse health, lower quality of life and increasing costs. In *Taming the Beloved Beast: How Medical Technology Costs are Destroying our Health Care System* (2009), Daniel Callahan states that the 'reigning model of limitless medical progress and technological innovation' has to be rejected (Callahan, 2009: 9).

All these studies show that there are no simple recipes to cure health care from problems of quality, limited access and rising costs. The main reason is that the present crisis is not caused by just a few identifiable problems, but is rooted in the behaviour of all stakeholders, the architecture of the system, and the values that drive individuals and that underlie the whole system. As a consequence, old paradigms, old ways of thinking, and old values cannot solve current problems. Perhaps more radical innovations are needed to develop a sustainable, accessible and high-quality health care.

In our view, the first question that has to be asked is: What is the specific nature of health care? In the terminology of the practice model (see Chapter 9) this question has to be reformulated: What rules of conduct define and constitute our health care system? Or: What is the qualifying function that characterises the primary process of the health care practice? One could argue that health care is about medical technology because progress is driven by innovations. One also could argue that health care is about markets and economics because an enormous amount of money is involved. However, health care is not *qualified* by the technological or economic aspect. First and foremost, doctors and nurses have to care for their patients. Consequently, the qualifying aspect of health care is the moral aspect: caring for patients (see Chapters 4 and 5). This idea drives health care from its early beginnings. For example, Ambroise Paré (1510–1590), one of the fathers of surgery and surgical techniques, defined the challenge of medical care as 'cure sometimes, relieve often and comfort always'. Therefore, the development of a sustainable, accessible and high-quality health care system has not to be guided by technological innovations or by the introduction of (more) market mechanisms but has to be led by the idea 'curing, relieving, and comforting'. In other words, it's all about the values that underlie the health care system. This fundamental point also has been recognised by Porter and Teisberg. They state that the present problems cannot be solved within the present health care system. They also state that effective leaders have to revisit the fundamental purpose of health care and have to develop different ways to attain it. They write: 'Ironically, the solution to the crisis lies in refocusing the health care system on health'. In their view, a new vision has to be developed in which 'everything in the system is realigned around its fundamental purpose – patient health' (Porter and Teisberg, 2006: 381).

2 Mayo Clinic

Is it possible to develop a medical practice that is characterised by caring for patients? Is it possible to develop a health care system that is aligned around its fundamental purpose: patient health? These questions have been answered affirmatively by Mayo Clinic, a foundation that is more than 100 years old. The history of this foundation can be described as 'from innovation to innovation' based on the primary value that 'the needs of the patients come first'. In *Breakthrough: from Innovation to Impact* the history of Mayo Clinic is described as follows:

> Mayo Clinic developed gradually from the medical practice of a pioneer doctor, Dr William Worrall Mayo, who settled in Rochester, Minnesota, in 1863. His dedication to medicine became a family tradition when his sons, Drs William James Mayo and Charles Horace Mayo, joined his practice in 1883 and 1888 respectively. A devastating tornado in 1883 brought the Mayo family and the Sisters of Saint Francis of Rochester together. The tornado destroyed one-third of the town, killing at least 37 people. A temporary hospital was set up to treat the injured. Afterwards, Mother Alfred Moes, OSF, founder of the Sisters of Saint Francis, asked Dr William W. Mayo to open a permanent hospital. The 27-bed Saint Mary's Hospital opened in September 1889. Attending staff were Dr Mayo and his sons. From the beginning, innovation was their standard, and they shared a pioneering zeal for medicine. As the demand for their services increased, the Mayos asked other doctors and basic science researchers to join them, forming the world's first private integrated group medical practice. The founders believed that the combined wisdom of peers was greater than that of any individual, and they worked with one another to solve practical problems for patients. In 1907, Henry Plummer, MD, an early partner of the Mayo brothers, developed a unified patient record. This shared record gave all providers easy access to a patient's critical health information and led to improved, integrated care for patients. The organized group practice was a revolutionary way to practice medicine and in retrospect, one of the most significant healthcare innovations of the twentieth century. In a 1910 commencement address at Rush University, Dr William J. Mayo detailed some of the circumstances that fostered this movement toward the group practice of medicine:

> 'As we grow in learning, we more justly appreciate our dependence upon each other … The best interest of the patient is the only interest to be considered, and in order that the sick may have the benefit of advancing knowledge, union of forces is necessary.'
>
> This statement forms the foundation of Mayo Clinic's primary value – the needs of the patient come first – and also reinforces the importance of team-based care.
>
> *(Breemen et al., 2014: 292)*

Breakthrough continues:

> In 1919, the Mayo brothers established Mayo Foundation as a humanitarian, not-for-profit organization to operate for the health and wellbeing of humanity. In partnership with administrators, the organization was led by physicians in order to ensure that the patient remained at the centre of decision-making. All physicians were salaried and all proprietary interests ceased. Today, that mission continues. Mayo Clinic remains physician-led with salaried staff and operates in six states (Minnesota, Florida, Arizona, Iowa, Wisconsin and Georgia), employing nearly 60,000 people and caring for 1 million people each year. It is a nearly $9 billion enterprise that reinvests about $400 million annually in medical education and research that will benefit patients. Mayo Clinic is governed by a 33-member Board of Trustees.
>
> *(Breemen et al., 2014: 292–293)*

3 Vision, mission and values

The history of Mayo Clinic clearly shows the shaping power of values. Recently, the vision and mission have been reformulated whereas the primary value has been maintained (Breemen et al., 2014: 299). The so-called *Mayo Clinic 2020* report gives the following definitions:

> **Vision**
>
> Mayo Clinic will provide an unparalleled experience as the most trusted partner for healthcare.
>
> **Mission**
>
> To inspire hope and contribute to health and wellbeing by providing the best care to every patient through integrated clinical practice, education and research.
>
> **Primary Value**
>
> The needs of the patient come first.

Based on these reformulations, the Mayo Clinic has defined its core business as 'Create, connect and apply integrated knowledge to deliver the best healthcare, health guidance and health information.'

The Mayo Clinic is one of the best hospitals in the world. In every ranking list, it takes a first, second or third position. It is believed that the outstanding performance is driven by its key value. Glenn S. Forbes, former CEO of Mayo Clinic in Rochester describes its role as follows:

> What makes Mayo Clinic distinct is that we have said, 'The needs of the patient come first' from the beginning. Over generations, we have driven the needs of the patient into our thinking about how policies were developed. We've driven it into our thinking about how we structure ourselves and our governance and how we allocate our resources. We've driven it into our thinking when we recruit people and form staffs. We've driven it so broadly and deeply in our management and operations that it becomes part of our culture. Thus, when we bring an issue forward, it's not a thin layer of, oh yes, that was the marketing mantra that somebody thought of last week. No, this is driven much more deeply into the fabric of the organization. That's what makes us different.
>
> *(Berry and Seltman, 2008: 20)*

4 Innovation and breakthrough

The *Mayo Clinic 2020* report presents the strategic outlook for the future. It is based on the idea that new ways of health care delivery – both physical and virtual – have to be developed. It takes as a starting point that the Mayo Clinic wants to have 'meaningful relationships' with patients by 'anticipating their healthcare needs before they are aware of them', and by 'keeping them well, helping them understand their risks and how we can diagnose and treat them earlier' (Breemen et al., 2014: 300). In their view, high-quality care has to be delivered to people regardless of location.

The main strategic highlights for innovation and breakthrough have been identified as:

- Continued thriving of Mayo Clinic campuses in Arizona, Florida and Minnesota. Mayo Clinic will continue to provide an unparalleled patient experience. The system will be fully integrated.
- A Mayo Clinic Health System (i.e., a wholly-owned provider network) will surround each of the Mayo Clinics. These will be fully integrated with each other and with the three clinics.
- Beyond the borders of the health system will be a national Affiliate Care Network. Mayo will provide support services that enhance patient-centred, integrated care outside Mayo Clinic regions.

- Mayo's activities will be a highly differentiated, personalised and consumer-driven integrated Web experience for consumers and patients worldwide. Components will include services for patients, referring physicians and specific groups such as benefactors, students, researchers and healthcare consumers in general.
- All of Mayo Clinic's sites, health systems and affiliates will provide appropriate levels of electronic visits (e-visits), e-consults and remote monitoring.
- Mayo Clinic will partner strategically with a few major organisations that will provide a broad network of distribution channels for Mayo Clinic's products and services to a large number of people. These relationships will enable Mayo Clinic to meet some of the healthcare needs of many people without requiring a physical visit.
- Mayo Clinic will have a more dynamic and active public presence, not only through its website but through partnerships with key, large media organisations. It will collaborate in the creation of content across the spectrum of media channels including film, television, gaming, documentaries, mobile applications and consumer products.
- Mayo Clinic will grow its international presence, using technology to deliver its expertise to people around the world.

(Breemen et al., 2014: 297)

These highlights clearly show the importance of (medical) innovation. New technologies have to be developed to deliver the best health care, health guidance and health information to patients. All these technologies have to be developed guided by the primary values of Mayo Clinic: the needs of the patients come first. Philosophically formulated, all these technologies have to be developed under the guidance of the qualifying function of health care: caring for people. In all these developments, technology will play an important role, economies of scale and business models are key, but their key characteristics will be moral: curing, relieving, and comforting.

References

Berry, L.L. and Seltman, K.D. (2008) *Management Lessons from Mayo Clinic. Inside One of the World's most Admired Service Organizations*, New York: McGrawhill.

Breemen, H. van den, Murray, D., Bilski, B. and Verkerk, M.J. (2014) *Breakthrough. From Innovation to Impact*, Lunteren: Owls Foundation.

Brownlee, S. (2007) *Overtreated. Why Too Much Medicine is Making us Sicker and Poorer*, New York: Bloomsbury.

Callahan, D. (2009) *Taming the Beloved Beast. How Medical Technology Costs are Destroying our Health Care System*, Princeton, NJ: Princeton University Press.

Porter, M.E. and Teisberg, E.O. (2006) *Redefining Health Care. Creating Value-Based Competition on Results*, Boston, MA: Harvard Business School Press.

Welch, H.G. (2011) *Overdiagnosed. Making People Sick in the Pursuit of Health*, Boston, MA: Beacon Press.

14

EXPECTATIONS FOR THE FUTURE

The secular sacred and the limits of technology

Summary

In this chapter the relation between technology and religion takes central place. At first sight it seems as if religion no longer has a place in our technological culture. People have taken their fate into their own hands and no longer need supernatural powers. However, one could ask whether this supposition can be substantiated. For in spite of the expectation that religion would quietly die, it seems that religion is once more prevalent. What is more, there are indications that technological development itself is being motivated to a great extent by religious driving forces. Could one speak of a secular form of the sacred?

14.1 Introduction

We live in a time of secularisation. God is dead, or at least has been declared dead. People have taken their fate into their own hands. Where formerly the gods and powers of nature governed, people now more and more determine their own lives. Science and technology have played a crucial role in this process. By means of a vastly increased insight into the functioning of the reality that surrounds us and the systematic use of this knowledge in technology and technological development human beings have gained a grip on nature. Almost all the utopian dreams of Francis Bacon in his *A New Atlantis* (1627) have in the meantime come true.

The prevailing image of modern times has been sketched above. It seems as if many people no longer need faith in God. Science has taken the place of traditional faith and the role of religion has declined. This has been partly the result of technological developments. People need no longer pray for their daily bread. They can now provide for themselves. Modern atheists like Daniel Dennett and

Richard Dawkins think that it is time to move past the childish superstitions of our ancestors and become mentally grown-up.

Yet there are cracks in this modern awareness. It is evident that there still is a lot of superstition in existence: horoscopes are eagerly consulted, alternative medicines are very popular and oriental mysticism is something commonly sought after. Moreover the decline in church membership, which is expected to continue for some time, has been proved not to lead to a decline in faith or religiosity. Rather, elsewhere in the world one could speak of a strengthening and sometimes also of a radicalisation of religiosity, for instance in Islam. So in a world in which people have more than ever taken their existence into their own hands, it is clear that there still is a limit to the human capacity of fully determining its own existence. As a result the question remains topical how we should deal with the limits of being human and what is beyond these limits.

The idea that human beings no longer need God has been dominant in the West for a long time. Secularization or breaking away from the church seems to be the inevitable consequence. This is expressed in what Peter Berger calls the 'secularization theory'. According to this theory 'modernization necessarily leads to a decline in religion, both in society and in the minds of individuals' (Berger, 1999: 2). However, this theory is gradually on the decline. Before dealing with the arguments currently brought against the secularization theory we would first like to investigate against which historic background it originated.

14.2 The origin of the modern age

In the Middle Ages the Roman Catholic Church was the ruling authority in huge parts of Europe. It kept the people in the grip of a tradition handed down with divine authority in combination with various popular forms of religiosity. Modernity started when people began taking leave of this ecclesiastical authority. In this the reformation and the Renaissance played a crucial part. According to the modern self-image the Renaissance was more radical in this than the reformation. For the latter professed to be a movement which attempted to return (reform) to the original sources of Christianity. The Renaissance, on the contrary, laid the foundations for humanism. The value of the human being took central position and the ideal of human autonomy, which gave great importance to reason, came to the fore.

This development did not mean a sudden radical break with the Christian faith. It was a more gradual process. Take for instance a philosopher like René Descartes who is generally seen as one of the founders of the modern self-image. Descartes wanted to uncover the indubitable foundation of human knowledge. But in contrast to the preceding ages of history he no longer sought this in ecclesiastical authority or in a faith based on divine revelation. The indubitable foundation had to be acceptable and in itself insightful to the subject and to human reason. For this Descartes devised the experiment of methodical doubt: he doubted all contents present in the consciousness and tested them for their reasonable tenability. Finally

Descartes retained *one* certainty: the 'I' that (while doubting) is conscious of itself. He called this the 'cogito' ('I think'). The first certainty to be deduced from this is the existence of consciousness: 'cogito, ergo sum' ('I think, therefore I am').

Descartes' argument would nowadays convince no-one. The most important element is its intention to look for the final foundation of certainty of our knowledge in the subject, the thinking and knowing 'I'. For that matter it seems that Descartes was less convinced of his argument than one would expect judging by his certainty. Based on the 'I' as consciousness, in order to reach certainty of the existence of the (material) world outside the 'I', it becomes evident that Descartes also has to prove the existence of God. For even if I have a 'clear and distinct' idea of the existence of a world outside myself, who will guarantee that the idea has not been given me by a God who deceives me or by an 'evil spirit'? This guarantee can only be given when it can be proved that there is a perfectly good God who is unable to deceive me on the grounds of his goodness and perfection.

Descartes' pursuit of the autonomy of the human subject, based on pure reason, set the philosophical tone for a whole period. But his elaboration of this idea soon met with fundamental critique. Philosophers like Ludwig Feuerbach, Karl Marx and Friedrich Nietzsche from the quest for the autonomy of the subject drew the conclusion that God is just an illusion, a projection, an entity generated by man's spirit. For instance, Nietzsche established his fame with the statement that God is dead.

With Feuerbach, Marx and Nietzsche (apart from some fore-runners like perhaps Baruch de Spinoza or David Hume) the era of atheism began. More and more philosophers and scientists explicitly broke with the Christian life view and religion in general. It became ever more apparent that religiosity and faith in the existence of a supernatural reality belonged to the intellectual past.

The founder of positivism and sociology, August Comte (1798–1857) was of the opinion that this was the inevitable course of history. In his famous *law of the three stages* he reconstructed world history in three episodes: the theological, the metaphysical and the positivist stage. In the first stage supernatural powers are called on to explain various phenomena. In the metaphysical stage people rather seek these explanations in abstract, metaphysical concepts like the spirit, reason or the soul. In the last stage phenomena are exclusively explained in an enlightened manner with the aid of the empirical practice of science. This is the stage of positive knowledge. Comte became so enthusiastic about his own positivist mission that he wanted to establish a *religion of humanity*, in which he did not hesitate to appoint himself in the position of high priest.

After ages in which the Christian faith had determined life in Western European societies, slowly but surely a new kind of obvious assumption took hold. Faith in a supernatural reality belongs to a definite past and is not something of this age. People who through personal opinion find it necessary to live by such a conviction of faith should be given room to do so, but at the same time they should realise that this is a matter of the private sphere with which they should bother other people as little as possible. The individual 'right' to perception of one's own religion or life view is immediately bound up with the 'duty' of seeing faith or religious conviction as a personal matter which is irrelevant to the public sphere of life.

This new obvious assumption at present comes under pressure. As far as organised, institutionally anchored religiosity is concerned, it appears that secularisation is advancing at an unrelenting pace. But if we look at the individual convictions of many people, it appears that an undiminished faith in a supernatural reality prevails. Many people believe in 'something'. And almost stronger than in ecclesiastical or evangelical circles it seems that at present even outside communities of faith there is an interest in 'miracles' and 'supernatural phenomena'.

This obvious assumption is heightened by the arrival of many people having a non-Western background in Western societies. Think for instance of people with a Muslim background who are acutely uncomfortable with the typically Western opinion that religion is a matter of the private life. They come from countries in which culture, society and state are formed to a high degree by the religious convictions of the majority of their inhabitants. This fundamental attitude progressively confronts the Western societies now inhabited by them with new problems and issues.

14.3 The myth of neutral science

In the 1930s there emerged in circles within the VU University Amsterdam a new philosophical school. The founders of this school, the professors Herman Dooyeweerd and Dirk van Vollenhoven, queried the *postulate of neutrality* made by modern science. This is the idea that theoretical knowledge is neutral regarding religion and worldview.

Dooyeweerd worked out his polemics against the postulate of neutrality in his so-called '*transcendental critique of theoretical thinking*'.[1] What Dooyeweerd attempts to show in this critique is that there are purely philosophical and theoretical reasons to doubt the self-image of modern philosophy and science. Believing and knowing are not in opposition as modern ideas of science postulate, but are intrinsically connected: without worldview or faith there is no science.

There is something paradoxical in this view of Dooyeweerd. For he claims that he can prove this connection on theoretical grounds. Supporters of the postulate of neutrality will bring against this that on the contrary it is a feature of science to set aside one's own life-viewish or religious convictions. These may have no role in scientific research. It is exactly against this idea of purity that Dooyeweerd aims his attack: the idea that the Archimedean point from which one can oversee the coherence of reality has to be sought in theoretical knowledge itself. Dooyeweerd states that the theoretical way of knowing offers no independent point of departure or *God's eye point of view*. Theoretical knowledge is limited and the boundary of this knowledge cannot be explained by theoretical insight. This is an opinion that inevitably leads to the conclusion that the scientific way of knowing can never be taken as a neutral point of departure for knowing reality.

Since Dooyeweerd claims that one can speak of an intrinsic connection between worldview and the practice of science, his opinion is not affected by the reproach that he mixes up science and faith or worldview. After all, they belong together. People think and act from the perspective of certain views about the world. These include views of the meaning and coherence of the world in which they live and the experiences they have in it. These views in various ways influence what people do and do not do and often give direction to their conduct. However, to Dooyeweerd acknowledging this does not mean that he is leaving the field of science and embarking on the road of theology or interpretation. On the contrary: neither can be had without the other. Therefore even in the theoretical way of knowing, worldviewish or religious controversies can occur. These cannot be solved by reproaching one another for worldviewish or theological dogmatism but solely by being, as a scientist, as explicit as possible about one's own worldviewish or religious points of departure.

14.4 The pursuit of perfection: David Noble

In *The Religion of Technology* (1997), David F. Noble defends the statement that religious motives are the most important driving forces behind the enormous technological developments of our age. In particular he means the Christian notions of redemption, perfection and the end of the world. He places the origin of this in two developments during the Middle Ages: (a) acceleration in the development of technology and (b) laying a connection between technology and higher religious values. This religious revaluation of technology is expressed in the term '*artes mechanicae*' which was probably first used by the medieval philosopher John Scotus Erigena. According to Erigena by means of these *artes* we can regain the lost 'image of God' and reach again the perfect state of Adam in paradise. Noble demonstrates that this idea was closely connected with the belief in an imminent advent of the thousand-year reign (millennium). During the Renaissance this development continued uninterruptedly. During the reformation in the sixteenth century the expectation of the millennium once more flared up powerfully.

Noble shows that religious motives played an important part in the breakthrough of the empirical approach in the natural sciences and the development of technology. Francis Bacon is a good example of someone who explicitly based himself on such motives. He, too, lived in the expectation of the advent of the thousand-year reign. Many well-known natural scientists who made important contributions to the rise of the engineering sciences also were Freemasons. For the American situation Noble particularly foregrounds the role of a movement like the 'Great Awakenings', a religious revival in eighteenth- and nineteenth-century America.

In spite of the fact that Noble bases his analysis largely on the meaning of the Christian religion for the development of natural science and technology he does use the concept 'religion' in a wide sense. This becomes evident for instance when he quotes the philosopher Augustus Comte as someone who encouraged the development of science and technology from religious motives. Noble shows that the objectives of positivist society remind one strikingly of the Christian expectations of the end of the world. The concepts 'redemption' and 'perfection' have, however, acquired a different connotation: they are no longer seen as things given by a higher hand, but on the contrary as something that human beings only achieve by banning all thought of something higher. Religion becomes a 'religion of humanity', an absolute faith in the human being.

This widening of the concept of religion that we have just pointed out in the analysis of Noble, however, seems to be more an exception. Mostly he speaks specifically about Christianity. This also becomes apparent in treating current subdivisions of technology like space travel (which he connects with the religious concept of 'ascension to heaven'), artificial intelligence (which he relates to the Biblical concepts 'resurrection' and 'eternal life') and genetic engineering (also sometimes called the 'eighth day of creation' which once again is interpreted by Noble as a more than metaphorical reference to the Christian faith). For that matter the analysis of Noble sometimes lacks subtle distinctions. Widely diverging forms of religiosity (and Christian religiosity in particular) are sometimes thrown together and are not always distinguished with the necessary qualification. So Noble has little acknowledgement for other religious motivations as, for instance, the desire to reach a deeper understanding of the greatness of God.

Noble concludes that technological developments were and still are stimulated by expectations of redemption originating in the Christian religion. In this he bases his opinions especially on marginal (Freemason), extreme (Anabaptist) and fundamentalist groups (chiliasm) in the Christian tradition and bypasses the more dominant tendencies within the Christian tradition as, for instance, Calvinism. Moreover Noble's effort is predominantly critical: he wants to point out the dangers attached to technological development driven by religious motives and expectations of redemption. Religion is mostly presented in a negative light as a potentially dangerous motive. What he does not discuss is the issue whether there is an intrinsic connection between worldview and religion on the one hand and science and technology on the other hand. Viewed thus, worldviewish and religious motives are more the exception than the rule. The idea that such motives *inevitably* play a role because they are always part of human life does not occur to Noble. These remarks, however, do not detract from Noble's merit, namely that he has shown that religious or worldviewish motives can play an important part in technological development.

14.5 The modern self-image doubted: Bruno Latour

The modern age has for a long time seen itself as an age separated from the previous period by a radical rift. We could say with Bruno Latour (1993: 10):

> The adjective 'modern' designates a new regime, an acceleration, a rupture, a revolution in time. When the word 'modern', 'modernisation' or 'modernity' appears, we are defining, by contrast, an archaic and stable past. Furthermore,

> the word is always being thrown into the middle of a fight, in a quarrel where there are winners and losers, Ancients and Moderns.

According to Latour (1993: 10–11) the self-image of the modern age is based on:

> two sets of entirely different practices which must remain distinct if they are to remain effective, but have recently begun to be confused. The first set of practices, by 'translation', creates mixtures between entirely new types of beings, hybrids of nature and culture. The second, by 'purification', creates two entirely distinct ontological zones: that of human beings on the one hand; that of non-humans on the other.

Latour states that both pre-modern and modern societies form hybrids or mixed forms of culture and nature. However, modern people think that – in contrast to their pre-modern ancestors – they have succeeded in adequately separating ('purifying') the world of things from that of human beings. For indeed they maintain a strict division between the world of objective facts (nature) and that of subjective opinions, political convictions and social interests (society). They have devised numerous practices to keep these zones separated as cleanly as possible. Science, the laboratory and technological research belong to the world of objective facts. In the university or research centre it is assumed that one keeps one's subjective convictions of faith and one's political opinions to oneself. Faith and politics do not belong in these practices. They have been banned to private life. In the modern age these domains must remain carefully separated and should not be mixed the one with the other.

According to Latour this 'purification' unintentionally facilitates an almost infinite 'translation' or 'mediation'. By this he means that the modern 'belief' that the world of nature and the world of human beings should be separated as cleanly as possible leads to the networks of human and non-human actors increasingly becoming globally encompassing. Through this it becomes ever less clear where exactly the possibilities and responsibilities are of directing and influencing scientific and technological developments.

Latour deals with the modern self-image. He wants to restore the symmetry between the pre-modern and modern worldview. For it is not easily to be distinguished which of both worldviews represents reality in the most adequate way. To give an example: our pre-modern ancestors often tried to placate the gods by bringing sacrifices or by prayers. Modern people no longer believe that such actions can influence the course of events. When the swine stock is threatened by swine fever it is decided exclusively on scientific-technological insights to put down hundreds of thousands of pigs. But is there a fundamental difference between 'sacrificing' to placate the gods and 'putting down' to get the stock healthy once more? Is the one a case of 'superstition' and the other of 'adequate insight'? Or do we here have two interpretations of the world of which one cannot say beforehand which is the better?

Both these convictions have their own standards for what is important and what is beyond any discussion. People who try to influence the gods think about the manner in which their lives are determined in a particular way. The gods rule the world, they can be influenced and can be induced by sacrifices to giving a different direction to fate. Whether this is an effective strategy of influencing fate is hard to say. For the bringing of sacrifices bears no guarantee of a good result. The readiness and will of the gods remain the determining factor and these seem to be elevated above direct influence. In a worldview dominated by a scientific-technological view matters are entirely different. On the basis of scientific insight people now believe that they know the cause of threatening phenomena. By doing something about the cause, one can influence the consequences. The actions flowing from these are regarded as rationally essential means. A pig only counts as a 'thing' that may be destroyed and given up to reach the goal – making the livestock economically expendable. The economic value of animals in this context is 'holy' in the sense of 'being beyond any discussion'.

Against the background of the above is it justified to speak of a modern or secular appearance of the 'sacred'? To consider this question let us look at another contemporary philosopher who researches the relationship between nature, technology and the sacred: Bronislaw Szerszynski.

14.6 The secular sacred: Bronislaw Szerszynski

According to Szerszynski modern thought is dominated by a certain view of the sacred and the secular. In this view the secular is seen as the term that needs no explanation. What has to be explained is religion and the religiosity of human beings. For it is seen as something special and as something which at first sight is not rational. Therefore the particular phenomenon that people attach value to faith and religion has to be explained. This view is an extension of the idea from the Enlightenment that science and technology have replaced religion by explaining and controlling reality in a rational manner. What Szerszynski wants to set against this is the suggestion that we should much rather see the secular as a specific product of religious and cultural history. He even calls the secular a religious phenomenon. Contemporary secular thought understands its own secularity as absolute. By this Szerszynski wants to discuss the established narrative of the disenchantment of the world. The sacred should therefore not so much be explained from the secular (during which the secular is considered as the not-problematic, obvious category) but the other way round: the secular is the specific product of history. And in history the sacred always plays a central part and will do so again: 'for on closer examination the narrative of disenchantment – the rendering of the world as totally profane and without spiritual significance – itself involves and calls forth new forms of enchantment' (Szerszynski, 2005a: 7). In our secular age the sacred has not so much disappeared as it has been newly ordered or reorganised.

Of course this evokes the question what exactly Szerszynski understands by the sacred. In his book he says very little about it. One of the reasons for this is that he wants to make a problem of the secular, not of the sacred. 'How did a cultural form emerge that understands itself not as engaging in heresy, idolatry, or apostasy but as *non*religious, to be understood in its own, immanent terms, with no need of a sacral reference point to make it intelligible?' (Szerszynski, 2005b: 815). The term 'religion' is less suitable to Szerszynski since in various aspects it already is a modern term and its meaning is determined by the contrast between the religious and the secular. Szerszynski is looking for a term that would be momentous enough to make a problem out of the secular. So we are here dealing with a relatively indeterminate use of the term 'holy' which is specifically focused on restoring the symmetry between the religious and the secular. The secular then can no longer be interpreted in an absolute sense, but should be seen as something that in some way or another is in a relationship with the (non-neutralisable) other-than-secular.

To achieve his goal Szerszynski needs a particularly open and broadly applicable concept of the sacred. To him there is no universal *homo religiosus* or a *transcendental reality* to which this concept would refer. On the contrary, he wants to abstract the sacred from all the particular forms and historical shapes with which it has been filled. Only then could it be used for modern philosophy.

It would be interesting to trace Szerszynski's opinions further with reference to his own examples. However, we now want specifically to draw a parallel between certain of Szerszynski's statements and the philosophy of Dooyeweerd, which we discussed earlier. Both authors want to restore the symmetry between the secular and the religious worldview. The former may no longer be taken as the obvious, unproblematic point of departure to explain the latter. So Dooyeweerd points out that science can take no meta-standpoint which transcends all other stances. Although in science its own rules are valid, it never can detach itself fully from the various worldviews from which it is practised.

Szerszynski has a different approach. He discusses in particular the implicit presupposition of secular thought in which religion is something to be explained in opposition to the secular view itself. The underlying thought is that the secular worldview forms the absolute and therefore obviously valid point of departure for an explanation of the sacred.

But if there are reasons for restoring the symmetry between the sacred and the secular – without taking *one* as the obvious point of departure for the other – then the question arises where the sacred is hiding in our secular age. Can we prove a 'transcendent motive' in our contemporary secular worldview? In our opinion such a transcendent motive becomes visible in the 'pursuit of perfection' which we have mentioned several times above. This is manifested in many domains in ever different appearances.[2] For example, women and sometimes even young girls may be so occupied with their outward appearance that they are prepared to undergo invasive and money-consuming operations to change it. In this the media play a crucial role with their images of women who look perfect. In addition the available medical technologies offer almost unlimited possibilities – for those who can pay – of perfecting their own bodies. More and more people experience a certain social pressure to avail themselves of such means and possibilities.

One can also experience a similar form of pressure in contemporary campaigns for a healthy lifestyle. In many ways people are confronted by the fact that their eating and lifestyle habits influence their health and that they can influence their own 'age' by paying attention to their lifestyle. By taking part in certain tests, one is supposed to be able to determine one's 'real age'. One can also get all kinds of advice on how, by changing one's lifestyle, one can bring down one's 'real age'. To many people this means living more and more constantly focusing on the scale, the tonometer (blood pressure meter) and the centimetres of their waistline.

The striking feature of all these examples is the connection between the high values people set for their safety and health on the one hand and the concern people have about their vulnerability in these areas on the other hand: Do I lead a safe and healthy life? Because of this connection one could speak of an extreme willingness to make sacrifices to improve their safety and health.

14.7 Dealing with risks

One can observe the connection between technology and the pursuit of perfection in modern secular life, amongst other things, in the way people deal with safety, risks and threats. In Chapter 11 we discussed Ulrich Beck's ideas about society as a 'risk society'. How do we in our age deal with the risks attached to life? How do we deal with the boundaries of our abilities to control reality and exclude risks as far as possible? And what does it tell us about the 'transcendent motive' which emerges from this pursuit?

Science and technology play an important part in making reality controllable and predictable. In this way people take control of reality and succeed in eliminating more and more threats and risks. Simultaneously by this approach new risks are created which again have to be controlled in the same way. This is connected with the mode of operation employed by science and technology. This mode of operation is characterised by what we would like to call a *strategy of distrust*. In the laboratory all kinds of variables are meticulously controlled to observe how certain materials or organisms react to certain interventions in exactly defined circumstances. And in technology all kinds of artefacts are made which work according to a carefully written script where upsetting factors are eliminated as far as possible or controlled. In this way people have succeeded in making reality ever more predictable and controllable.

This strategy of distrust does not mean that modern science and technology have disturbed a relationship between human beings and nature which originally was a harmonious one. The contrary is true. All through the ages nature always formed a threat to human existence. In reaction to this human beings have often approached nature in a violent manner. This was purely a form of self-preservation. Science and technology have diminished our dependence on capricious nature to an enormous extent and thereby in many respects have given a friendlier appearance to the relationship between human beings and nature. At the same time the scale of intervention in nature has grown tremendously. Destructive forces in nature have emerged which we will have difficulty in controlling: environmental pollution, depletion of raw materials and nuclear waste. Human beings themselves have become an actual threat to nature and to the survival of the environment in which they live.

The *strategy of distrust* is in sharp contrast with the religious way of dealing with life's unpredictability. In religion the evil that threatens human beings is in one way or another connected with a 'supernatural' reality. In Christian faith people think of the involvement of God or his Son, Jesus Christ, in what happens on earth. What people experience in their lives is not detached from the will of God or from the gods and has an intention. And even though often this intention remains hidden to people, still religions offer ways of making contact with God or the gods. In other words: religion offers ways of dealing with the experience of evil and suffering, not by controlling them, but by entering into a relationship with what they believe is hidden behind the actual course of events. By worshipping God a person can come into a relationship with the intention hidden behind the evil that happens to them. In a different approach than in technology people do not derive the security of their existence so much from control as from their surrender to and dependence on divine forces and the extent to which they trust them.[3]

One could illustrate the difference between science and technology on the one hand and religion on the other hand with the difference between two kinds of certainty: *certitudo* and *securitas* (Grethlein, 1999; Gregersen, 2004). This pair of concepts was used as long ago as by the reformer Maarten Luther. In his work it relates to the way in which believers search for certainty in their faith. Do they try to gain the favour of God (*securitas*) on the grounds of their own merit (good deeds) or do they commit themselves to God's grace (*certitudo*). *Securitas* is connected with control and holding in check the circumstances. With *certitudo*, on the contrary, the issue is trusting in something that one does not control or cannot control. In this regard one can understand the call often heard in the Christian tradition to trust in the promises of God. While the certainty of everyday existence is shaky, Christians find their certainty in God's promise of salvation. In such a case one surrenders oneself in trust to someone whom one does not keep in check or control.

Since in our age technology plays a central role in the way in which we engage with reality it is probable that the certainty we nowadays look for is of the type *securitas*. It appears that the *certitudo* which is more connected with the religious sphere, the certainty of an intimate relation with the transcendent or God, increasingly takes a lesser role in our daily lives.[4] We allow ourselves to be led by an ever more one-sided perspective. As a result of this one-sided orientation to *securitas* the ways at our disposal of reacting to the shattered condition of human existence are severely restricted. Nowadays when people are confronted with an unforeseen disaster the first step is to look for possibilities to get better control of risks. Immediately the focus is on who erred, how such a disaster could have been prevented and why the security measures were inadequate or not followed properly. The better human beings become in controlling reality by means of technology, the more embarrassing the question about how people should deal with the boundaries of the human ability to control becomes.

14.8 Restricted repertoire of actions

In our age, for example, the possibilities of preventing pregnancies or of promoting them with supportive aids have increased tremendously. If a woman becomes pregnant it is usually the case that she explicitly wished to have a child. And where a wish is concerned the expectations are often intense. The prospective parents are very curious, and, even at an early stage can join the obstetrician in looking on the sonar at what is happening in the uterus. It is a wonderful moment when people first recognise the foetus and see it moving: the child in a rudimentary form is visible on a screen. Well ahead of time preparations are made to give the child a perfect reception.

In some exceptional situations it happens, however, that the sonar pictures cause anxiety. Sometimes the treating professionals may at first keep this to themselves. But it can also be a reason to have an interview with the parents about the possibilities for closer investigation. Of course the people involved have the right to refuse this. But if one can get certainty, one often wants it. Is there anything serious afoot? And would timely intervention perhaps be able to set right what is needed? The expectant parents will not hesitate long to accept the offer of closer investigation.

Say, for instance, that after a few visits to the hospital it becomes clear that something is wrong with the child. Exactly what is hard to tell. But it seems that certain functions are not going well. And furthermore there are indications that the brain is not fully developed. Because, as time goes by it becomes increasingly more uncertain whether the child will live to the confinement, it is decided to bring on the birth prematurely. Many weeks too soon the child is born. And quite soon it appears that what was feared is true: the child has some serious handicaps. Then a very difficult time starts for the parents. However much they had wanted the child, this was the last thing they had reckoned with. Where happy expectation had become silent hope, this hope after the birth turns into the knowledge that they bear a heavy burden. During the first months the daily rhythm of the parents is completely determined by the baby. Soon they realize that they will have to change their lives drastically. What follows is emotionally painful for everyone concerned. A number of examinations and operations begin. The child survives the operations but the results of the examination are far from joyful. Although it is very hard to tell what exactly is wrong and what the life expectancy of the child will be, it is in any case clear the child will be living with a serious handicap.

This example demonstrates an ambivalent feature of contemporary technology. On the one hand we have a tremendous increase in technological possibilities. Even children who are born many weeks early can be kept alive. On the other hand in the case of such troubles the serious handicaps that are the consequence of, for instance, genetic abnormalities can only partly be prevented or cured. Much can be done, but by far not everything. And sometimes diagnostic clarity cannot be reached in the prognosis of a child with serious abnormalities. The baby has been successfully kept alive, but no one knows for what kind of life and for how long.

The parents become attached to the baby. The doctors want to do their utmost to keep the child alive. Repeatedly decisions have to be made. In every situation where a choice has to be made there are many uncertainties. And although we call them situations of choice the words 'freedom of choice' in this context do not enter one's mind. On the contrary, everyone in certain respects has their backs to the wall.

The growth in what is technologically possible, therefore, does not mean that all limits have been conquered. If the technological possibilities had not been there the child would inevitably have died. Then the process of mourning or of acceptance can start directly. But as long as there still are possibilities of having a favourable influence on the fate of the child, acceptance of the fate becomes more difficult. Parents can sometimes be so busy with the technological possibilities existing for their child that they hardly have time to accept or to 'enjoy' the given situation. Sometimes the possibilities of influencing seem never-ending although one feels at the same time that the marginal results of these are decreasing. For what would further intervention add to the quality of life? However much our possibilities of influencing sometimes look infinite, there always is a point at which further intervention seems senseless.

The example also shows that when we in our everyday – in many respects secular – world come up against problems we are very soon inclined to look for technological solutions. This not only applies to all kinds of economic and organisational problems but also to problems in our personal lives. Technological alternatives for action have increasingly replaced the non-technological ways of dealing with problems. No matter how far we have pushed back the frontiers of our technological possibilities there always remain limits to these possibilities. And it is precisely at these limits where we experience a kind of embarrassment about how to act because the alternatives to a technological approach seem to get stranger all the time.[5]

No matter how much technological possibilities have expanded, there always will be limits. The pursuit of perfection which seems to be inherent to the way people deal with modern technology has not led to fulfilment of the promise given by the technological utopian dream: the technological paradise. Therefore it is essential that people also have other means at their disposal for dealing with the boundaries of the technologically achievable. As we have shown for this there is more awareness within religion than within a secular perspective on our existence. It seems as if in our age there is a growing sensitivity to the fact that the infinite pursuit of *securitas* is in the final instance unsatisfactory. The need to partake in communal forms of commemoration in times of disaster and the observing of 'silent marches' at the consequences of senseless violence, can be seen as an expression of such a development.

14.9 Conclusion

In this chapter we have queried the way in which religion is regarded in our age. According to many people religion is an outdated phenomenon. The insights of science and the possibilities of technology have made faith in God or gods superfluous. We have seen that this dominant outlook is based on the presupposition that the secular worldview is obvious and needs no explanation. The religious, on the contrary, would have to be explained in terms of secularised thought. In line with Dooyeweerd one could protest against the apparent neutrality of this secular worldview which is based on so-called scientific insight. In line with Szerszynski one could show that it rather is the secular worldview itself that demands explanation. This is a very specific result of a history that was for the greater part dominated by the sacred.

In former chapters we have already seen that worldviews are a factor contributing to the various directions in which technological possibilities can be disclosed. In modern technology there is an inherent *pursuit of perfection*. We analysed this pursuit in terms of the *secular sacred*. Human beings try to transcend the human condition with the help of science and technology. They want to perfect the imperfect as far as possible and banish the threats and risks of life. This one-sided orientation towards technological solutions to the problems of modern people leads to a restricted repertoire of actions, a 'locking out' of possibilities.

Notes

1 With this designation Dooyeweerd refers to the view of the German philosopher of the enlightenment, Immanuel Kant, who also indicated his own philosophy as a transcendental critique. The three main works by Kant all have the word *critique* in their titles: *Kritik der reinen Vernunft* (*Critique of Pure Reason*), *Kritik der praktischen Vernunft* (*Critique of Practical Reason*) and *Kritik der Urteilskraft* (*Critique of the Power of Judgment*).

2 Goudzwaard in his book *Capitalism and Progress* (1979: 41) points out that during the Enlightenment 'the barrier of the lost paradise' was taken so that the perfection of the human being could be pursued: 'Now the third barrier is razed by the Enlightenment philosophers, who place rational self-destination in the context of an attainable perfect future and of the guarantee that the lost paradise can be regained by man's own activities.'

3 For that matter it is not really possible to speak about religion here in a generalised way. In religions, too, strategies of distrust occur, probably the most explicit in the phenomenon of magic. Magic is in fact a manipulative engagement of a human being with the gods.

4 Here we should note that intimate engagement with the transcendent is not necessarily representative of every religion. It does apply most emphatically, however, to the Jewish-Christian tradition (both Old and New Testament).

5 Here we can think for instance of what the American philosopher of technology Borgmann designated as 'focal activities'. See Section 12.5.

References

Bacon, F. (1627) *A New Atlantis*, reprinted in *New Atlantis and The Greet Instauration*, editor J. Weinberger, Wheeling, IL: Harland Davidson, 1989.

Berger, P. (1999) *The Secularization of the World: Resurgent Religion and World Politics*, Grand Rapids, MI: Eerdmans.
Goudzwaard, B. (1979) *Capitalism and Progress*, Grand Rapids, MI: Eerdmans. Translation of *Kapitalisme en voortuitgang* (1976).
Gregersen, N.H. (2004) 'Risk and religious certainty: conflict or coalition', *Tidsskriftet Politik*, 8(1): 22–32.
Grethlein, C. (1999) 'Die Verantwortung der Kirche für Bildung in einer pluralistischen Gesellschaft', in Ammer, C. and Karpinski, H. (eds), *Die Zukunft lieben. Herausforderung zum verantwortlichen Handeln*, Leipzig: Thomas Verlag, pp. 81–92.
Latour, B. (1993) *We Have Never Been Modern*, Cambridge, MA: Harvard University Press.
Noble, D.F. (1997) *The Religion of Technology. The Divinity of Man and the Spirit of Invention*, New York: Knopf.
Szerszynski, B. (2005a) *Nature, Technology and the Sacred*, Oxford: Blackwell.
Szerszynski, B. (2005b) 'Science, religion, and secularity in a technological society. Rethinking the secular: science, technology and religion today', *Zygon*, 40: 813–822.

ANDREW FEENBERG

(1943)

PORTRAIT 14 Andrew Feenberg

Andrew Feenberg is one of the most well-known philosophers of technology in our times. He studied philosophy at Johns Hopkins University, the University of Paris and the University of California. From 1969 to 2003 Feenberg was professor in philosophy of technology at the San Diego State University. At the moment he holds the Canada Research Chair in Philosophy of Technology at Simon Fraser University. He has also been a guest lecturer at more than ten universities in Europe, Japan and in the USA.

During the course of time Feenberg has made a change in course from classical philosophy of technology to a more empirically oriented philosophy of technology. At the onset of his career he was mainly inspired by the philosopher Herbert Marcuse. Marcuse was of the *Frankfurter Schule,* a group of well-known neo-Marxist philosophers from the previous century. Although he actually never discarded this orientation, the way in which he articulated and based his criticism of technology was increasingly inspired by empirical studies. It began in his first important publication on philosophy of technology, *Critical Theory of Technology* (1991). In it he did mention some concrete new developments in technology but these did not yet fill a definite role in his argumentation. This changed in his second book on philosophy of technology, *Alternative Modernity* (1995). In this he based his most important claims on a number of case studies. In his most recent book, *Questioning Technology* (1999) he takes a constructivist approach (see Chapter 12) and enters into a discussion with other philosophers of technology like Jacques Ellul, Martin Heidegger and Albert Borgmann.

What is significant in Feenberg's thinking on technology is that the development of a new technology is not only determined by technological-scientific considerations. For technology always has something indeterminate: one cannot by using 'logic' deduce from technological-scientific arguments what the best design is. Other, social considerations also come into play. Therefore technical developments are partly determined by social actors, like authorities and consumers. Technology can also become a means of realising social objectives. To Feenberg this means particularly the democratisation of society. But then there has to be a broad interest and motivation within society for influencing technology. On this point Feenberg criticises both Marcuse and the constructivists. Marcuse places the will to influence technology too much on the minorities in society (e.g., students) and the constructivists limit themselves too much to disconnected case studies. With reference to examples like the French Minitel computer system and the contraceptive pill Feenberg shows that consumers in the past were able to give a drastic turn to a technological development. According to him this should take place with all technological developments. He pleads for a 'process of instrumentalisation' consisting of two steps. In the first step, the primary instrumentalisation, a technological-social issue is lifted from its context to reach a first solution. In this mainly technological-scientific arguments play a role. Then follows the second step, the secondary instrumentalisation, in which the initial solution is placed back into its context and can undergo significant changes under the influence of considerations of social actors who have influence in this context.

Feenberg himself articulated the essence of his philosophy of technology in an unpublished article. Below are two quotes:

> Technology is not the product of a unique technical rationality but of a combination of technical and social factors. The study of these factors must include not only the empirical methods of social science but also the interpretive methods of the humanities in order to get at the underlying meaning of technical objects and activities for participants. Meaning is critically important insofar as technical objects are socially defined. What it is to be an automobile or a television is settled by social processes that establish definitions of these objects and grant them specific social roles. Technology itself cannot determine the outcome of these processes. Rather, conflicting worldviews shape alternatives and so the results cannot be understood without a grasp of how worldviews come to be concretized in technology. These problems of meaning have been insufficiently considered in technology studies. It has been my contribution to emphasize them as the specific contribution of the humanities.
>
> *(Feenberg, n.d.)*

> There have been many attempts in philosophy to define the essence of technology and to distinguish the specific difference of modern and premodern technologies. I have argued that these various theories are all unilateral and fail to grasp the full complexity of their object. I distinguish two levels of technical 'instrumentalization' and define eight different aspects of these two levels. At the primary level technology reifies its objects, i.e. decontextualizes them and manipulates them. At the secondary level various compensations are introduced to recontextualize technical objects once again, for example, by providing them with ethical and aesthetic dimensions. In premodern technology these two levels are distinguishable only analytically. They become institutionally separate in the development of modern technology. Various different configurations of the levels and their constituents define distinct technological rationalities. On these terms it is possible to develop a constructive critique of the culture of technology in our society and to imagine an alternative.
>
> *(Feenberg, n.d.)*

References

Feenberg, A. (n.d.) 'Summary remarks on my approach to the philosophical study of technology', unpublished paper, www.rohan.sdsu.edu/faculty/feenberg/method1.html (retrieved 29 December 2014).

Feenberg, A. (1991) *Critical Theory of Technology*, New York: Oxford University Press.

Feenberg, A. (1995) *Alternative Modernity. The Technical Turn in Philosophy and Social Theory*, Chicago, IL: University of Chicago Press.

Feenberg, A. (1999) *Questioning Technology*, London: Routledge.

INDEX

abstraction 27, 56, 119, 122–23, 126, 140, 174
accountability 281, 288, 299
Achterhuis, H. 300–1
actor-network theory 231
actualisation *see* disclosure
aesthetics 11
alterglobalists 242–246, 250–251
alterity relation (between human beings and technology) 278
anabaptism 320
analogy 77
analytical function of philosophy 3, 4
anticipation 78
anticulture 280–1
applied science (technology as) 4, 5, 7
aspects (of reality) 66–9, 70–6, 111–12, 120–23

background relation (between human beings and technology) 278
Bacon, F. 315
Beck, U. 246–250, 324
Berger, P. 316
bicycle 231
Bijker, W.E. 231
Borgmann, A. 258–260, 279–80
Bourdieu, P. 296
Buckyball 110
Bunge, M. 9
Bush, V. 4
Business Process Reengineering 176

Calvin, J. 23
Calvinism 320
Capek, K. 63
car navigation system 57
Castells, M. 246–250
cathedral (Gothic) 93
certitudo 325
Chaplin, Ch. 165–66
chiliasm 320
clock 271
codified knowledge 119–120, 128, 159
community 240–246
complexity 41, 42
component 95–9
component-material relationship 103–04
compound component 97
Comte, A. 317, 319
concretisation 81, 107
constitutive rules 205–206
constructivism 331
context (of an artifact) 92–4
contextual side (of normative practices) 265
convergence 81
conveyor belt 100, 161–63
corporate social responsibility 303
cosmopolitanism 246, 249–251
critical function of philosophy 3, 5, 6
critical management studies 164
culture and technology 107–08
cultural development 240–241, 255
cultural diversity 242–243, 249–251
cultural identity 246, 249–250

cultural rights 246
cybernetics 281
cyborg 274–5

democratic tradition *see* Socio-Technical Systems Design
Descartes, R. 316–7
design: context of 139; direction of 139; and market 142–50; and meaning 137; methodology of 136; model of 139–142; normal 137; proper function 136–39; and quality 147–50; revolutionary 137; structure of 139
Dessauer, F. 8
determinism (technological determinism) 281
Device paradigm 259, 279–80
Dilthey, W. 32
directional function of philosophy 3, 6
disclosure 91, 100, 104, 128, 240–5, 251, 255; and meaning 137
disenchantment 322
disengagement 279–80
Dooyeweerd, H. 63, 66–9, 73–5, 80, 318
Drexler, E. 111
Dreyfus, H. 278–9
dual nature of technical artefacts 8, 80, 82, 120, 138

Ellul, J. 9, 79, 212, 220–222, 226, 233, 236–238, 243, 300
embodiment relation (between human beings and technology) 271, 277, 287
empirical turn 12, 223, 227, 231, 233, 287
encaptic enlacement 101, 104, 265
Enlightenment 71
eotechnology 271
Engelmeier, P.K. 8
ensemble 81
epistemology 10
essence, essentialism 274–5
ethical dilemma 13
ethics: deontology 288, 290–1; consequentionalism 288, 291–4; utilitarianism 293–4; virtue ethics 288, 294; new (forms of) ethics 298–9
eudaimonia 294
external goods (of a social practice) 88, 199, 204, 294–5

facilitating rules 206
faith 255
Failure Mode and Effect Analysis (FMEA) 128
feedback 82
Feenberg, A. 300–1, 330–1.
floppy disk 19–20
focal activities 259, 279
Ford, H. 162–166, 168, 173–75
Foucault, M. 176–178
foundational function *see* technological artifact
foundational rules 205
Frankfurter Schule 331
freedom 280–1
Freemasons 319, 320
functional characteristics (artefact) 82

game 196–197, 203–204, 231–232
Geertsema , H. 282
Gehlen, A. 268–9
Gemeinschaft 245
general assemblers 111
Gesellschaft 245
globalisation 242, 245, 246, 249, 251, 252
good life 294

Haraway, D. 274–5
Harding, S. 253–254
Hart, H. 72
health care 309–11
Heidegger, M. 9, 24, 27, 28, 35–7, 217–9, 223, 226, 233
Heidenhain, M. 101
hermeneutic relation (between human beings and technology) 277
Hertz, N. 243
Hickman, L. 182–84
Holst, G. 46
'homo technicus' 268–9
hot air engine 5
Human Relations Movement 166–67, 173
hybrid 274–5, 321

identity (artefact) 90
Ihde, D. 277
individuation 107
induction 224
interlaced structures 99–104
Intermediary (role of technology) 277
internal goods (of a social practice) 88, 198–199, 202, 204, 294–5.
internal structure *see* technological artefact

Jonas, H. 298–9.

Kant, I. 32
Kapp, E. 7, 267–8
knowledge *see* technological knowledge
Kuhn , Th. 224–225, 226, 232

Kurzweil, R. 274–5
Kymlicka, W. 246

lab-on-chip 110
language game 197
Laszlo, E. 281
Latour, B. 54, 139, 153–55, 320–1
lean production 175
life cycle 147–49
Luddites 281

machine 271
MacIntyre, A. 86–8, 197–9, 201. 204, 294–5
manufacturability 145–46
Marcuse, H. 331
marketing 142–45
Marx, K. 7
mathematics 27
Mayo, E. 166
Mayo Clinic 311–2
meaning 18, 22, 28, 137
megamachine 272–3
metaphysics 11
methodology 11
Mini-Company concept 172–73, 176, 188
Mitcham, C. 7, 10, 15–17
modalities (modal aspects) 68, 75
module 95–9
Monsanto 53–5
Moral decision making 266
moralising (of apparatus) 300
Mumford, L. 8, 59–61, 271

Nanobots 276
nanotechnology 109–114, 124
National Science Foundation 113
neoconservatism 254–5
network logic 248
network society 247–8
neutralisation 98–9, 102–03
neutrality (postulate of, for science) 318
NBIC convergence (convergence of Nanotechnology, Biotechnology, Information technology, Cognitive sciences) 275
neotechnology 271
New-Age 281
Noble, D. 33, 319–20

object function 69–70, 91, 111, 142
OncomouseTM 274
ontology 10
open innovation 150–51
operating plan 138–39
operational function *see* technological artifact
organised irresponsibility 302
Ortega y Gasset, J. 8
over-determination 81

paleotechnology 271
Paré, A. 310
part-whole relation 265
participatory tradition *see* Socio-Technical Systems Design
perfection 323
personal computer 19
Philips (company) 5, 43, 145
Philips Natuurkundig Laboratorium ("Natlab") 45
Pirsig, R. 279–80
Plumbicon 43–7
plurality (of lifestyles) 249–252, 256
Polanyi, M. 119–20, 124, 128
Popper, K. R. 223–225
Porter, M. 309–10
postcolonial science and technology studies 252–6
post-industrial society 246–8
power (in organizations) 176–78
practice (normative), constitutive/structural side 204–206, 263, 264, 266, 296; context 204–7; directional/regulative side 204–205, 207–208, 263–4, 266
predator view effect 264
projection 271
proper function *see* design of technological artefact

qualifying function see technological artifact
qualifying rules 205–6
Quality Function Deployment (QFD) 124, 149
quality technologies 178–79

radu 25–6
Rapp, F. 9
Rathenau Institute (The Netherlands) 279
Redner 242–4
reductionism 79
Reich, C. 280
religion 315–6
responsibility 281, 295–6, 301–2
retrocipation 77
Reuleaux, F. 68
reverse-engineering 276
Riessen, H. van 8
risk 129, 324
risk society 247–8, 324

robot 63, 77, 94
Roszak, T. 280
RoundUp 53
rules 195–9, 203–7, 209
rules of engagement 263, 265, 266

sacred 299, 315–6
safety 324
Schön, D. 4
Schuurman, E. 9, 126, 215–6, 281
scientific management 157–61, 164
script *see* operating plan
Second Life 279
secularisation 315
secularization theory' 316
securitas 325
Simon, H. 125–27, 131–33
Simondon, G. 8, 80, 106–8
Smith, A. 159
social constructivism 212–3, 232–3, 254
Social Construction Of Technology (SCOT) 231–3
socio-technical systems 188; design 168–72, 173–76; Australian approach 174; Dutch approach 169–172; Scandinavian approach 174
sociotechnology *see* Socio-Technical Systems Design
sphere universality 79
standardisation: 98–9, 141, 159, 162–63
standards of excellence 199, 201–2, 203, 206
Stirling, R. 5
strategy of distrust 324–5
Strong Artificial Intelligence 276
structural characteristics (artefact) 82
subject function 69–70, 91, 111, 142
Strijbos, S. 250–1
soya bean modification 52–7
Strong Programme 226, 229
superstructure 270
Suzaki, K. 172–173
Szerszynski, B. 322–3

tacit knowledge 119–120, 128, 159
Taylor, F.W. 157–66, 168, 173–75,186,188
Tavistock Institute of Human Relations 167–68
technicism 216, 241
technicity 81
technological artefact: analysis of 107–8; extensions 267–8; foundational function 91, 96; identity 90, 94; internal structure 93; operational function 94–6; qualifying function 92–4, 96, 301; specific destination 93; technological code 300
technological fix 53
technological knowledge: dual nature 120; kinds of 116–23; limited nature 128–29; modal aspects 120–23, norms in 127–28
technological object *see* technological artefact
technological script 324
technology assessment (TA) 150
technology road map 150
technological science 125–27
technoscience 54, 298
Teisberg, E.O. 309–10
television camera 43
traditional societies 244–5, 252
transcendental criticism 318
transcendental direction 80
transhumanism 275, 282
Total Quality Management (TQM) 147–50, 178–79
Toyota Production System *see* Lean Production
Turkle, S. 279

Vijverdal (hospital) 47–52
Vincenti W. 116–19, 134–35
virtue ethics 86, 198, 294–5
Vollenhoven, D. van 68–9, 75, 80, 318
Vries, P. de 57

Whitbeck, C. 12
whole-component relationship 102–03
whole-whole relationship 101–03
Wiener, N. 281
Wiener Kreis 223–224
Winner, L. 112–3, 211–3, 222–3
Wittgenstein, L. 195–197
worldview 30, 271, 274, , 282, 318–9
World Wide Web 99

CPSIA information can be obtained
at www.ICGtesting.com
Printed in the USA
BVHW050253061118
532257BV00003B/14/P